Handbook of Statistics
Volume 44

Data Science: Theory and Applications

Handbook of Statistics

Series Editors

C.R. Rao
C.R. Rao AIMSCS, University of Hyderabad Campus,
Hyderabad, India

Arni S.R. Srinivasa Rao
Medical College of Georgia, Augusta University, United States

Handbook of Statistics

Volume 44

Data Science: Theory and Applications

Edited by

Arni S.R. Srinivasa Rao
Medical College of Georgia, Augusta, Georgia, United States

C.R. Rao
AIMSCS, University of Hyderabad Campus, Hyderabad, India

North-Holland

An imprint of Elsevier

North-Holland is an imprint of Elsevier
Radarweg 29, PO Box 211, 1000 AE Amsterdam, Netherlands
The Boulevard, Langford Lane, Kidlington, Oxford OX5 1GB, United Kingdom

Notices
Knowledge and best practice in this field are constantly changing. As new research and experience broaden our understanding, changes in research methods, professional
practices, or medical treatment may become necessary.

Practitioners and researchers must always rely on their own experience and knowledge in evaluating and using any information, methods, compounds, or experiments described herein. In using such information or methods they should be mindful of their own safety and the safety of others, including parties for whom they have a professional responsibility.

To the fullest extent of the law, neither the Publisher nor the authors, contributors, or editors, assume any liability for any injury and/or damage to persons or property as a matter of products liability, negligence or otherwise, or from any use or operation of any methods, products, instructions, or ideas contained in the material herein.

ISBN: 978-0-323-85200-5
ISSN: 0169-7161

For information on all North-Holland publications
visit our website at https://www.elsevier.com/books-and-journals

Publisher: Zoe Kruze
Acquisitions Editor: Sam Mahfoudh
Editorial Project Manager: Chris Hockaday
Production Project Manager: Abdulla Sait
Cover Designer: Greg Harris

Typeset by SPi Global, India

Contents

Section II
Engineering sciences data

Section III
Statistical estimation designs: Fractional fields, biostatistics and non-parametrics

Section IV
Network models and COVID-19 modeling

Contributors

Numbers in Parentheses indicate the pages on which the author's contributions begin.

Amparo Baíllo (3), Departamento de Matemáticas, Universidad Autónoma de Madrid, Madrid, Spain

B. Munwar Basha (105), Department of Civil Engineering, Indian Institute of Technology Hyderabad, Kandi, Sangareddy, India

José Enrique Chacón (3), Departamento de Matemáticas, Universidad de Extremadura, Badajoz, Spain

Chung-Chou H. Chang (155), Department of Biostatistics; Department of Medicine, University of Pittsburgh, Pittsburgh, PA, United States

Srikanth Cherukupally (75), Indian Institute of Science, Bengaluru, India

Somak Dutta (131), Iowa State University, Ames, IA, United States

Masayuki Kakehashi (235), Department of Health Informatics, Graduate School of Biomedical and Health Sciences, Hiroshima University, Hiroshima, Japan

C.C. Kerr (291), Institute for Disease Modeling, Global Health Division, Bill & Melinda Gates Foundation, Seattle, WA, United States; School of Physics, University of Sydney, Sydney, NSW, Australia

Tae Jin Lee (235), Center for Biotechnology and Genomic Medicine, Medical College of Georgia, Augusta University, Augusta, GA, United States

Sunil Mathur (201), College of Science and Engineering, Texas A&M University-Corpus Christi, Corpus Christi, TX, United States

Debashis Mondal (131), Oregon State University, Corvallis, OR, United States

J. Panovska-Griffiths (291), Department of Applied Health Research; Institute for Global Health, University College London, London; The Wolfson Centre for Mathematical Biology and The Queen's College, University of Oxford, Oxford, United Kingdom

Pranav R.T. Peddinti (105), Department of Civil Engineering, Indian Institute of Technology Hyderabad, Kandi, Sangareddy, India

Arni S.R. Srinivasa Rao (235), Laboratory for Theory and Mathematical Modeling, Division of Infectious Diseases, Medical College of Georgia, Augusta, GA, United States

Diana Rypkema (39), Department of Natural Resources, Cornell University, Ithaca; The Nature Conservancy, 652 NY-299, Highland, NY, United States

Sireesh Saride (105), Department of Civil Engineering, Indian Institute of Technology Hyderabad, Kandi, Sangareddy, India

Brijesh P. Singh (257), Department of Statistics, Institute of Science, Banaras Hindu University, Varanasi, India

R.M. Stuart (291), Department of Mathematical Sciences, University of Copenhagen, Copenhagen, Denmark; Disease Elimination Program, Burnet Institute, Melbourne, VIC, Australia

Victor B. Talisa (155), Department of Biostatistics, University of Pittsburgh, Pittsburgh, PA, United States

Shripad Tuljapurkar (39), Department of Biology, Stanford University, Stanford, CA, United States

W. Waites (291), School of Informatics, University of Edinburgh, Edinburgh; Centre for the Mathematical Modelling of Infectious Diseases, London School of Hygiene and Tropical Medicine, London, United Kingdom

Preface

Data Science: Theory and Applications, Volume 44 can be treated both as a sequel and complementary to Principles and Methods for Data Science, Volume 43 in the *Handbook of Statistics* series. Both these volumes highlight new advances in the field. Volume 44 presenting interesting and timely topics, including animal models and ecology, network models, large data parameter estimation procedures, theoretical advancements and graphs, fractional Gaussian fields to COVID-19 models, application for the transportation and road construction engineering to nonparametric statistics applications of data science.

Data Science: Theory and Applications has been developed with brilliantly written chapters by authors from various aspects of data science. All the authors took utmost care in making their chapters available either for a data-oriented scientist or theoretician or method developer whose work interfaces with societal issues using data. Authors have experience of one or more of the skills in handling the data, namely, designing a course on data science, developing theory for data ensemble techniques, methods on high-dimensional data designing for biotechnology companies, teachers of data science to computer graduates or purely developers of theory to be applied by others, etc. Utmost care is taken to make chapters are assessable to the students and equally entertain subject experts. Students can use our volume for learning a new topic that has not been taught in their curriculums or using some of the chapters as additional reading material. Faculty can use our volume as a quick reference book on the subject matter.

The authors' expertise ranges from statistics, engineering sciences, physics to mathematics, pure data science to computational epidemiology. We strongly feel that we have brought compelling material for data science lovers in applied and purely theoretical. We have divided 10 chapters of volume 44 into four sections:

Section I: Animal models and Ecological Large Data Methods
Section II: Engineering Sciences Data
Section III: Statistical Estimation Designs: fractional fields, biostatistics, and non-parametrics
Section IV: Network Models and COVID-19 modeling

Section I contains two chapters: the first chapter by A. Baillo and J.E. Chacon brings statistical estimations involved in animal models and patches of the home range of animals. They also discuss "optimal" home range modeling and newer methodologies; the second chapter by D. Rypkema and S. Tuljapurkar wrote a detailed theoretical account on generalized extreme value distributions and their role in climate events. They also describe the modern implications of such methods and quantitatively estimating their impacts on ecosystems.

Section II contains two chapters: the first chapter by S. Cherukupally writes a comprehensive lecture note on blockchain technologies and technicalities by keeping a graduate student to an advanced engineer who wants to apply such techniques. Blockchain technology is an emerging field in various scientific and engineering disciplines. The second chapter by S. Saride, P. R. T. Peddinti, and B. M. Basha bring practical data collection methods and modern techniques of sustainability of pavements. They also explain practical modeling of environmentally smart futuristic road constructions.

Section III contains three chapters: the first chapter by S. Dutta and D. Mondal write a detailed chapter on lattice approximations for fractional Gaussian fields. The chapter also brings an overview of theoretical developments about the continuum fractional Gaussian fields and how they can be derived from the lattice approximations; the second chapter by V. Talisa and C-C. H. Chang is for the biostatistical data scientists who work on treatment effects, related modeling approaches. They provide detailed analysis techniques and measuring causal effects; in the third chapter, S. Mathur provides a comprehensive account of nonparametric large complex data-related statistical inference with applications in data analytics. The chapters bring a review on cutting-edge analytical tools, general data-mining novel approaches, and a practical guide for data scientists.

Section IV contains three chapters: the first chapter by T.J. Lee, M. Kakehashi, and A. S. R. Srinivasa Rao write a review on network models in epidemiology from traditional and modern perspectives. They provide the argument for applicability of network structures where epidemic data can be modeled and interpreted for emerging viruses; the second chapter by B. P. Singh writes various modeling approaches for modeling Indian COVID-19 growth approximations, comparisons, and advantages of statistical vs mathematical approaches. The ideas of this chapter are intended for scientists and modelers who require to handle necessary precautions while modeling the pandemics; the third chapter by J. Panovska-Griffiths, C.C. Kerr, W. Waites, and R.M. Stuart provide comprehensive perspectives on mathematical modeling of COVID-19, data availability, newer approaches of modeling with limited data. They describe models by their own experiences and perspectives from other such modelers. A useful guide for collaborations and providing timely pandemic assessment techniques and mathematics for policy developments are discussed.

Our sincere thanks to Mr. Sam Mahfoudh, Acquisition Editor (Elsevier and North-Holland) for his overall administrative support throughout the preparation of this volume. His valuable involvement in time to time handling of authors' queries is highly appreciated. We also thank Ms. Naiza Mendoza, Developmental Editor (Elsevier), and Ms. Hilal Johnson, Mr. Chris who worked during the initial stage of the development of this volume as the Editorial Project Manager (Elsevier) for their excellent assisting of editors and authors in various technical editorial aspects throughout the preparation. Ms. Mendoza provided valuable assistance toward the proofs and production. Our thanks are also to Md. Sait Abdulla, the Project Manager, Book Production, Chennai, India, RELX India Private Limited for leading the production and printing activities and assistance to the authors. Our sincere thanks and gratitude to all the authors for writing brilliant chapters by keeping our requirements of the volume.

We genuinely believe that this volume on data science has addressed both theory and applications in a well-balanced content. We are convinced that this collection will be resourceful for data scientists across several disciplines.

Arni S.R. Srinivasa Rao
C.R. Rao

Section I

Animal models and ecological large data methods

Chapter 1

Statistical outline of animal home ranges: An application of set estimation

Amparo Baíllo[a],* and José Enrique Chacón[b]

[a]*Departamento de Matemáticas, Universidad Autónoma de Madrid, Madrid, Spain*
[b]*Departamento de Matemáticas, Universidad de Extremadura, Badajoz, Spain*
Corresponding author: e-mail: amparo.baillo@uam.es

Abstract

The home range of an individual animal describes the geographic area where it spends most of the time while carrying out its common activities (eating, resting, reproduction, etc.). Motivated by the abundance of home range estimators, our aim is to review and categorize the statistical methodologies proposed in the literature to approximate the home range of an animal, based on a sample of observed locations. Further, we highlight that any statistical home range estimator is derived using a technique from the theory of set estimation, which aims to reconstruct a set on the basis of a multivariate random sample of points. We also address the open question of choosing the "optimal" home range from a collection of them constructed on the same sample. For clarity of exposition, we have applied all the estimation procedures to a set of real animal locations, using R as the statistical software. As a by-product, this review contains a thorough revision of the implementation of home range estimators in the R language.

Keywords: Animal space use, Tracking locations, Utilization distribution, Nonparametric, Density level set

1 Introduction to home range estimation

1.1 Problem statement

The scientific interest in identifying different characteristics (degrees of use, geographical boundaries, environmental descriptors, etc.) of space use by whole animal species or just an individual animal has more than a century of history. The space use can be described by various (occasionally interrelated) concepts. In this work we specifically focus on the concept of *home range*. As Seton (1909, p. 26) pointed out when speaking about the home range of

Handbook of Statistics, Vol. 44. https://doi.org/10.1016/bs.host.2020.10.002

an individual, "no wild animal roams at random over the country; each has a home-region, even if it has not an actual home. The size of this home region corresponds somewhat with the size of the animal. [...] In the idea of a home-region is the germ of territorial rights."

Burt (1943) is considered to be the first in formalizing the idea of home range, as "that area traversed by the individual in its normal activities of food gathering, mating and caring for young. Occasional sallies outside the area, perhaps exploratory in nature, should not be considered as in part of the home range." Water sources, resting and shelter zones, and the routes traveled between all these vital locations may also be part of the home range. The home range of an animal should not be confused with its territory, which is the defended portion of the home range. Consequently, home ranges of different individuals can overlap (for instance, watering holes might be shared).

In everyday practice, conservation biologists and spatial ecologists, among others, have a keen interest in estimating space use maps from animal tracking data, for example, for monitoring threatened species and for conservation planning. Further, the widespread use of geolocated smartphones has lead to massive amounts of analogous tracking data for billions of human individuals (Meekan et al., 2017). Analysis of space use is, thus, more appealing than ever, not only from a conservational point of view (analysis of interactions between animals and humans), but also from an economical or even anthropological perspective (Walsh et al., 2010).

Marking the home range perimeter is a popular and simple way of describing the space use of a monitored individual (see Péron, 2019 for considerations on the time frame of a tracking study and its implication on the conclusions about space use). This problem has been tackled in a variety of ways of increasing complexity. For instance, individual space use can also be described by the *utilization distribution* (UD), the probability distribution of the animal locations over a period of time (van Winkle, 1975). Nowadays, *statistical* procedures aiming to obtain the home range usually define it as a level set, with high probability content (by default 95%), of the utilization density. Another appealing possibility is to use *mechanistic* home range techniques (see Moorcroft and Lewis, 2006), where first the trajectory of the animal is modeled via a stochastic process and then the utilization level set is obtained. Stochastic dependence among tracking locations is indeed a current matter of concern and interest. Fleming et al. (2015) stress that animal tracking data are, by nature, autocorrelated, an obvious consequence of animal movement being a continuous process. Similarly to the mechanistic procedures, these authors define a home range as a region with a prespecified probability content (usually 95%) of the *range distribution*, which is the "probability distribution of all possible locations as determined from the distribution of all possible paths." Fleming et al. (2015) consider that an autocorrelated sample contains less geometrical information about the density level set contour than a sample of independent observations of the same size. However, the existence of autocorrelation between observations implies that we can take advantage of information from the past to predict the future movements of the animal and, thus, it should be used for home range estimation.

The purpose of this work is to provide a thorough review of the statistical procedures proposed in the literature to estimate the home range of a specific animal (Section 2). At the same time, along with the conceptual review, the libraries in the statistical software R (R Core Team, 2020) available for the different procedures will be also surveyed (see also Tétreault and Franke, 2017; Joo et al., 2020). These techniques are mostly of a descriptive nature, and based on set estimation approaches (see Sections 1.2 and 2.1), either through direct geometrical procedures to estimate a set or through density estimation of the UD and posterior obtention of its level set. Finally, there is a natural question stemming precisely from this review. Given such a wealth of possible estimators, which is the "best" one? To our knowledge, this is an interesting, still-open question. In Section 3.3 we review the proposals available in the literature and point out their drawbacks. More generally, Section 3 is devoted to some open problems and possible future scope for research.

1.2   Connection to the set estimation problem

In this section we introduce the statistical problem of set estimation and explain its connection to the home range approximation problem. In its content there are more mathematical technicalities than in the rest of the paper, but readers from a less theoretical (e.g., biological) background can skip Section 1.2.

In mathematical statistics, the problem of set estimation consists in approximating an unknown compact set $S \subset \mathbb{R}^d$, for $d \geq 1$, using the information of a random sample of n points generated by a probability distribution $P_\mathbf{X}$ (the population distribution), which is linked to S in some way. The reader interested in delving into set estimation is referred to two exhaustive surveys by Cuevas (2009) and Cuevas and Fraiman (2010). There are two well-known and important problems usually considered in set estimation that are closely related to home range estimation. The first problem is that of approximating the support S of $P_\mathbf{X}$, that is, the smallest closed subset of $\mathbb{R}^d$ having probability 1. The second problem concerns the estimation of the level set $\{f \geq c\}$, for some fixed level $c > 0$, of the density f corresponding to the distribution $P_\mathbf{X}$ (when $P_\mathbf{X}$ is absolutely continuous with respect to the Lebesgue measure μ_L in $\mathbb{R}^d$).

In the context of home range estimation, the sample points correspond to the recorded locations of the animal (with dimension $d = 2$ if only the planar position is considered and with $d = 3$ if the altitude is also included), the probability measure $P_\mathbf{X}$ is the utilization distribution (see Section 1.1) and f is the utilization density. In general, under the assumption that the locations of the tracked animal are independent, any of the home range estimators proposed in the literature (see Section 2.1) is in fact an approximation either of the support or of a level set of the utilization density.

The rest of Section 1.2 is devoted to comment more details and results on the estimation of the support (Section 1.2.1) and level sets (Section 1.2.2). To study the performance of an estimator S_n of the set S, we introduce two commonly used discrepancy measures. The first one is the Hausdorff distance

$$\rho_H(S, S_n) = \inf \{\epsilon > 0 : S_n \subset S^\epsilon \text{ and } S \subset S_n^\epsilon\},$$

where $A^\epsilon = \bigcup_{x \in A} B(x, \epsilon)$ defines the ϵ-parallel set of a set A and $B(x, \epsilon)$ denotes the open ball centered at x and with radius ϵ. The second discrepancy is the distance in measure $d_\mu(S, S_n) = \mu(S \Delta S_n)$, where μ is a reference measure (the Lebesgue measure, for instance) and $S \Delta A = (S \cap A^c) \cup (S^c \cap A)$ denotes the symmetric difference between two sets S and A.

1.2.1 Support estimation

When no parametric model is assumed on the probability distribution P_X, it is common to impose some geometric conditions on S in order to avoid "strange" shapes that complicate the estimation process. The most used and deeply studied geometrical restriction on the support is convexity. If S is convex, the natural estimator of S is the convex hull of the sample (see Brunel, 2018 for a recent review on convex set estimation). Thus, home range estimators such as the convex hull (see Section 2.1) are examples of support approximations. Rényi and Sulanke (1963, 1964), some of the earliest and best known references on the convex hull, established the consistency in probability of this estimator with respect to d_{μ_L} in the two-dimensional case. Dümbgen and Walther (1996) obtained a.s. rates of convergence to 0 of the Hausdorff distance between the convex hull and the corresponding convex target S.

The convexity assumption is quite restrictive in practice (see Section 2.1.1). In the Computational Geometry literature, the notion of α-convexity (first employed by Perkal, 1956) is introduced as a generalized convexity concept (see also Walther, 1997). The set S is said to be α-convex if $S = C_\alpha(S)$, where

$$C_\alpha(S) = \bigcap_{\{x \in \mathbb{R}^d : B(x, \alpha) \subseteq S^c\}} (B(x, \alpha))^c$$

denotes the α-convex hull of S; i.e., the intersection of the complements of all the balls of radius α that are fully contained in S^c. If S is α-convex, the natural estimator is the α-convex hull of the sample, for which Rodríguez-Casal (2007) obtained a.s. convergence rates with respect to d_H and d_{μ_L}. Rodríguez-Casal and Saavedra-Nieves (2016) provide an automatic, data-driven method to select the smoothing parameter α.

In absence of geometric information on the support S, Devroye and Wise (1980) suggested using the union of closed balls centered at the sample points and with a fixed radius $r > 0$ as a simple and intuitive support estimator. The resulting estimator has a shape similar to the PPA home range (see Section 2.2.1 and Fig. 12). Biau et al. (2009) obtained convergence rates and asymptotic normality, as $n \to \infty$ and $r = r_n \to 0$, for this support estimator with respect to the distance in measure d_{μ_L}. From a completely different perspective, Schölkopf et al. (2001) introduced an algorithm, based on support vector machines and valid even for a high dimension d, to estimate a general support S. Using similar, kernel-based tools de Vito et al. (2010) introduce the regularized spectral support estimator. Rudi et al. (2017) prove that a set estimator defined via kernel principal component analysis is a.s. consistent with respect to d_H.

1.2.2   Level set estimation

In this setting, geometrical assumptions are sometimes placed on the unknown, population level set to be estimated. However, the most common, direct, and intuitive way of approximating $\{f \geq c\}$ is provided by the plug-in methodology, where the unknown f is replaced by a consistent density estimator $\hat{f}$, thus yielding the set estimator $\{\hat{f} > c\}$ (see Chen et al., 2017 for the asymptotic distribution of the Hausdorff distance for this plug-in estimator and a review of previous references). Other procedures for estimating level sets have also been defined; for instance, the excess mass approach in Polonik (1995), or the partition-based method in Willett and Nowak (2007) and in Singh et al. (2009) (see also Klemelä, 2018 for a review of other proposals).

A particular instance of the level set estimation problem refers to the case where the level set is imposed to have a prespecified probability content $1 - \alpha$, for some $0 < \alpha < 1$. In this case, the problem is to approximate the population set $\{f \geq c_\alpha\}$, where c_α is such that

$$1 - \alpha = \int_{\{f \geq c_\alpha\}} f.$$

When the value α is close to 0, the compact level set $\{f \geq c_\alpha\}$ is considered as a sort of "effective" support of $P_{\mathbf{X}}$, more informative than the actual support S (that might be too large or even unbounded). The plug-in methodology estimates the set $\{f \geq c_\alpha\}$ by $\{\hat{f} \geq \hat{c}_\alpha\}$, where $\hat{c}_\alpha$ satisfies

$$1 - \alpha = \int_{\{\hat{f} \geq \hat{c}_\alpha\}} \hat{f}.$$

When $\hat{f}$ is a kernel density estimator, the plug-in approximation $\{\hat{f} \geq \hat{c}_\alpha\}$ gives rise to one of the most popular home ranges, namely, the kernel home range (see Section 2.1). The rates of convergence of the set estimator $\{\hat{f} \geq \hat{c}_\alpha\}$ with respect to d_{μ_L}, when $\hat{f}$ is a kernel estimator, have been studied by Cadre et al. (2013), and the problem of bandwidth selection has been considered in Doss and Weng (2018) and Qiao (2020).

1.3   Characteristics of animal location data

As mentioned in Sections 1.1 and 1.2, the sample information used in home range estimation consists of animal tracking data. Originally, animal locations were obtained using live traps (Blair, 1940) or continuous observation of the animal in short periods of time (Odum and Kuenzler, 1955). In the 1960s these techniques were substituted by radio telemetry, that is, attaching a transmitter to the animal and recording its position via the transmission of radio signals. The resulting observations, quite spaced apart in time, were considered independent and several home range estimators (see Section 2), derived from set estimation techniques, were designed under this independence assumption. These estimators are still in wide use today.

Recent advances in wildlife tracking and biotelemetry devices (e.g., satellite positioning, see MacCurdy et al., 2019) provide animal location data with an ever increasing frequency and accuracy (Kie et al., 2010), together with physiological aspects of the animal (heart rate, body temperature, etc.) and details about its external environment (such as temperature, salinity, depth, or altitude; see Cooper et al., 2014, Wilmers et al., 2015). This, together with improvement in computing resources, has originated the use of new estimation techniques (Tracey et al., 2014), the analysis of time evolution and autocorrelation effects on home range estimation (Noonan et al., 2019) and an increasing generalized interest in the analysis of space use based on tracked individuals.

1.4 A real data set: Mongolian wolves

Throughout this work we will review a large number of techniques for home range estimation proposed in the literature. The procedures will be illustrated via the analysis of the real data set of relocations with ID 14291019 downloaded from Movebank (www.movebank.org), an animal movement database coordinated by the Max Planck Institute for Ornithology, the North Carolina Museum of Natural Sciences, and the University of Konstanz. This data set contains the locations of a pair of Mongolian wolves, Boroo (female) and Zimzik (male), observed with GPS technology during a period starting May 2003 (Boroo) or March 2004 (Zimzik) and ending September 2005 (both) (Kaczensky et al., 2006, 2008). Signal transmission took place between one to three times a day, at irregularly-spaced times. In this work we focus on the female wolf, Boroo, with 2226 observed locations as depicted in the map of Fig. 1. The trajectory seems to correspond to a home range with interesting mathematical features (possibly more than one connected component, nonconvex components, etc.). Previously, Baíllo and Chacón (2018, 2020) analyzed in detail the home range estimators derived from the locations and trajectory of the male wolf, Zimzik.

The map comprises part of the provinces of Hovd and Govi-Altay in southwestern Mongolia and China (the Mongolia–China border is in light gray at the bottom of the map). We can see that a high proportion of the locations of Boroo fell inside the Great Gobi B Strictly Protected Area, an International Biosphere Reserve in the Gobi desert (for more detailed information, see Kaczensky et al., 2008). In general, it seems reasonable to believe that there are usually explanatory variables related (to some extent) to the

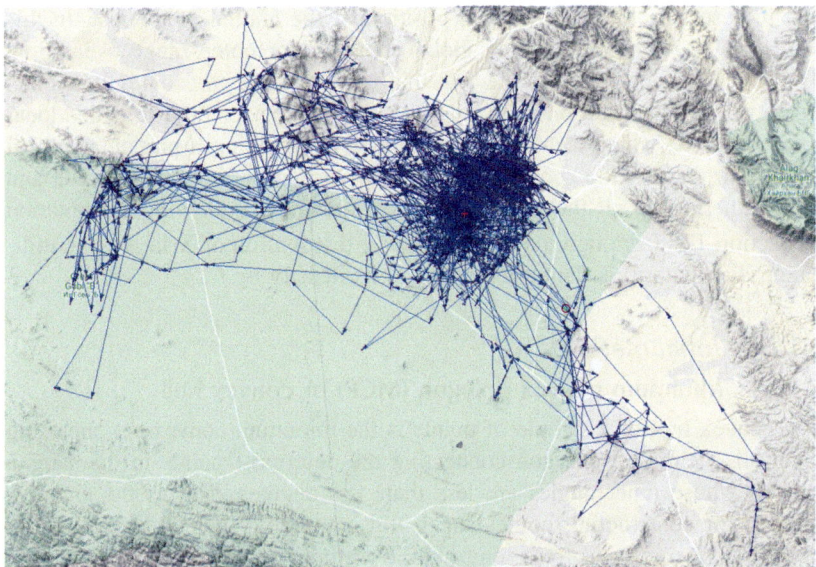

FIG. 1 Physical map showing $n = 2226$ observed locations of the Mongolian wolf Boroo, along with the trajectories between them.

utilization distribution and the home range of an animal, for instance, geographical elevation, presence/absence or degree of abundance of preys, presence/absence of hunters, etc. The problem of how to incorporate this information onto the home range estimator has been little analyzed yet (see Section 3.2).

2 Statistical techniques for home range estimation

The statistical procedures to estimate home ranges are of a descriptive and nonparametric nature. They either estimate the home range directly, using geometric-type procedures, or they estimate first the utilization density and then compute the 95% level set of the density estimator. The earliest estimation techniques were based on scarce location data, which were assumed to be independent. However, the high-frequency observation in current tracking samples has recently fostered the incorporation of time dependence into the analysis of positional data. To distinguish between these two setups, in Section 2.1 we review the simpler home range estimators proposed under the independence assumption, and in Section 2.2 we introduce the more recent proposals dealing with time-dependent locations.

2.1 Assuming location data are independent

In this section we denote the animal relocations by $\mathbf{x}_1, ..., \mathbf{x}_n$ and we assume they are independent realizations of a random vector $\mathbf{X}$ taking values in $\mathbb{R}^2$

and representing the geographical position of the animal of interest. In this setup, we begin by introducing global methods for home range estimation. The term "global" refers to the fact that each of such procedures is applied to the whole data set at once. In contrast, later we focus on the so-called localized methods, which stem from the former by applying a particular global method to each of the subsets of a suitably chosen partition of the data and then gathering up all the local pieces thus obtained. Following the notation of Section 1.2, from now on the probability distribution of $\mathbf{X}$ in $\mathbb{R}^2$ (the utilization distribution) is denoted by $P_\mathbf{X}$ and its density by f.

2.1.1   Global methods

2.1.1.1   Minimum convex polygon (MCP) or convex hull

The convex hull of a sample of points is the minimum convex set enclosing them all, yielding a polygon connecting the outermost points in the sample and all whose inner angles are less than 180 degrees. This is the simplest method for constructing home ranges and computing their areas and it has been widely employed for a long time, even until recently. The convex hull of the whole sample is a natural, consistent estimator of the support of $P_\mathbf{X}$ if this latter set is convex (see, e.g., Schneider, 1988; Majumdar et al., 2010 and references therein).

A variant of the MCP home range estimator is obtained by removing the proportion α of the sample points farthest from the sample centroid. More generally, Bath et al. (2006) compare the MCP home ranges obtained after applying different criteria to "peel" individual observations from the sample of locations: they remove points farthest from the centroid or from the harmonic mean or those with a great influence on the area of the home range.

References using the minimum convex polygon (alone or in comparison with other methods) to estimate the home range of an individual are very abundant. Let us just mention a few: Mohr (1947), Odum and Kuenzler (1955), Worton (1987), Carey et al. (1990), Harris et al. (1990), Pavey et al. (2003), List and Macdonald (2003), Nilsen et al. (2008), Signer et al. (2015), and Mammen et al. (2017).

Obviously, the convexity restriction has serious drawbacks, among them home range overestimation, as illustrated in Fig. 2, showing the MCP of Boroo locations. The MCP was computed using the R package adehabitatHR (Calenge, 2006). This home range does not adapt to the "croissant-shaped" territory to which this wolf usually circumscribed its movements. To overcome this drawback, Harvey and Barbour (1965) proposed the minimum area method (also called by Kenward et al., 2001 the concave polygon). First, they computed the *range length*, the distance between the two locations furthest apart. Then, one outermost point (that is, in the boundary of the MCP) is connected with the next outermost one lying inside a ball of radius one quarter of the range length. The resulting minimum area home range is not necessarily convex,

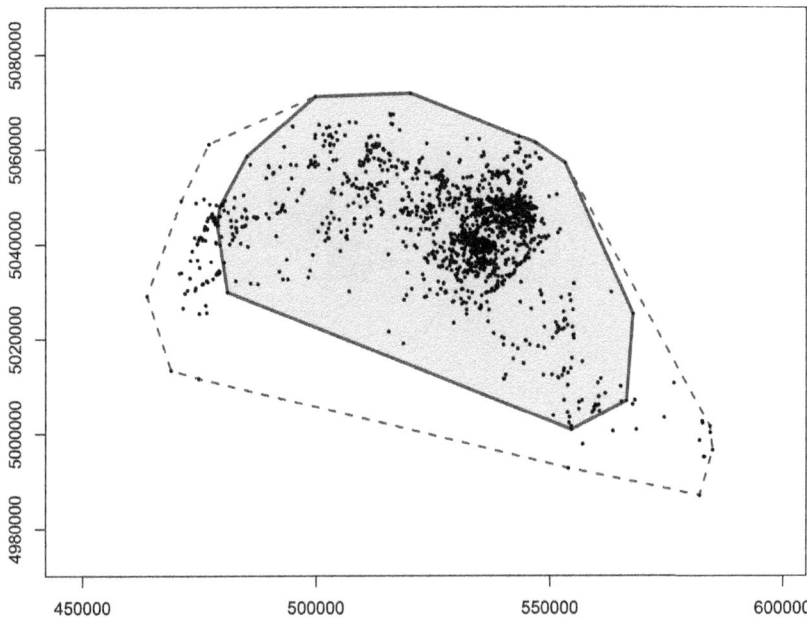

FIG. 2 Minimum convex polygons (MCPs) for Boroo location data. The MCP containing 95% of the sample points is *colored* in *gray*, with its boundary in *solid lines*. The *dashed lines* correspond to the MCP computed on the basis of the whole sample.

but it is still a polygon and adapts better to the underlying sample shape. With the same objective of determining the "envelope" of a set of points, Moreira and Santos (2007) and Park and Oh (2012) introduced a "concave hull" algorithm that digs into the original MCP by identifying the nearest inner points from the convex hull edge.

2.1.1.2 α-Convex hull

As already mentioned in Section 1.2.1, there have been other proposals in the literature to relax the assumption of convexity, which is particularly restrictive in the problem of home range estimation. Defined in simple terms, the α-convex hull of the sample is the intersection of the complements of all the open balls of radius α which do not contain any sample points. Fig. 3 shows the α-convex hull of Boroo locations, with a value of $\alpha = 16,000$, computed using the R package alphahull (Pateiro-López and Rodríguez-Casal, 2019) after removing 5% of the data points farthest from the data centroid (see also Pateiro-López and Rodríguez-Casal, 2010). The boundary of this home range estimator clearly shows the arcs of some of the balls of radius α that form the complement of the estimator. The improvement over the convex hull is clear, since the α-convex hull does not contain some of the large portions of never visited terrain that the convex hull enclosed.

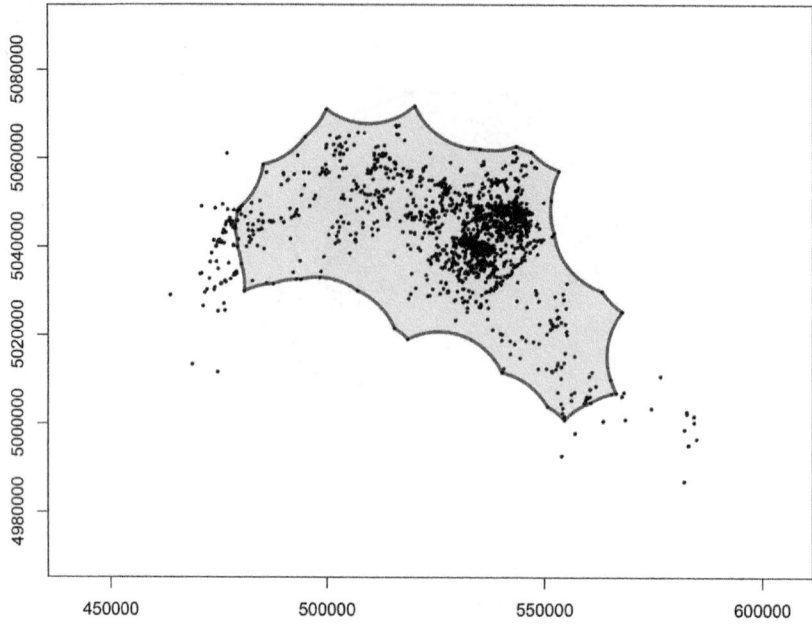

FIG. 3 Home range estimation by means of the α-convex hull for Boroo location data, computed after removing 5% of the data points farthest from the data centroid, using a value of $\alpha = 16,000$.

The radius α controls the regularity of the resulting estimator: large values of α make the α-convex hull very similar to the ordinary convex hull, and small values yield quite a fragmented home range estimator. The suggested value of $\alpha = 16,000$ was chosen here by visual inspection.

Let us point out that there are other existing home range estimation techniques that receive a name similar to "α-convex hull," although they refer to a completely different concept. To avoid confusion, it is worth mentioning the procedure of Burgman and Fox (2003), called α-hull, which simply consists in obtaining the Delaunay triangulation of the data points and eliminating those line segments longer than a multiple α of the average line length. Although this latter α-hull was initially introduced to estimate a species range, it has also been used for home range estimation (see, e.g., Foley et al., 2014).

In fact, Burgman and Fox's α-hull is closer to another related object in the Computational Geometry literature: the α-shape of the data points. This is another generalization of the convex hull, defined as the polygon whose edges are those segments connecting two data points if there is an open ball of radius α having those two data points in its boundary and no other data point in its interior (Edelsbrunner et al., 1983). The α-shape corresponding to Boroo locations, computed with the R package alphahull, is depicted in Fig. 4, for the same value of $\alpha = 16,000$ as before and after removing the 5% of the data points farthest from the data centroid. Its appearance is very similar to the

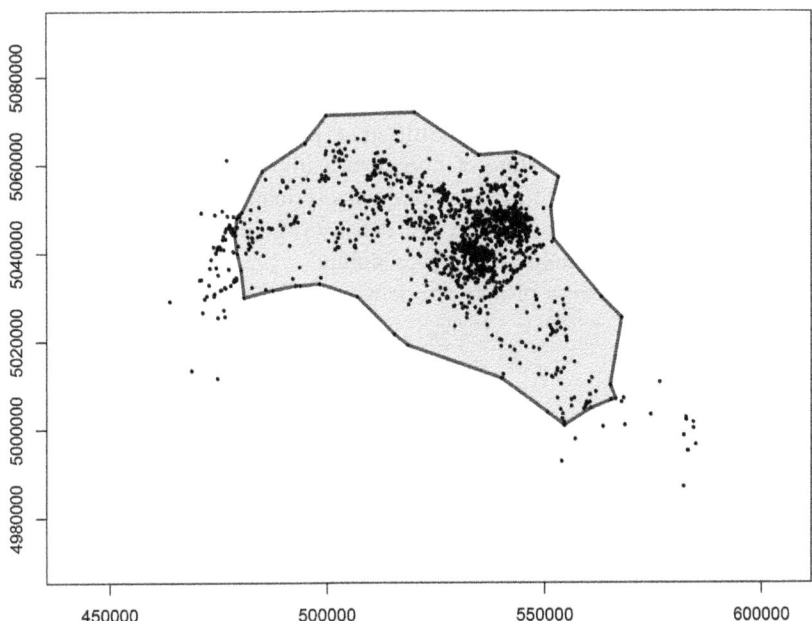

FIG. 4 Home range estimation by means of the α-shape for Boroo location data, computed after removing 5% of the data points farthest from the data centroid, using a value of $\alpha = 16,000$.

α-convex hull in Fig. 3, with the notable difference that its boundary is formed by straight lines, instead of circular arcs.

Still other recent generalizations of the convex hull are available for independent relocations, as for instance the cone-convex hull of Cholaquidis et al. (2014). In the case of dependent locations, Cholaquidis et al. (2016) propose to estimate the home range by the α-convex hull of the trajectory or by the reflected Brownian motion (RBM) sausage (the outer parallel set of the trajectory). Here we will not explore these many variants.

2.1.1.3 Kernel density estimation (KDE)

Let us recall that, in Section 1, we introduced the utilization density f as the probability density of the geographical position of the animal during a time period. We also mentioned that the home range of an animal is frequently defined as the level set of the utilization density attaining a 95% probability content, that is, the set $\{f \geq c_{0.05}\}$ such that

$$0.95 = \int_{\{f \geq c_{0.05}\}} f(\mathbf{x}) \, d\mathbf{x}.$$

A plug-in estimator of the home range is the analogous level set $\{\hat{f} \geq \hat{c}_{0.05}\}$ of an estimator $\hat{f}$ of f, where

$$0.95 = \int_{\{\hat{f} \geq \hat{c}_{0.05}\}} \hat{f}(\mathbf{x}) \, d\mathbf{x}.$$

The first proposals along these lines assumed a bivariate normal model for the UD (Calhoun and Casby, 1958). Still in a parametric setting, Don and Rennolls (1983) used a mixture of normal densities to estimate f. However, noting that such parametric models were usually inappropriate, in a seminal paper Worton (1989) introduced kernel density estimators as a natural non-parametric approach in home range procedures.

The general expression of a kernel density estimator is

$$\hat{f}(\mathbf{x}) = \frac{1}{n} \sum_{i=1}^{n} K_{\mathbf{H}}(\mathbf{x} - \mathbf{x}_i), \tag{1}$$

where $K : \mathbb{R}^2 \rightarrow [0, \infty)$ is the kernel function (in this case, a probability density in $\mathbb{R}^2$), $\mathbf{H} = (h_{ij})$ is a symmetric, positive-definite 2×2 bandwidth matrix, and the scaling notation $K_{\mathbf{H}}(\mathbf{x}) := |\mathbf{H}|^{-1/2} K(\mathbf{H}^{-1/2}\mathbf{x})$ has been used, with $\mathbf{H}^{-1/2}$ standing for the inverse of the matrix square root of $\mathbf{H}$ (see Chacón and Duong, 2018, for a recent monograph on multivariate kernel smoothing).

It is well known that the choice of the kernel K has little effect on the accuracy of the estimator $\hat{f}$, compared to that of the bandwidth $\mathbf{H}$. Worton (1989) chose a constrained bandwidth matrix $h^2\mathbf{I}$ (where $\mathbf{I}$ denotes the identity matrix) depending on a single smoothing parameter $h > 0$, which was proposed to be selected either via the so-called "ad hoc" method (which refers to the optimal choice for the Gaussian distribution) or via least-squares cross-validation. Worton (1989) additionally considered an adaptive KDE, that is, a kernel-type estimator where the bandwidth is allowed to vary from one observation $\mathbf{x}_i$ to the other.

The KDE home range for the wolf data is displayed in Fig. 5 for the one-dimensional ad hoc bandwidth $h = 4163.9$ (as computed using the R package adehabitatHR) and the bidimensional unconstrained plug-in bandwidth matrix

$$\mathbf{H} = \begin{pmatrix} 2805936.6 & -184535.9 \\ -184535.9 & 1525691.0 \end{pmatrix} \tag{2}$$

of Chacón and Duong (2010), obtained with the package ks (Duong, 2007; [Duong, 2019]). As already noted by Bauder et al. (2015), the use of this unconstrained plug-in bandwidth matrix outperforms the single smoothing parameter approaches. It is worth mentioning that, due to the very large values of the UTM coordinates, it was necessary to prespherify the location data in order for the numerical routine to be able to perform the optimization correctly. This means that the data were pre-multiplied by $\mathbf{S}^{-1/2}$, the inverse of the square root of their sample variance matrix S, so that the sample variance matrix of the transformed data became the identity matrix. A plug-in bandwidth was obtained

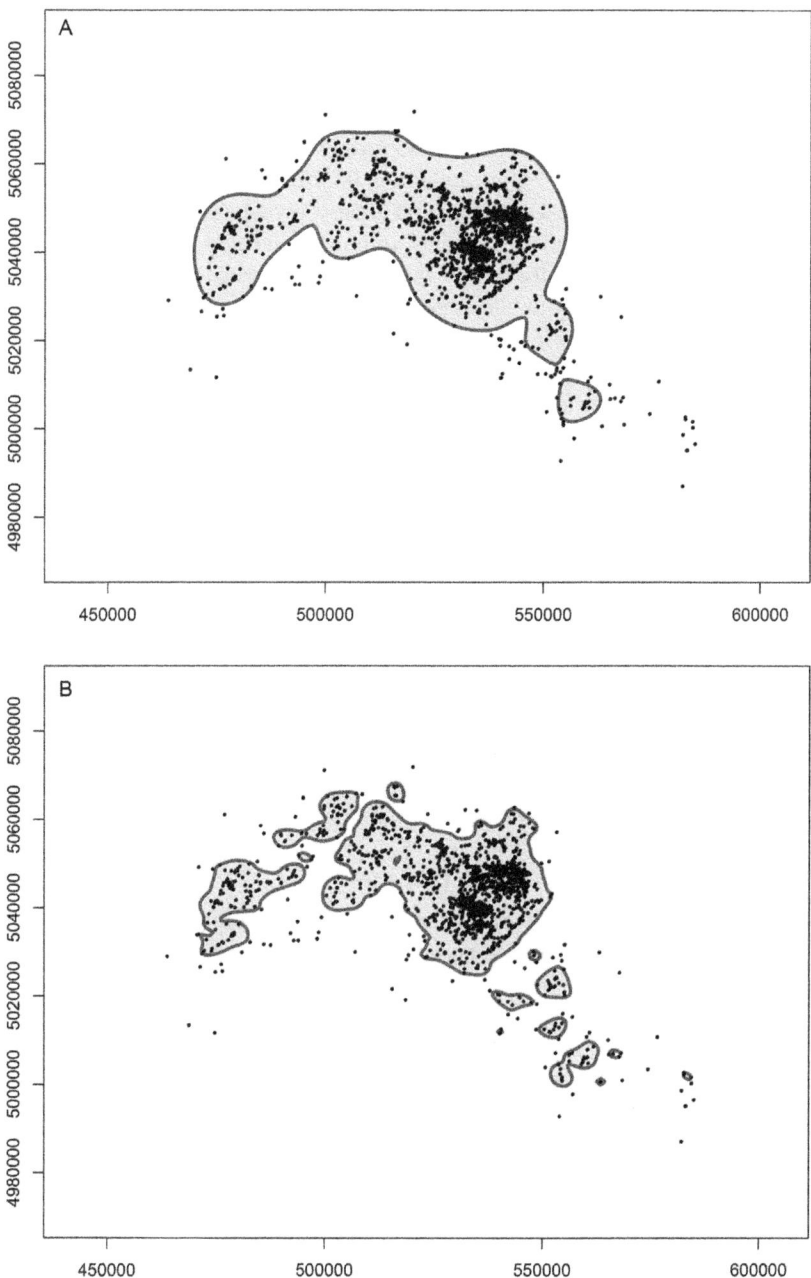

FIG. 5   KDE home ranges for Boroo location data using (A) the one-dimensional ad hoc bandwidth $h = 4163.9$, and (B) the unconstrained plug-in bandwidth matrix $\mathbf{H}$ given in (2).

from these sphered data and was then scaled back by pre- and postmultiplying it by $\mathbf{S}^{1/2}$ to finally obtain the above plug-in matrix for the original data (see Duong and Hazelton, 2003, section 4).

It must be noted at this point that the preliminary step involved in the previous hull-related methods, where points furthest away from the data centroid were trimmed to obtain a region containing 95% of the data points, might not be a good idea, since the UD estimators clearly reveal that those points may not necessarily correspond to locations within low density zones.

Since its introduction, the use of the KDE home range has become widespread and a great deal of publications make use of it, even if the density estimator given in (1) is originally thought for independent observations (see Section 2.2). To cite a few case studies using this KDE home range, Bertrand et al. (1996), Kie and Boroski (1996), Girard et al. (2002), Hemson et al. (2005), Berger and Gese (2007), Pellerin et al. (2008), Jacques et al. (2009). Seaman and Powell (1996) and Seaman et al. (1999) analyze the KDE home range performance via simulations. Matthiopoulos (2003) presents a modification of the KDE to incorporate additional information on the animal.

Another ingenious modification of the KDE of a home range, namely the permissible home range estimator (PHRE), is given by Tarjan and Tinker (2016). These authors transform the original geographical sighting coordinates into coordinates with respect to relevant landscape features influencing animal space use. A KDE is constructed on the new coordinates and, afterwards, the corresponding estimated UD is backtransformed to geographical coordinates. The PHRE makes full sense in the context considered by Tarjan and Tinker (2016), but it is not clear that, in general, relevant features for the animal can always give rise to a new coordinate system.

2.1.1.4 Harmonic mean home range

Dixon and Chapman (1980) define the areal sample moment of order -1 with respect to a point $\mathbf{x} \in \mathbb{R}^2$ as the harmonic mean of the distances of $\mathbf{x}$ to the observed locations:

$$\mathrm{HM}(\mathbf{x}) = \frac{1}{n^{-1} \sum_{i=1}^{n} \| \mathbf{x} - \mathbf{x}_i \|^{-1}},$$

where $\| \mathbf{x} - \mathbf{x}_i \|$ is the Euclidean distance between $\mathbf{x}$ and $\mathbf{x}_i$. Then, as pointed out in Worton (1989), $\mathrm{HM}(\mathbf{x})^{-1}$ can be considered as a kernel-type estimator of the UD, so that the harmonic mean home range is a level set of HM^{-1} containing 95% of the locations. In fact, Devroye and Krzyżak (1999) showed that $\mathrm{HM}(\mathbf{x})^{-1}$ is not a consistent estimator of the density, although this flaw is easily amended after a simple normalization, which consists of dividing $\mathrm{HM}(\mathbf{x})^{-1}$ by $V_d \log n$, where d is the dimension of $\mathbf{x}$ and V_d is the volume of the Euclidean unit ball in $\mathbb{R}^d$. This normalized form of HM^{-1} is called

Hilbert kernel density estimate. In any case, note that the 95% level set is unaffected by that normalization. We display in Fig. 6 the resulting home range, which obviously overestimates the real region, sharing thus the main drawback of the MCP.

2.1.2 Localized methods

The previous global methods share the common feature that they employ the whole data set (or a slightly trimmed subsample) simultaneously to construct the home range estimate, either by applying some geometric construction to the sample points (convex hull and its variants) or by previously obtaining an estimate of the utilization density from them and then computing the density level set with 95% probability content (kernel methods).

In contrast, localized methods proceed in three steps: first, a local neighborhood of points is selected according to some criterion (preclustering, nearest neighbors, points lying on a certain ball); then, one of the previous global methods is applied only to the selected points to obtain a local home range estimate, and, finally, many of these local home ranges are gathered together by taking their union. This way, many variants can be considered, depending on two choices: the way in which local neighborhoods are constructed and the global method applied to each of these neighborhoods. Next we describe the most popular ones.

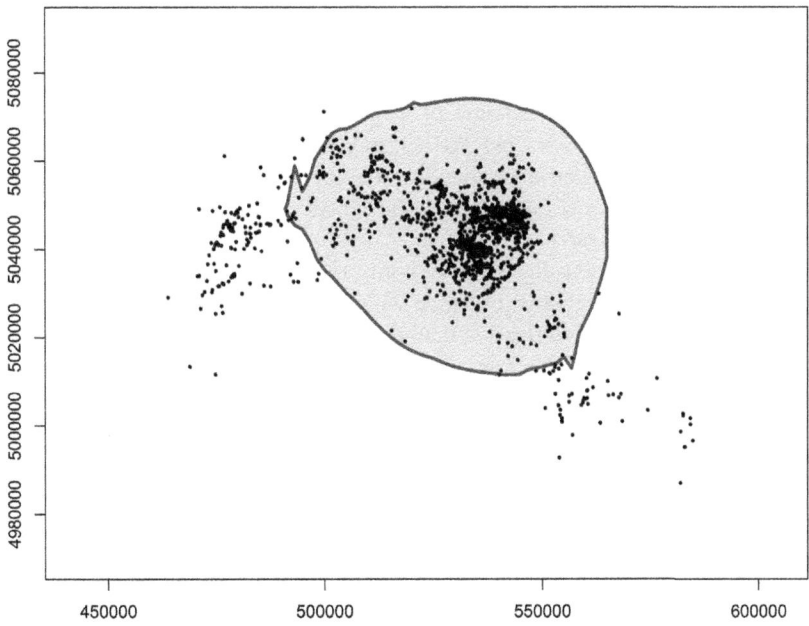

FIG. 6 Harmonic mean home range for Boroo location data.

2.1.2.1   Local convex hull (LoCoH) or k-nearest neighbor convex hulls (k-NNCH)

This is a localized version of the MCP. For a fixed integer $k > 0$, Getz and Wilmers (2004) construct the convex hull, k-NNCH$_i$, of each sample point x_i and its $k - 1$ nearest neighbors (NN) (with respect to the Euclidean metric, although other metrics could be employed). Then these hulls are ordered according to their area, from smallest to largest. The LoCoH home range estimate is the isopleth that results of progressively taking the union of the hulls from the smallest upwards, until a specific percentage (e.g., 95%) of sample points is included. Getz et al. (2007) extend the original LoCoH procedure (called k-LoCoH) to the r-LoCoH (where a fixed sphere of influence was used instead of the k nearest neighbors) and to the a-LoCoH (the a standing for adaptive sphere of influence). A detailed study of the theoretical properties and finite sample performance of the r-LoCoH can be found in Aaron and Bodart (2016).

The optimal choice of the number of neighbors, k, depends on the topological features of the home range. In particular, one possibility is to choose the minimal k resulting in a prefixed genus (number of holes), as long as this information is known. Alternatively, Getz and Wilmers (2004) suggest guessing the genus after inspection of the physical characteristics of the territory. Another possibility is to examine the areas of the final isopleths as a function of the values of k (see the documentation of the R package tlocoh, Lyons, 2019; Lyons et al., 2013).

The idea of incorporating topological information on the estimator has been little explored in the home range estimation literature. However, it represents a very active area in statistical research recently, encompassing a variety of methodologies under the name of Topological Data Analysis, which are nicely reviewed in Wasserman (2018) and implemented in the R package TDA (Fasy et al., 2019). Undoubtedly, the application of these new techniques to the field of home range estimation shows promise as a very interesting venue for further research.

In Fig. 7 we display the LoCoH home range, based on Boroo data, for $k = 10$ and $k = 50$ neighbors, obtained with the tlocoh package. It is clear that this procedure is far more flexible than the convex hull, although it retains the simplicity of the latter. We can also see the effect of selecting too small a value of k, which produces an estimator full of holes, slivers and thin passages. In contrast, a larger value of k obviously increases the similarity of the k-LoCoH to the convex hull.

2.1.2.2   Characteristic hull polygon (CHP)

As the LoCoH, the characteristic hull polygon proposed by Downs and Horner (2009) is a union of convex sets and a generalization of the MCP. First, the Delaunay triangulation of the sample of locations is constructed (see Fig. 8).

FIG. 7 LoCoH isopleths with probability content 95% based on Boroo locations for (A) $k = 10$ and (B) $k = 50$ neighbors.

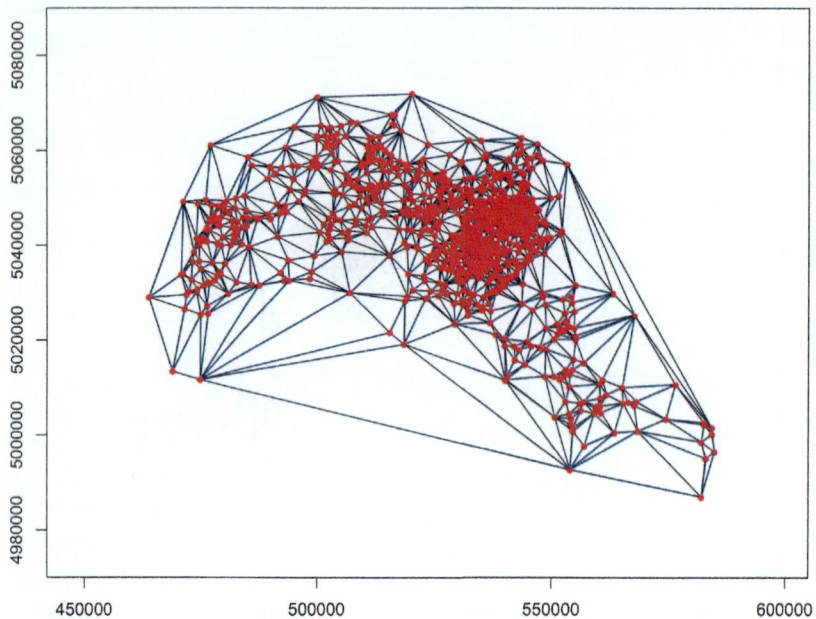

FIG. 8 Delaunay triangulation of Boroo location data, with no triangle removed.

Then, the largest 5% of the triangles are removed, where the size of the triangle is measured by its perimeter. As Downs and Horner (2009) point out, it would be interesting that the proportion of triangles to remove should somehow depend on the actual shape of the real home range (for instance, this proportion should be 0 for convex domains).

Any method for constructing a characteristic hull of a set of points consists in removing "large triangles." For example, the χ (chi) algorithm in Duckham et al. (2008) iteratively removes the longest exterior edge from the triangulation, such that the edge to be removed is longer than a prefixed parameter and the exterior edges of the resulting triangulation form the boundary of a simple polygon.

The package adehabitatHR is the only one implementing the CHP in R, but it measures the size of the triangles by their area. As a home range, the CHP is unsatisfactory, since it may contain slivers corresponding to seldom visited locations. In particular, in the case of Boroo locations, its CHP is depicted in Fig. 9.

2.1.2.3 Single-linkage cluster

Kenward et al. (2001) proposed yet another localized variant of the MCP. The difference with the LoCoH lies in the way that local neighborhoods are constructed. Their proposal starts from the nearest-neighbor distances between locations. These distances are clustered using single-linkage (see Everitt, 2011), aiming to minimize the mean joining distance inside each

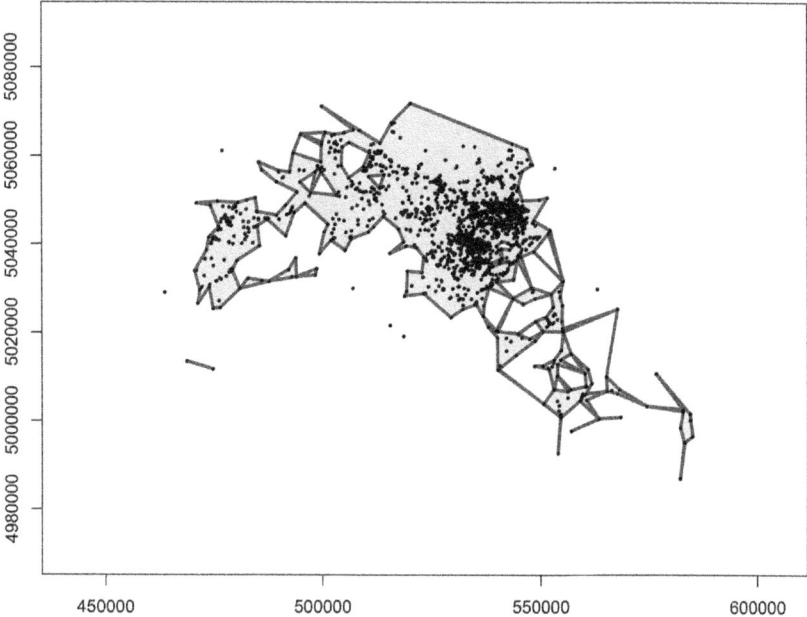

FIG. 9 Characteristic hull polygon for Boroo location data, constructed by removing the triangles corresponding to the 5% largest perimeters from the Delaunay triangulation.

group and imposing a certain minimum cluster size. Once the locations are grouped in clusters, the convex hull of each of them is computed and the final home range is defined as the union of these polygons.

The single-linkage cluster home range of Boroo locations, produced using the package adehabitatHR, appears in Fig. 10. The method identified a large cluster of locations, four medium-size clusters and a myriad of small or tiny clusters. The final home range estimate is obtained by gathering together the convex hulls of each of the clusters. This home range is not realistic or practical due, first, to the presence of a large amount of small, negligible hulls and, second, to using the little-adaptable convex hull to construct each connected component. In Baíllo and Chacón (2018) we can see that the single-linkage cluster of Zimzik locations had exactly the same drawbacks.

2.2 Incorporating time dependency

As noted in Section 1.3, the advent of satellite technologies to the animal tracking context has posed new interesting challenges. The high frequency in the signal transmission originates stochastic dependence among locations which are close in time. This dependence can somehow be incorporated into the home range estimator and is also a source of information on the behavioral mechanisms of the animal. Noonan et al. (2019) study the effect of autocorrelation on several home range estimates.

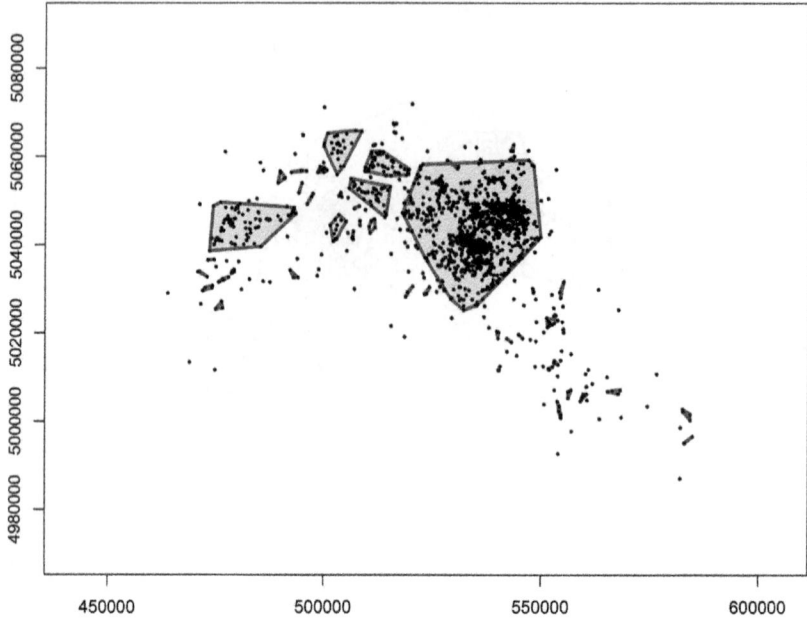

FIG. 10   Single-linkage cluster home range for Boroo location data, built by gathering together the convex hulls of each of the clusters obtained after a preliminary single-linkage clustering step.

In this section we review some proposals that incorporate the time dependency into the home range estimator. The enlarged sample including the observation times is denoted by $(t_1, \mathbf{x}_1), \ldots, (t_n, \mathbf{x}_n)$, with $t_1 < \cdots < t_n$. We note that for most studies the observation times are fixed by the researcher, so they should not be treated as random.

We want to remind that static home range estimators are usually defined as a high probability contour of the utilization distribution. Most dynamic home range estimation methodologies seem to be also oriented toward this natural population goal, but some caution is required regarding recently proposed dynamic techniques, since some of them seem to lose track of the target that they intend to estimate.

2.2.1   Global methods

2.2.1.1   Time geography density estimation (TGDE)

Time geography is a perspective of social geography introduced by Hägerstrand (1970). It aims to analyze human phenomena (transportation, socio-economic systems, etc.) in a geographic space taking into account the restrictions and trade-offs in human spatial behavior (the paths followed by individuals) imposed by different types of constraints (capability, coupling or authority); see Miller (2005).

An important concept in time geography is the space-time prism, which bounds all possible space-time paths of an individual between two observed locations $\mathbf{x}_i$ and $\mathbf{x}_{i+1}$ at consecutive time points, taking into account the accessibility constraints of the individual (see Fig. 11). The *geo-ellipse* g_i or *potential path area* (PPA) of these two consecutive locations delineates in the geographic space all points that the individual can potentially reach during the time interval $[t_i, t_{i+1}]$:

$$g_i = \{\mathbf{x} : \| \mathbf{x} - \mathbf{x}_i \| + \| \mathbf{x} - \mathbf{x}_{i+1} \| \le (t_{i+1} - t_i)v_i\},$$

where v_i denotes the maximum velocity in the time interval.

Downs (2010) and Downs et al. (2011) propose to integrate KDE and time geography methodologies by using the geo-ellipses derived from consecutive location pairs as surrogates for the KDE level sets. Downs (2010) claims that a drawback of KDE home range estimation is that it includes areas where the

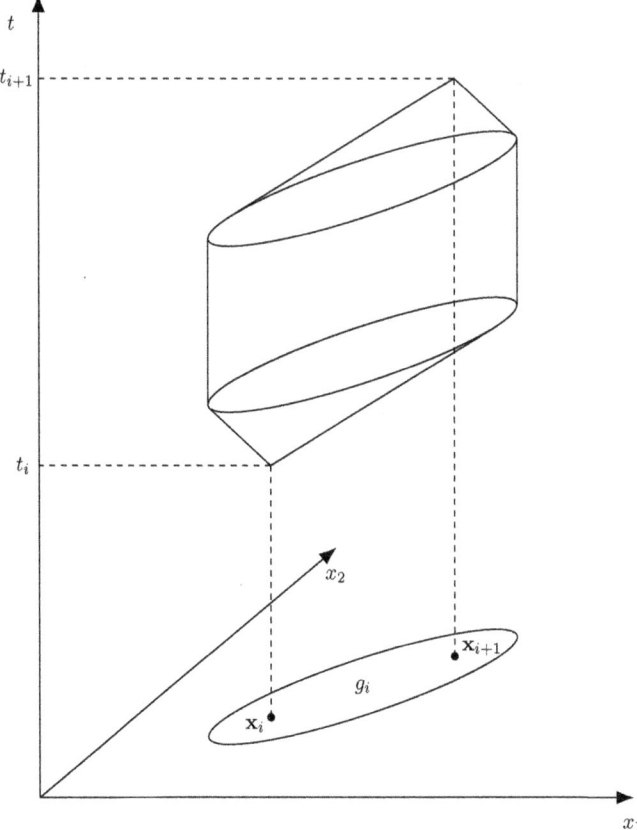

FIG. 11　Space-time prism and geo-ellipse g_i corresponding to consecutive locations $\mathbf{x}_i$ and $\mathbf{x}_{i+1}$, observed at times t_i and t_{i+1}, respectively.

individual could not have been located, given the spatial and temporal constraints established by the observed locations and the maximum potential velocity. As an alternative, if the maximum interval velocities v_i are all assumed to be equal to some global maximum velocity v, then the time-geographic density estimate at a point $\mathbf{x}$ is defined as

$$\hat{f}_{\text{TG}}(\mathbf{x}) = \frac{1}{(n-1)[(t_n - t_1)v]^2} \sum_{i=1}^{n-1} G\left(\frac{\|\mathbf{x} - \mathbf{x}_i\| + \|\mathbf{x} - \mathbf{x}_{i+1}\|}{(t_{i+1} - t_i)v}\right), \qquad (3)$$

where G is a decreasing function, defined on $[0, \infty)$, and playing the role of the kernel. Given that G is maximal at zero, the individual summands of the estimator (3) assign the highest probability to the points along the straight path between the two consecutive observations $\mathbf{x}_i$ and $\mathbf{x}_{i+1}$ and spread out the remaining probability mass in ellipsoidal contours having the observed locations as their foci. The velocity v plays the role of the smoothing parameter and, as such, it is reasonable to derive its value both from biological knowledge about the animal or the species and from the animal locations and their corresponding observation times (see Long and Nelson, 2012). The TGDE home range is the 95% level set, $\{\hat{f}_{\text{TG}} \geq \hat{c}_{0.05}\}$, of the density estimate given in (3), where

$$0.95 = \int_{\{\hat{f}_{\text{TG}} \geq \hat{c}_{0.05}\}} \hat{f}_{\text{TG}}(\mathbf{x}) \, d\mathbf{x}.$$

In parallel, Long and Nelson (2012) defined the PPA home range as the union of the $n - 1$ geo-ellipses obtained from all the pairs of consecutive locations in the sample:

$$\text{PPA}_{\text{HR}} = \bigcup_{i=1}^{n-1} g_i, \qquad (4)$$

with $v_i = v$, for all $i = 1, \dots, n - 1$. The estimator (4) is a particular case of the TGDE home range $\{\hat{f}_{\text{TG}} \geq \hat{c}_{0.05}\}$ with a uniform kernel G. Regarding the choice of the parameter v, once the segment velocities $v_{i,i+1} = \|x_{i+1} - x_i\|/(t_{i+1} - t_i)$, $i = 1, \dots, n - 1$, are computed, Long and Nelson (2012) suggest several estimates $\hat{v}$ based on the order statistics of $v_{i,i+1}$, $i = 1, \dots, n - 1$, and following the works of Robson and Whitlock (1964) and van der Watt (1980). More generally, Long and Nelson (2015) introduce the dynamic PPA home range (dynPPA), an adaptation of the PPA estimator to possibly different motion states. It is obtained by allowing the mobility parameter v to change over time. In practice, these authors divide the trajectory of the animal into dynamic phases and obtain an estimator of v in each of them.

The PPA and the TGDE home ranges are implemented in the R package wildlifeTG (Long, 2020). In Fig. 12 we show both home range estimators

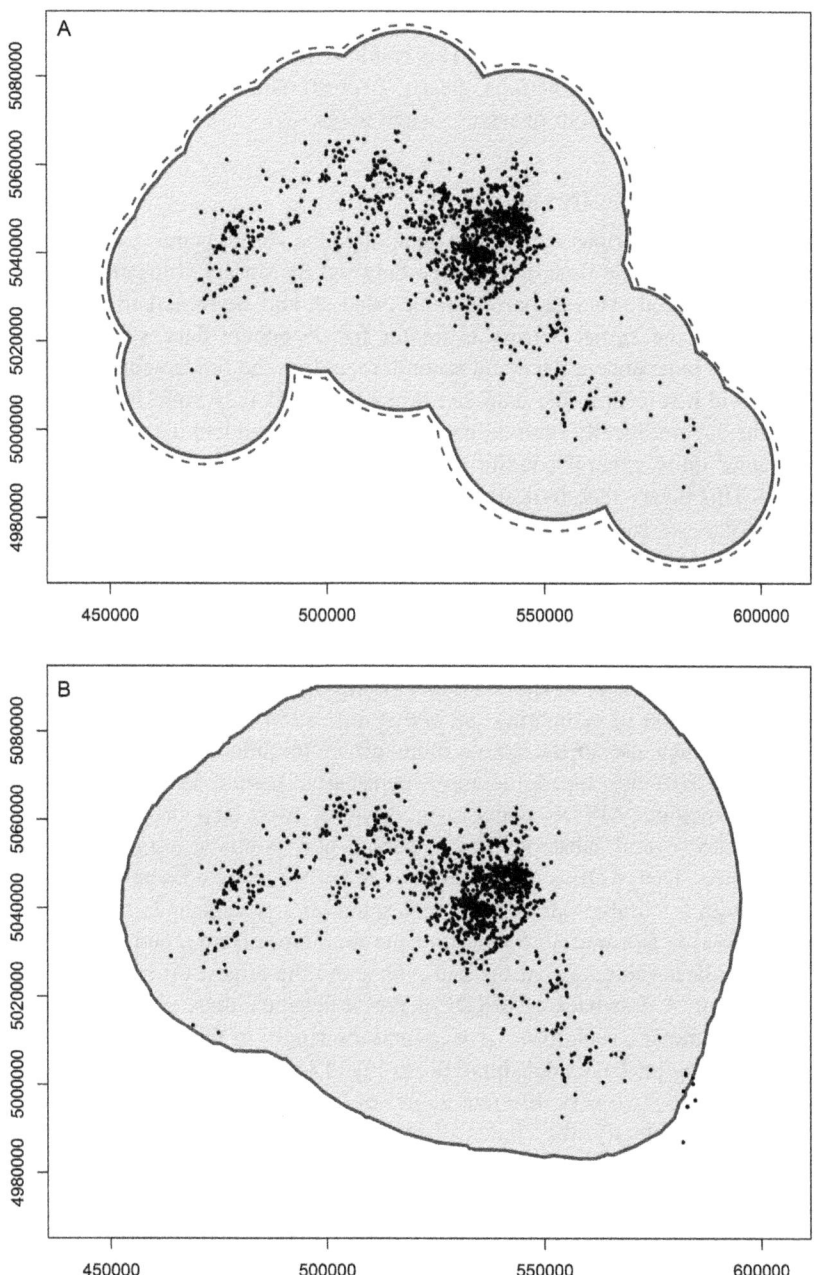

FIG. 12   Home ranges for Boroo location data using (A) the potential path area, or PPA (in *gray* for a estimated velocity parameter $\hat{v}$ chosen as suggested in Robson and Whitlock (1964), and with a *dashed boundary* for a velocity parameter estimated as suggested in van der Watt, 1980) and (B) TGDE with a Gaussian kernel.

for Boroo locations, where the function G is Gaussian-like (i.e., proportional to $e^{-x^2/2}$) for the TGDE in (B). The result is not satisfactory, as the PPA and the TGDE approximations clearly overestimate the population home range and include seldom or never visited areas.

2.2.1.2 Kernel density estimation

There have been various attempts to generalize the kernel home range estimator to incorporate the time dependence between the observed locations. Nevertheless, there are two important issues that should be remarked: first, the definition of the kernel density estimator for dependent data is exactly the same as for independent data, and second, regarding the fundamental problem of bandwidth selection, the data can be treated as if they were independent, since the asymptotically optimal bandwidth for independent data is also optimal under quite general conditions of dependence, as shown in Hall et al. (1995). This means that, to design methods to estimate the utilization distribution density, we can proceed exactly the same as for independent data.

Keating and Cherry (2009) suggested a product kernel density estimator where time was incorporated as an extra variable to the two-dimensional location vector, thus yielding three-dimensional observations. This approach does not seem appropriate, since time is not a random variable whose frequency we want to analyze, as we noted at the beginning of Section 2.2.

In the context of estimating the *active* utilization distribution (describing space frequency use in the active moments of the animal), Benhamou and Cornelis (2010) developed the movement-based kernel density estimation (MKDE) method. MKDE consists in dividing each step or time interval $[t_i, t_{i+1}]$ into several substeps, that is, adding new points at regular intervals on each step. Then KDE is carried out on the known and the interpolated relocations with a variable one-dimensional smoothing parameter $h_i(t)$. For each time interval h_i is a smooth function of the time lapse from t_i and to t_{i+1}, taking its smallest value h_{min} at the end points and the largest (at most h_{max}) at the midpoint. A drawback of MKDE is that it depends, thus, on the choice of several parameters, such as h_{min}, h_{max} and the length of the subintervals. For instance, using package adehabitatHR, in Fig. 13 we have plotted the MKDE home ranges for two very different values of h_{min} but equal values of the rest of parameters: clearly, the choice of this smoothing parameter can substantially alter the resulting home range. Optimal simultaneous selection of all the parameters of MKDE with respect to some criterion seems computationally unfeasible even for moderate sample sizes. Mitchell et al. (2019) analyze the influence of fix rate and tracking duration on the home ranges obtained with MKDE and KDE, thus providing also a comparison between the performance of the two methods on a specific set of locations.

There have been extensions to the original MKDE proposal. For instance, to incorporate knowledge of boundaries that the animal does not go through

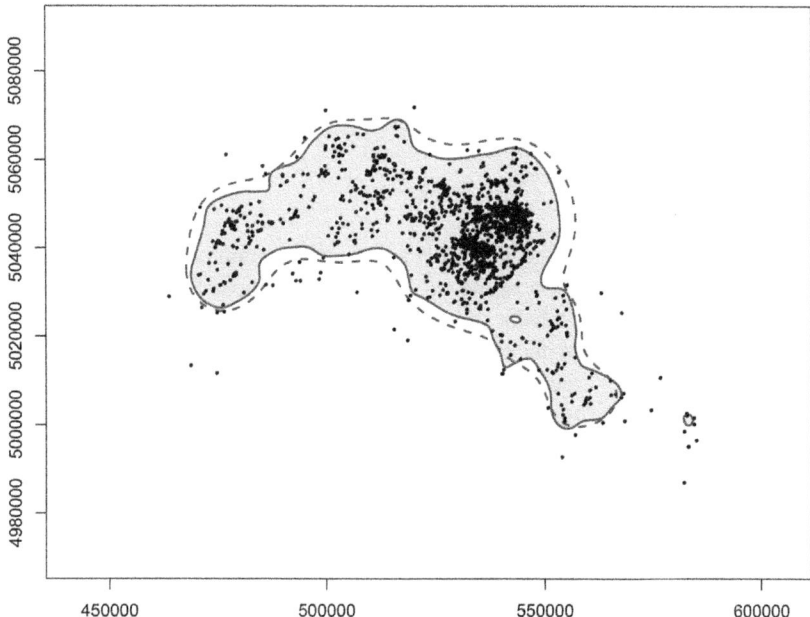

FIG. 13 MKDE home range with $h_{min} = 1$ (*solid line*) and $h_{min} = 4000$ (*dashed line*).

(rivers, cliffs, etc.), Benhamou and Cornelis (2010) suggest to reset to 0 beyond the boundary and reflect with respect to the boundary the resulting estimate of the utilization density. Also, Tracey et al. (2014) use the MKDE on 3-dimensional location data.

Steiniger and Hunter (2013) propose the *scaled line kernel home range* estimator (SLKDE), which is similar to the MKDE of Benhamou and Cornelis (2010). Each line segment connecting two consecutive relocations, x_i and x_{i+1}, is divided into subsegments and a KDE is constructed on the line segment based on observed and interpolated relocations. This estimator is called the *raster* $r_{i, i+1}$. Then all the rasters are pooled to construct an unnormalized estimator of the UD (it does not integrate to 1). The main difference with MKDE is that the smoothing parameter of the SLKDE is selected in such a way that it "inflates" the home range estimator on the observed relocations and thins it in the lines connecting them: the rasters have what Steiniger and Hunter (2013) call a "bone-like" shape. This procedure is not implemented in R.

One further generalization of the KDE home range to the time-dependent context is given by Fleming et al. (2015), who propose the *autocorrelated kernel density estimator* (AKDE). As usual, the probability density f of the UD is estimated by a bivariate KDE $\hat{f}$ with a Gaussian kernel K and an unconstrained smoothing parameter **H**. As in the ad hoc bandwidth proposal of Worton (1989), the AKDE uses a bandwidth which is derived from the assumption of

Gaussianity of the underlying distribution (which naturally leads to over-smoothing in most practical scenarios), with the only difference that instead of assuming that the location data are independent, they are supposed to be observations from a Gaussian stochastic process.

2.2.2 Localized methods

2.2.2.1 T-LoCoH

Lyons et al. (2013) define the T-LoCoH as a generalization of the LoCoH home range that incorporates the observation times of the relocations. T-LoCoH uses the time associated to each location in two phases of the LoCoH algorithm: nearest neighbor selection and the ranking of hulls. First, the NN selection relies on the so-called time-scaled distance (TSD), which transforms the time coordinate into a third one of Euclidean space $\mathbb{R}^3$. Specifically, the TSD between two sample points, $(t_i, \mathbf{x}_i)$ and $(t_j, \mathbf{x}_j)$, is defined as

$$\mathrm{TSD}_{ij}^2 = \| \mathbf{x}_i - \mathbf{x}_j \|^2 + s^2 v_{\max}^2 (t_i - t_j)^2, \tag{5}$$

where $s \geq 0$ is a scaling factor of the maximum theoretical velocity $v_{\max}$. It is also possible to define an alternative equation for the time-scaled distance based on a Gaussian diffusion model (as if the animal had moved during the time interval in a random walk):

$$\mathrm{TSD}_{ij}^2 = \| \mathbf{x}_i - \mathbf{x}_j \|^2 + c(t_i - t_j), \tag{6}$$

where c is a parameter. Lyons et al. (2013) point out that using (5) or (6) does not make a great difference in the resulting nearest neighbor ranking. Finally, to construct the isopleths, local hulls are sorted according to a hull metric, chosen to reflect the spatial or time information we might want to use (see Table 1 in Lyons et al., 2013 for a list of many possible hull metrics). Then, as in LoCoH, the T-LoCoH is the union of the smallest hulls that covers 95% of the sample points.

Following the indications in Lyons et al. (2013), with the aid of the graphical procedures available in the package tlocoh, we have chosen the values of $s = 0.05$ and $k = 50$ as T-LoCoH parameters for Boroo locations. Alternatively, Dougherty et al. (2017) suggest a cross-validation procedure to select the tuning parameters k and s for the T-LoCoH. The value of $v_{\max}$ was internally chosen by R. Fig. 14 displays the resulting T-LoCoH home range.

3 Developing trends, open problems, and future work

The aim of this final section is to mention some possible paths for future research and sum up active areas of interest in home range estimation or their corresponding counterparts in set estimation.

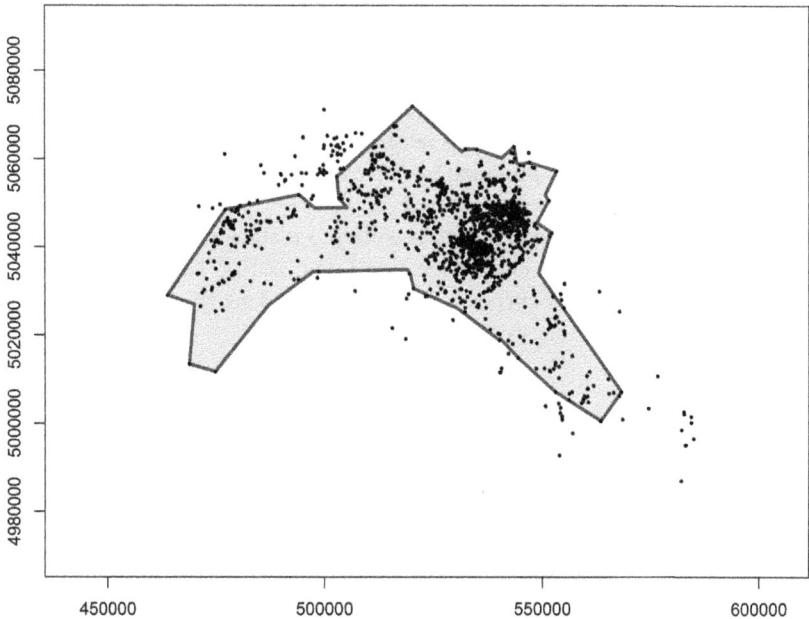

FIG. 14 T-LoCoH isopleth with probability content 95% for Boroo location data, computed with parameters $s = 0.05$ and $k = 50$.

3.1 Incorporating time dependence into the home range estimator

As Noonan et al. (2019) rightly pointed out, failure to incorporate autocorrelation into the home range estimator can produce biased home range estimates. We believe that there is still room to propose further nonparametric statistical procedures to approximate the home range that do not ignore the time dependence or, specifically, the fact that the observed locations are realizations of a two- or three-dimensional continuous stochastic process. Mechanistic home range analysis has the drawback of having to specify a model which might not be flexible enough to adapt to the real, observed locations of the individual animal. Thus, it is not clear that the home range derived from the model is a good approximation to that of the individual of interest. Procedures such as the Brownian bridge movement model and its variants (see, e.g., Horne et al., 2007; Kranstauber et al., 2012) are actually conditional on the observed sample and their aim is to give a sort of confidence region for the complete, continuous realization of the path that the animal followed. Consequently, the resulting "home range" derived from these techniques is likely to have a negative bias.

3.2 Taking advantage of explanatory variables

In Section 1.4 (see also Baíllo and Chacón, 2018, 2020) we pointed out that the tracked locations of the Mongolian wolves and most other animals are likely to depend on the values of certain explanatory variables (environmental covariates) at those locations, in the sense that presence of an animal is more probable in sites with resources and without threats. Indeed, animal space use (in particular, the home range) is of interest for various reasons, among others, the need to preserve critical habitats necessary for species conservation. Thus, the additional information provided by these covariates could be useful to better describe the space use of an animal.

However, there does not exist yet a unified, satisfactory approach to make use of this complementary knowledge (see Long, 2019). The synoptic model of Horne et al. (2008) expresses the utilization density as a function of k environmental predictors as follows:

$$f(x) = \frac{f_0(x) \prod_{i=1}^{k}(1 + \beta_i H_i(x))}{\int f_0(x) \prod_{i=1}^{k}(1 + \beta_i H_i(x))dx},$$

where f_0 is a density representing the null model of space use in the absence of any effect from environmental covariates, H_i is a function representing the effect of the i-th predictor and the value of β_i serves to give more or less weight to that covariate. For instance, if the i-th covariate is binary, then H_i takes the value 0 or 1 depending on the value of the covariate at the location x. Slaught et al. (2013) also use the synoptic model to account for constraints imposed by linear features in the environment (such as rivers and shorelines).

The randomized shortest paths proposed by Long (2019) incorporate hard and soft barriers into the space use measures. This technique establishes a raster or grid on the landscape which acts as a weighted directional network: it is, thus, a discretized model. Environmental factors that limit or promote animal movement are integrated in the model via an affinity matrix associated to the raster and describing the ability to move from one cell to the other. A randomized shortest path is then a cell-by-cell movement along the grid subject to the values defined by the affinity matrix.

Finally, Guo et al. (2019) propose to substitute, in the mathematical expression of certain home ranges (such as the synoptic model of Horne et al., 2008), the classical Euclidean distance by a cost distance that takes into account obstacles and terrain characteristics.

3.3 Selecting the "optimal" home range

Based on a single sample of positional data, it is possible to construct several of the home range estimates revised in Section 2. The problem is then how to select the most adequate or realistic one. Usually the ecologist (or, in general, an expert) chooses a reasonable result according to previous information on the animal or its species. The possibility of making an automatized home range selection, based on objective statistical criteria, remains an open question (see Cumming and Cornélis, 2012).

It is true, however, that there have been some proposals to compare home ranges using different procedures. In this work we mention the ideas valid for real samples (methods designed only for simulated data are not included due to their limited interest in practice). One possibility, especially if locations are treated as independent observations, is to separate the sample into two subsets: a training one to construct the home range estimator and a test sample to check its predictive accuracy. This idea has been employed, for example, in Kranstauber et al. (2012) and in Tarjan and Tinker (2016). Similarly, Fleming et al. (2015) compare the areas of the home range estimates obtained with the complete sample and with the first half of the data (chronologically speaking).

Cumming and Cornélis (2012) and Walter et al. (2015) compare home range estimates using the area-under-the-curve (AUC) corresponding to receiver operating characteristic (ROC) curves. The ROC curve (one for each HR) crucially depends on the choice of a certain raster containing all the home ranges. The curve is computed on the basis of the probability assigned by the estimated UD (derived from the home range) to each cell in the raster and of the label indicating whether there is any sighting in the cell or not. As a concern, observe that ROC curves are graphical devices designed for a binary classification problem, although in the home range setting the location data are not naturally separated into two classes: there is only one class of locations corresponding to where the individual was observed. Recently, Baíllo and Chacón (2020) proposed the penalized overestimation ratio, a quantitative indicator to grade a collection of home ranges constructed using the same sample of locations.

It also seems reasonable to quantify the adequacy of the constructed home range by connecting it with the characteristics of the terrain comprised in the set perimeter. As we already mentioned, the location of the observed animal will depend on geographical features, food, water, or shelter availability, predators presence or absence, etc. To mention some proposals along this line, Long and Nelson (2015) compare different home range estimates by computing their areas and their habitat composition (land cover). Kenward et al. (2001) study the relationship between the logarithm of the home range area and, for instance, environmental factors known to influence the animal behavior (such as food availability or population density). Apart from the area, Steiniger and Hunter (2013) use shape complexity as given, for example, by holes and patches and the presence or not of corridors in the home range. Nevertheless, a single quantitative optimality measure of the home range based on the many possible available predictors relating the animal and the estimated set is still lacking.

Acknowledgments

The authors are grateful to Antonio Cuevas for drawing their attention to the home range estimation problem and for some insightful discussions.

References

Aaron, C., Bodart, O., 2016. Local convex hull support and boundary estimation. J. Multivar. Anal. 147, 82–101.

Baíllo, A., Chacón, J.E., 2018. A survey and a new selection criterion for statistical home range estimation. ArXiv:1804.05129.

Baíllo, A., Chacón, J.E., 2020. A new selection criterion for statistical home range estimation. J. Appl. Stat. https:/doi.org/10.1080/02664763.2020.1822302.

Bath, S.K., Hayter, A.J., Cairns, D.A., Anderson, C., 2006. Characterization of home range using point peeling algorithms. J. Wildl. Manage. 70, 422–434.

Bauder, J.M., Breininger, D.R., Bolt, M.R., Legare, M.L., Jenkins, C.L., McGarigal, K., 2015. The role of the bandwidth matrix in influencing kernel home range estimates for snakes using VHF telemetry data. Wildl. Res. 42, 437–453.

Benhamou, S., Cornelis, D., 2010. Incorporating movement behavior and barriers to improve biological relevance of kernel home range space use estimates. J. Wildl. Manage. 74, 1353–1360.

Berger, K.M., Gese, E.M., 2007. Does interference competition with wolves limit the distribution and abundance of coyotes? J. Anim. Ecol. 76, 1075–1085.

Bertrand, M.R., DeNicola, A.J., Beissinger, S.R., Swihart, R.K., 1996. Effects of parturition on home ranges and social affiliations of female white-tailed deer. J. Wildl. Manage. 60, 899–909.

Biau, G., Cadre, B., Mason, D.M., Pelletier, B., 2009. Asymptotic normality in density support estimation. Electron. J. Probab. 14, 2617–2635.

Blair, W.F., 1940. Notes on home ranges and populations of the short-tailed shrew. Ecology 21, 284–288.

Brunel, V.E., 2018. Methods for estimation of convex sets. Stat. Sci. 33, 615–632.

Burgman, M.A., Fox, J.C., 2003. Bias in species range estimates from minimum convex polygons: implications for conservation and options for improved planning. Anim. Conserv. 6, 19–28.

Burt, W.H., 1943. Territoriality and home range concepts as applied to mammals. J. Mammal. 24, 346–352.

Cadre, B., Pelletier, B., Pudlo, P., 2013. Estimation of density level sets with a given probability content. J. Nonparametr. Stat. 25, 261–272.

Calenge, C., 2006. The package "adehabitat" for the R software: a tool for the analysis of space and habitat use by animals. Ecol. Model. 197, 516–519.

Calhoun, J.B., Casby, J.U., 1958. Calculation of home range and density of small mammals. Public Health Monogr. 55, 1–24.

Carey, A.B., Reid, J.A., Horton, S.P., 1990. Spotted owl home range and habitat use in Southern Oregon coast ranges. J. Wildl. Manage. 54, 11–17.

Chacón, J.E., Duong, T., 2010. Multivariate plug-in bandwidth selection with unconstrained pilot bandwidth matrices. Test 19, 375–398.

Chacón, J.E., Duong, T., 2018. Multivariate Kernel Smoothing and Its Applications. Chapman & Hall.

Chen, Y.-C., Genovese, C.R., Wasserman, L., 2017. Density level sets: asymptotics, inference, and visualization. J. Am. Stat. Assoc. 112, 1684–1696.

Cholaquidis, A., Cuevas, A., Fraiman, R., 2014. On poincaré cone property. Ann. Stat. 42, 255–284.

Cholaquidis, A., Fraiman, R., Lugosi, G., Pateiro-López, B., 2016. Set estimation from reflected Brownian motion. J. R. Statist. Soc. B 78, 1057–1078.

Cooper, N.W., Sherry, T.W., Marra, P.P., 2014. Modeling three-dimensional space use and overlap in birds. Auk 131, 681–693.

Cuevas, A., 2009. Set estimation, another bridge between statistics and geometry. BEIO 25, 71–85.

Cuevas, A., Fraiman, R., 2010. Set estimation. In: Kendall, W.S., Molchanov, I. (Eds.), New Perspectives on Stochastic Geometry. Oxford University Press, pp. 374–397.

Cumming, G.S., Cornélis, D., 2012. Quantitative comparison and selection of home range metrics for telemetry data. Divers. Distrib. 18, 1057–1065.

de Vito, E., Rosasco, L., Toigo, A., 2010. Spectral regularization for support estimation. In: Lafferty, J.D., Williams, C.K.I., Shawe-Taylor, J., Zemel, Culotta, A. (Eds.), NIPS'10: Proceedings of the 23rd International Conference on Neural Information Processing Systems–Volume 1, pp. 487–495.

Devroye, L., KrzyŻak, A., 1999. On the hilbert kernel density estimate. Stat. Probab. Lett. 44, 209–308.

Devroye, L., Wise, G.L., 1980. Detection of abnormal behavior via nonparametric estimation of the support. SIAM J. Appl. Math. 38, 480–488.

Dixon, K.R., Chapman, J.A., 1980. Harmonic mean measure of animal activity areas. Ecology 61, 1040–1044.

Don, B.A.C., Rennolls, K., 1983. A home range model incorporating biological attraction points. J. Anim. Ecol. 52, 69–81.

Doss, C.R., Weng, G., 2018. Bandwidth selection for kernel density estimators of multivariate level sets and highest density regions. Electron. J. Stat. 12, 4313–4376.

Dougherty, E.R., Carlson, C.J., Blackburn, J.K., Getz, W.M., 2017. A cross-validation-based approach for delimiting reliable home range estimates. Mov. Ecol. 5, 19.

Downs, J.A., 2010. Time-geographic density estimation for moving point objects. In: Fabrikant, S.I., Reichenbacher, T., VanKreveld, M., Schlieder, C. (Eds.), Geographic Information Science. Lecture Notes in Computer Science, vol. 6292, pp. 16–26.

Downs, J.A., Horner, M.W., 2009. A characteristic-hull based method for home range estimation. Trans. GIS 13, 527–537.

Downs, J.A., Horner, M.W., Tucker, A.D., 2011. Time-geographic density estimation for home range analysis. Ann. GIS 17, 163–171.

Duckham, M., Kulik, L., Worboys, M.F., Galton, A., 2008. Efficient generation of simple polygons for characterizing the shape of a set of points in the plane. Pattern Recogn. 41, 3224–3236.

Dümbgen, L., Walther, G., 1996. Rates of convergence for random approximations of convex sets. Adv. Appl. Probab. 28, 384–393.

Duong, T., 2007. ks: kernel density estimation and kernel discriminant analysis for multivariate data in R. J. Stat. Softw. 21, 1–16.

Duong, T., 2019. ks: Kernel smoothing. R package version 1.11.6. https:/CRAN.R-project.org/package=ks.

Duong, T., Hazelton, M.L., 2003. Plug-in bandwidth matrices for bivariate kernel density estimation. J. Nonparametr. Stat. 15, 17–30.

Edelsbrunner, H., Kirkpatrick, D.G., Seidel, R., 1983. On the shape of a set of points in the plane. IEEE Trans. Inf. Theory 29, 551–559.

Everitt, B., Landau, S., Leese, M., Stahl, D., 2011. Cluster Analysis, fifth ed. Wiley.

Fasy, B.T., Kim, J., Lecci, F., Maria, C., Millman, D.L., Rouvreau, V., 2019. TDA: statistical tools for topological data analysis. The included GUDHI is authored by Clement Maria, Dionysus by Dmitriy Morozov, PHAT by Ulrich Bauer, Michael Kerber and Jan Reininghaus. R package version 1.6.5. https:/CRAN.R-project.org/package=TDA.

Fleming, C.H., Fagan, W.F., Mueller, T., Olson, K.A., Leimgruber, P., Calabrese, J.M., 2015. Rigorous home range estimation with movement data: a new autocorrelated kernel density estimator. Ecology 96, 1182–1188.

Foley, A.M., Schroeder, B.A., Hardy, R., MacPherson, S.L., Nicholas, M., 2014. Long-term behavior at foraging sites of adult female loggerhead sea turtles (Caretta caretta) from three Florida rookeries. Marine Biol. 161, 1251–1262.

Getz, W.M., Wilmers, C.C., 2004. A local nearest-neighbor convex-hull construction of home ranges and utilization distributions. Ecography 27, 489–505.

Getz, W.M., Fortmann-Roe, S., Cross, P.C., Lyons, A.J., Ryan, S.J., Wilmers, C.C., 2007. Locoh: nonparametric kernel methods for constructing home ranges and utilization distributions. PLoS ONE 2.

Girard, I., Ouellet, J.P., Courtois, R., Dussault, C., Breton, L., 2002. Effects of sampling effort based on GPS telemetry on home-range size estimations. J. Wildl. Manage. 66, 1290–1300.

Guo, J., Du, S., Ma, Z., Huo, H., Peng, G., 2019. A model for animal home range estimation based on the active learning method. ISPRS Int. J. Geo Inf. 8, 490. https://doi.org/10.3390/ijgi8110490.

Hägerstrand, T., 1970. What about people in regional science? Pap. Reg. Sci. Assoc. 24, 6–21.

Hall, P., Lahiri, S.N., Truong, Y.K., 1995. On bandwidth choice for density estimation with dependent data. Ann. Stat. 23, 2241–2263.

Harris, S., Cresswell, W.J., Forde, P.G., Trewhella, W.J., Woollard, T., Wray, S., 1990. Home-range analysis using radio-tracking data—a review of problems and techniques particularly as applied to the study of mammals. Mammal Rev. 20, 97–123.

Harvey, M.J., Barbour, R.W., 1965. Home range of Microtus ochrogaster as determined by a modified minimum area method. J. Mammal. 46, 398–402.

Hemson, G., Johnson, P., South, A., Kenward, R., Ripley, R., Macdonald, D., 2005. Are kernels the mustard? Data from global positioning system (GPS) collars suggests problems for Kernel homerange analyses with least-squares cross-validation. J. Anim. Ecol. 74, 455–463.

Horne, J.S., Garton, E.O., Krone, S.M., Lewis, J.S., 2007. Analyzing animal movements using Brownian bridges. Ecology 88, 2354–2363.

Horne, J.S., Garton, E.O., Rachlow, J.L., 2008. A synoptic model of animal space use: Simultaneous estimation of home range, habitat selection, and inter/intra-specific relationships. Ecol. Model. 214, 338–348.

Jacques, C.N., Jenks, J.A., Klaver, R.W., 2009. Seasonal movements and home-range use by female pronghorns in sagebrush-steppe communities of western South Dakota. J. Mammal. 90, 433–441.

Joo, R., Boone, M.E., Clay, T.A., Patrick, S.C., Clusella-Trullas, S., Basille, M., 2020. Navigating through the R packages for movement. J. Anim. Ecol. 89, 248–267.

Kaczensky, P., Ganbaatar, O., Enksaikhaan, N., Walzer, C., 2006. Wolves in Great Gobi B SPA GPS tracking study 2003–2005 dataset. Movebank data repository. www.movebank.org).

Kaczensky, P., Enkhsaikhan, N., Ganbaatar, O., Walzer, C., 2008. The Great Gobi B strictly protected area in Mongolia—refuge or sink for wolves Canis lupus in the Gobi. Wildl. Biol. 14, 444–456.

Keating, K.A., Cherry, S., 2009. Modeling utilization distributions in space and time. Ecology 90, 1971–1980.

Kenward, R.E., Clarke, R.T., Hodder, K.H., Walls, S.S., 2001. Density and linkage estimators of home range: nearest-neighbor clustering defines multinuclear cores. Ecology 82, 1905–1920.

Kie, J.G., Boroski, B.B., 1996. Cattle distribution, habitats, and diets in the Sierra Nevada of California. J. Range Manage. 49, 482–488.

Kie, J.G., Matthiopoulos, J., Fieberg, J., Powell, R.A., Cagnacci, F., Mitchell, M.S., Gaillard, J.-M., Moorcroft, P.R., 2010. The home-range concept: are traditional estimators still relevant with modern telemetry technology? Philos. Trans. R. Soc. B 365, 2221–2231.

Klemelä, J., 2018. Level set tree methods. WIREs Comput. Stat. 10, e1436.

Kranstauber, B., Kays, R., LaPoint, S.D., Wikelski, M., Safi, K., 2012. A dynamic Brownian bridge movement model to estimate utilization distributions for heterogeneous animal movement. J. Anim. Ecol. 81, 738–746.

List, R., Macdonald, B.W., 2003. Home range and habitat use of the kit fox (*Vulpes macrotis*) in a prairie dog (*Cynomys ludovicianus*) complex. J. Zool. 259, 1–5.

Long, J.A., 2019. Estimating wildlife utilization distributions using randomized shortest paths. Landscape Ecol. 34, 2509–2521.

Long, J.A., 2020. wildlifeTG: Time Geograhic Analysis of Wildlife Telemetry Data. R Package Version 0.4. http:/jedalong.github.io/wildlifeTG.

Long, J., Nelson, T., 2012. Time geography and wildlife home range delineation. J. Wildl. Manage. 76, 407–413.

Long, J., Nelson, T., 2015. Home range and habitat analysis using dynamic time geography. J. Wildl. Manage. 79, 481–490.

Lyons, A., 2019. T-locoh. R package version 1.40.07. http:/tlocoh.r-forge.r-project.org/.

Lyons, A.J., Turner, W.C., Getz, W.M., 2013. Home range plus: a space-time characterization of movement over real landscapes. Movement Ecol. 1, 2.

MacCurdy, R.B., Bijleveld, A.I., Gabrielson, R.M., Cortopassi, K.A., 2019. Automated wildlife radio tracking. In: Zekavat, R., Buehrer, R.M. (Eds.), Handbook of Position Location: Theory, Practice, and Advances. IEEE Press and Wiley, pp. 1219–1262.

Majumdar, S.N., Comtet, A., Randon-Furling, J., 2010. Random convex hulls and extreme value statistics. J. Statist. Phys. 138, 955–1009.

Mammen, K., Mammen, U., Resetaritz, A., 2017. Red kite. In: Hötker, H., Krone, O., Nehls, G. (Eds.), Birds of Prey and Wind Farms: Analysis of Problems and Possible Solutions. Springer, pp. 13–96.

Matthiopoulos, J., 2003. Model-supervised kernel smoothing for the estimation of spatial usage. Oikos 102, 367–377.

Meekan, M.G., Duarte, C.M., Fernández-Gracia, J., Thums, M., Sequeira, A.M.M., Harcourt, R., Eguiluz, V.M., 2017. The ecology of human mobility. Trends Ecol. Evol. 32, 198–210.

Miller, H.J., 2005. A measurement theory for time geography. Geogr. Anal. 37, 17–45.

Mitchell, L.J., White, P.C.L., Arnold, K.E., 2019. The trade-off between fix rate and tracking duration on estimates of home range size and habitat selection for small vertebrates. PLoS ONE 14 (7), e0219357.

Mohr, C.O., 1947. Table of equivalent populations of North American small mammals. Am. Midl. Nat. 37, 223–449.

Moorcroft, P.R., Lewis, M.A., 2006. Mechanistic Home Range Analysis. Princeton University Press.

Moreira, A.J.C., Santos, Y.M., 2007. Concave hull: a *k*-nearest neighbors approach for the computation of the region occupied by a set of points. In: Braz, J., Vázquez, P.-P., Pereira, J.M. (Eds.), GRAPP 2007—Proceedings of the Second International Conference on Computer Graphics Theory and Applications. INSTICC, pp. 61–68.

Nilsen, E.B., Pedersen, S., Linnell, J.D.C., 2008. Can minimum convex polygon home ranges be used to draw biologically meaningful conclusions? Ecol. Res. 23, 635–639.

Noonan, M.J., Tucker, M.A., Fleming, C.H., Alberts, S.C., Ali, A.H., Altmann, J., Antunes, P.C., Belant, J.L., Berens, D., Beyer, D., Blaum, N., Böhning-Gaese, K., LauryCullen, Jr, de Paula, R.C., Dekker, J., Farwig, N., Fichtel, C., Fischer, C., Ford, A., Goheen, J.R., Janssen, R., Jeltsch, F., Kappeler, P., Koch, F., LaPoint, S., Markham, A.C., Medici, E.P., Morato, R.G., Nathan, R., Oliveira-Santos, L.G.R., Patterson, B.D., Paviolo, A., Ramalho, E.-E., Roesner, S., Selva, N., Sergiel, A., Silva, M.X., Spiegel, O., Ullmann, W., Zieba, F., Zwijacz-Kozica, T., Fagan, W.F., Mueller, T., Calabrese, J.M., 2019. A comprehensive analysis of autocorrelation and bias in home range estimation. Ecol. Monogr. 89 (2), e01344.

Odum, E.P., Kuenzler, E., 1955. Measurement of territory and home range size in birds. Auk 72, 128–137.

Park, J.S., Oh, S.J., 2012. A new concave hull algorithm and concaveness measure for n--dimensional datasets. J. Inf. Sci. Eng. 28, 587–600.

Pateiro-López, B., Rodríguez-Casal, A., 2010. Generalizing the convex hull of a sample: the R package alphahull. J. Stat. Softw. 34, 1–28.

Pateiro-López, B., Rodríguez-Casal, A., 2019. Generalizing the convex hull of a sample: the R package alphahull. R package version 2.2. https:/cran.r-project.org/package=alphahull.

Pavey, C.R., Goodship, N., Geiser, F., 2003. Home range and spatial organisation of rock-dwelling carnivorous marsupial, Pseudantechinus macdonnellensis. Wildl. Res. 30, 135–142.

Pellerin, M., Saïd, S., Gaillard, M., 2008. Roe deer Capreolus capreolus home-range sizes estimated from VHF and GPS data. Wildl. Biol. 14, 101–110.

Perkal, J., 1956. Sur les ensembles ε-convexes. Colloq. Math. 4, 1–10.

Péron, G., 2019. The time frame of home-range studies: from function to utilization. Biol. Rev. 94, 1974–1982.

Polonik, W., 1995. Measuring mass concentrations and estimating density contour clusters an excess mass approach. Ann. Stat. 23, 855–881.

Qiao, W., 2020. Asymptotics and optimal bandwidth selection for nonparametric estimation of density level sets. Electron. J. Stat. 14, 302–344.

Core Team, R., 2020. R: A Language and Environment for Statistical Computing. R Foundation for Statistical Computing, Vienna, Austria. www.R-project.org/.

Rényi, A., Sulanke, R., 1963. Über die konvexe Hülle von n zufällig gewählten Punkten. Z. Wahrsch. Verw. Gebiete 2, 75–84.

Rényi, A., Sulanke, R., 1964. Über die konvexe Hülle von n zufällig gewählten Punkten II. Z. Wahrsch. verw. Gebiete 3, 138–147.

Robson, D.S., Whitlock, J.H., 1964. Estimation of a truncation point. Biometrika 51, 33–39.

Rodríguez-Casal, A., 2007. Set estimation under convexity type assumptions. Ann. I. H. P. Probab. Stat. 43, 763–774.

Rodríguez-Casal, A., Saavedra-Nieves, P., 2016. A fully data-driven method for estimating the shape of a point cloud. ESAIM: Probab. Stat. 20, 332–348.

Rudi, A., de Vito, E., Verri, A., Odone, F., 2017. Regularized kernel algorithms for support estimation. Front. Appl. Math. Stat. 3, 23.

Schneider, R., 1988. Random approximation of convex sets. J. Microsc. 151, 211–227.

Schölkopf, B., Platt, J.C., Shawe-Taylor, J., Smola, A.J., Williamson, R.C., 2001. Estimating the support of a high-dimensional distribution. Neural Comput. 13, 1443–1471.

Seaman, D.E., Powell, R.A., 1996. An evaluation of the accuracy of kernel density estimators for home range analysis. Ecology 77, 2075–2085.

Seaman, D.E., Millspaugh, J.J., Kernohan, B.J., Brundige, G.C., Raedeke, K.J., Gitzen, R.A., 1999. Effects of sample size on kernel home range estimates. J. Wildl. Manage. 63, 739–747.

Seton, E.T., 1909. Life-Histories of Northern Animals: An Account of the Mammals of Manitoba. vol. I Charles Scribner's Sons (Grass-Eaters).

Signer, J., Balkenhol, N., Ditmer, M., Fieberg, J., 2015. Does estimator choice influence our ability to detect changes in home-range size? Anim. Biotelemetry 3, 16.

Singh, A., Scott, C., Nowak, R., 2009. Adaptive Hausdorff estimation of density level sets. Ann. Stat. 37, 2760–2782.

Slaught, J.C., Horne, J.S., Surmach, S.G., Gutiérrez, R.J., 2013. Home range and resource selection by animals constrained by linear habitat features: an example of Blakiston's fish owl. J. Appl. Ecol. 50, 1350–1357.

Steiniger, S., Hunter, A.J.S., 2013. A scaled line-based density estimator for the retrieval of utilization distributions and home ranges from GPS movement tracks. Eco. Inform. 13, 1–8.

Tarjan, L.M., Tinker, M.T., 2016. Permissible home range estimation (PHRE) in restricted habitats: a new algorithm and an evaluation for sea otters. PLoS ONE 11 (3), e0150547.

Tétreault, M., Franke, A., 2017. Home range estimation: examples of estimator effects. In: Anderson, D.L., McClure, C.J.W., Franke, A. (Eds.), Applied Raptor Ecology: Essentials From Gyrfalcon Research. The Peregrine Fund, pp. 207–242.

Tracey, J.A., Sheppard, J., Zhu, J., Wei, F., Swaisgood, R.R., Fisher, R.N., 2014. Movement-based estimation and visualization of space use in 3D for wildlife ecology and conservation. PLoS ONE 9 (7), e101205.

van der Watt, P., 1980. A note on estimation bounds of random variables. Biometrika 67, 712–714.

van Winkle, W., 1975. Comparison of several probabilistic home-range models. J. Wildl. Manage. 39, 118–123.

Walsh, P.D., Boyer, D., Crofoot, M.C., 2010. Monkey and cell-phone-user mobilities scale similarly. Nat. Phys. 6, 929–930.

Walter, W.D., Onorato, D.P., Fischer, J.W., 2015. Is there a single best estimator? Selection of home range estimators using area-under-the-curve. Mov. Ecol. 3, 10.

Walther, G., 1997. Granulometric smoothing. Ann. Stat. 25, 2273–2299.

Wasserman, L., 2018. Topological data analysis. Ann. Rev. Stat. Appl. 5, 501–532.

Willett, R., Nowak, R., 2007. Minimax optimal level set estimation. IEEE Trans. Image Process. 16, 2965–2979.

Wilmers, C.C., Nickel, B., Bryce, C.M., Smith, J.A., Wheat, R.E., Yovovich, V., 2015. The golden age of bio-logging: how animal-borne sensors are advancing the frontiers of ecology. Ecology 96, 1741–1753.

Worton, B.J., 1987. A review of models of home range for animal movement. Ecol. Model. 38, 277–298.

Worton, B.J., 1989. Kernel methods for estimating the utilization distribution in home-range studies. Ecology 70, 164–168.

Chapter 2

Modeling extreme climatic events using the generalized extreme value (GEV) distribution

Diana Rypkema[a,b] and Shripad Tuljapurkar[c,*]

[a]*Department of Natural Resources, Cornell University, Ithaca, NY, United States*
[b]*The Nature Conservancy, 652 NY-299, Highland, NY, United States*
[c]*Department of Biology, Stanford University, Stanford, CA, United States*
[*]*Corresponding author: e-mail: tulja@stanford.edu*

Abstract

Extreme climatic events (ECEs) are increasing in frequency, intensity, and/or duration due to climate change. These large deviations from average environmental conditions have strong impacts on ecosystems, exposing organisms to more severe conditions for longer periods of time than previously seen before. Many ECEs are well described by the generalized extreme value (GEV) distribution. The GEV has a long history as an accurate model across disciplines, including hydrology, finance, civil engineering, and climate science. Although the GEV is often used by climate scientists to characterize ECEs, such as hurricanes and floods, it is rarely included in ecological models. Furthermore, climate scientists typically model ECEs on large spatial scales ($\gg 10^5$ km^2) whereas ecologically relevant spatial scales are often much smaller (e.g., $< 10^3$ km^2). Here, we demonstrate how to estimate a GEV for ECEs and "downscale" the model to an ecologically relevant spatial scale. We then show how to shift the GEV to account for climate change predictions and incorporate the GEV into ecological population models. The GEV is a flexible tool to model ECEs across event types and ecosystems. We discuss the implications and challenges of accurately modeling ECEs and quantitatively estimating their impacts on ecosystems.

Keywords: Extreme climatic events, Ecology, Climate change, Generalized extreme value distribution

1 Introduction

Extreme climate events (ECEs) are large deviations from average conditions and are, by definition, rare (Coles, 2001). Regions experience "typical" conditions throughout the year, such as seasonal precipitation, temperature, and wind (IPCC, 2012). An ECE is a large deviation from these "average" conditions, such as a heatwave—a period of time with very high temperatures compared to the long-term average (Robinson, 2001; Xu et al., 2016). Other examples of ECEs include wildfires, floods, droughts, and hurricanes. With climate change, the frequency, intensity, duration, and/or the timing of ECEs is shifting (Easterling et al., 2000; IPCC, 2013); these changes will continue into the foreseeable future, exposing ecosystems to previously unseen conditions (Harris et al., 2018; IPCC, 2012).

Extreme events strongly impact humans and ecosystems worldwide (Easterling et al., 2000; IPCC, 2012; Parmesan et al., 2000). ECEs can damage or destroy buildings and infrastructure, cause mortality or injury, and displace people temporarily or permanently (Below et al., 2009). Consequences of ECEs can be direct or indirect, e.g., reduced crop yields have a direct economic impact on farmers and their communities, but also indirectly affects consumers by decreasing food availability and increasing prices (IPCC, 2012; IPCC, 2019; Zampieri et al., 2017). Economically, ECEs can result in large costs, from insurance payouts to rebuilding expenses (Below et al., 2009; Dinan, 2017; IPCC, 2012). Disasters may also have a psychological toll on affected communities, a consequence which is difficult to measure (Neria et al., 2008). Exposure to ECEs is higher as a result of increasing human population size and development (Dinan, 2017; IPCC, 2012). With climate change, the damage caused by ECEs is further increasing due to shifts in event frequency, severity, and/or duration (IPCC, 2012; IPCC, 2019).

Ecologically, ECEs may have negative or positive effects (Gutschick and BassiriRad, 2010; Parmesan et al., 2000). Some ecosystems are uniquely adapted for extreme events, such as giant sequoias (*Sequoiadendron giganteum*) which need fire to reproduce (Harvey et al., 1980) or *Hypseleotris* species which outcompete an invasive species when there is regular flooding (Ho et al., 2013). However, as the climate changes and alters ECE patterns, ecosystems are experiencing never-before-seen conditions (Harris et al., 2018; IPCC, 2012), even harming systems that are ECE-adapted. For example, wildfires are now more damaging than in the past due to climate change, compounding effects of longer and more severe droughts, and human efforts to prevent wildfires which leave more forest debris on the ground as fuel when a fire does occur (Westerling and Bryant, 2008; Westerling et al., 2006). Linked to hotter and drier conditions, bark beetle infections of giant sequoias are more severe than in the past, increasing stress on sequoia individuals (Greenfield and Lampcov, 2020; Nydick et al., 2017). These combined phenomena lead to higher giant sequoia mortality when wildfires do occur, having a negative effect on resilience and recovery despite the species depending on fire to reproduce (Greenfield and Lampcov, 2020; Nydick et al., 2017).

Furthermore, there are socio-ecological impacts of ECEs, such as deterioration of ecosystem services (van der Geest et al., 2019). Ecosystem services are the free benefits humans reap from the environment, including water supply, water purification and disease control (Corvalan et al., 2005; Daily, 1997). Thus, agricultural systems in West Africa depend on rain as the main water source for their crops (IPCC, 2012). However, as precipitation is becoming more variable due to climate change, crop yields are decreasing due to droughts and/or extreme rains that wash away crops, leading to increasing food insecurity in a region with a rapidly growing population (IPCC, 2012). In some of the same areas, episodes of increased flooding lead to more standing pools of water and create breeding ponds for disease vectors (e.g., mosquitoes) leading to higher spread and incidence of infectious disease (e.g., malaria) (Okoth and Odhiambo, 2018; Snow et al., 2015).

Climate change is altering the frequency, intensity, duration, spatial extent, spatial location, and/or timing of ECEs (Easterling et al., 2000; IPCC, 2012). For example, heatwaves and droughts are likely increasing in frequency, intensity and duration across most of the globe (IPCC, 2012). In other cases, the magnitude and direction of change is region-specific (Alexander et al., 2006; IPCC, 2012, 2013); there was a significant increase in the number of cold days globally from 1951 to 2003 but a significant increase in the number of cold days per year in the central United States (Alexander et al., 2006). In some regions, flooding may happen earlier in the spring as peak flows from snowmelt and glaciers occurs earlier in the season (IPCC, 2012). Other examples of observed and projected changes are included in Table 1.

As a result of these changes in ECE patterns, return periods for events of any given intensity are decreasing, so more intense events are occurring more often and providing less time for ecosystem recovery between events (Dreesen et al., 2014; Schwager et al., 2006); the Intergovernmental Panel on Climate Change (IPCC) predicts that daily maximum temperatures that were reached once every 20 years in the recent past (1981–2000) will likely occur once every 5 years in the near future (2046–2065) and once every 1–2 years at the end of the century (2081–2100) (IPCC, 2012). This decrease in return period for ECEs, coupled with human population growth and development, will amplify the consequences of ECEs for people and nature (IPCC, 2012).

Since ECEs have such diverse consequences, we need to be able to quantify their impact. The probability distribution of any kind of extreme event is described mathematically by a generalized extreme value (GEV) distribution, via a powerful argument similar to that underlying the limit of a normal distribution (the Fisher–Tippett–Gnedenko theorem, Fisher and Tippett, 1928 and Gnedenko, 1943). For this reason, the GEV provides a natural, quantitative description of extreme events of all kinds (Coles, 2001), and has a long history in many disciplines, including finance (Coles, 2001), civil engineering (Gumbel, 1958), hydrology (Martins and Stedinger, 2000), and climatology (Coles, 2001; Gumbel, 1958). Although extreme events are naturally expected to be important ecosystem drivers and sources of disturbance in ecology, to

TABLE 1 Examples of observed and predicted changes in ECE characteristics with climate change.

Characteristic	Observed change	Predicted change
Frequency	Decreased number of cold waves in the United States (USGCRP, 2017)	Increasing frequency of tornadoes in Northern Eurasia (Chernokulsky et al., 2017)
Intensity	Hotter heatwaves across the African continent (Russo et al., 2016)	More severe droughts in the Brazilian Cerrado (Rodrigues et al., 2020)
Duration	Longer heatwaves in the Southeast Asia (Li, 2020)	Droughts lasting up to 90% longer in the Middle East (Tabari and Willems, 2018)
Spatial Extent	Increasing extent of hypoxic zones in the Baltic Sea (Carstensen and Conley, 2019)	Marine heatwaves are predicted to increase in size by 25% globally (Frölicher et al., 2018)
Spatial Location	Tropical cyclones are reaching maximum intensity farther away from the Equator and (closer to the Poles) (Kossin et al., 2014)	Wildfires are moving to higher elevations in the Sierra Nevada mountains, USA (Schwartz et al., 2015)
Timing	Wildfire season is starting earlier (and lasting longer) in Australia (Clarke et al., 2013)	Flooding predicted to occur earlier from snowmelt- and glacier-fed streams (IPCC, 2012)

Predicted conditions are generally for 2100. This is a small set of examples; there are many more examples in the literature (e.g., IPCC, 2012, 2013).

our knowledge, there are few examples of GEV models proposed or applied in ecology (examples of such applications include Meehl et al. (2000), Denny et al. (2009), and Easterling et al. (2000)). As yet, we lack a realistic method to systematically incorporate climate-driven changes in ECEs into ecological models, and so do not know how ecosystems are affected by climate-driven increases in severity or rate of extreme events. Here, we describe how to (1) model ECEs on an ecologically relevant spatial scale using the GEV; (2) shift the GEV's parameters to include climate change predictions; (3) incorporate the GEV into empirically based ecological population models; and (4) use these models to examine the repercussions of ECEs on ecosystem damage levels and recovery dynamics. We begin by briefly reviewing previous models of ECEs in ecology and end with a discussion of some of the challenges of realistically modeling ECEs and quantifying their effects on ecosystems.

1.1   Previous models of extreme events in ecology

Previously, ecological studies mainly focused on biological responses to mean environmental conditions; some studies included environmental variability, but did not explicitly include extremes (for more information, see Jentsch et al., 2007; Smith, 2011; van de Pol et al., 2017; Vasseur et al., 2014). van de Pol et al. (2017) asserts that the most interesting biological responses to ECEs are nonlinear because linear biological responses to extreme conditions are already captured by studies looking at shifting means and increasing environmental variability. However, studies that only examine changes in means and variability will not accurately capture a nonlinear biological response only seen under extreme conditions (see figure 1 from van de Pol et al., 2017 for an illustration of potential biological response to climate variable relationships and Bailey and van de Pol, 2016 for more details).

Recently, there has been a sharp spike in ECE-related ecological research (Bailey and van de Pol, 2016). However, there are several opportunities for improvement in this rapidly developing field. Currently, ecologists lack universally accepted terminology and a consensus definition of what constitutes an ECE. For example, definitions may be purely climatological (e.g., a heatwave is a series of consecutive days above a certain percentile compared to historical conditions in that region) or they may be impact-related (e.g., the biological response *and* the climatological conditions must both be extreme). Studies on ECEs are often short-term or focused on a single event occurrence, missing the long-term effects of ECEs on the ecosystem (van de Pol et al., 2017). To understand the impacts of an ECE on a system, we need sufficient records of conditions prior to the event and enough observations after the event to capture lagged effects. Single event studies also make it difficult to parse out the effects of multiple drivers on a system, potentially combining additive or interacting responses related to the ECE and other forces (e.g., anthropogenic stressors). See van de Pol et al. (2017) for a more detailed discussion of these issues.

Related frameworks to the study of ECEs in ecology include disturbance ecology and "catastrophes." Disturbance ecology studies biological responses to disturbances, discrete events that disrupt ecosystem, community, and/or population structure (White and Pickett, 1985). These disturbances are not necessarily due to extreme meteorological conditions and do not necessarily have to elicit an extreme biological response. In contrast to ECE ecology, which must have a climatological component, disturbances may be biological (e.g., species invasions) or anthropogenic (e.g., land conversion, oil spills) in nature. Other previous studies have described ECEs as "catastrophes" (e.g., in paleo-ecology, Grant et al. (2017), Katz et al. (2005) or Lande (1993)). For example, Lande (1993) modeled the response of a population to catastrophic events, which essentially devastated the population, causing extinction most of the time. However, many ecosystems are adapted to historical ECE conditions and recover after the event, and some ECEs may even have positive effects on certain species (such as the hurricane-disturbed understory shrub in our case study). Also, the catastrophe model often only includes event frequency, lacking other important ECE aspects, such as event magnitude.

Similarly, models that explicitly incorporate ECEs often only include event frequency, missing other key ECE components, such as ECE duration, timing, and intensity. Models that do include ECE intensity may use distributions that do not describe the unique qualities of extremes (e.g., Pardo et al. (2017) which models sea surface temperature extremes using a normal distribution). Using the statistics of extremes employed in other fields (see below) is a natural way to accurately model ECE magnitude.

However, ECE models used in other fields have often been developed on too broad a spatial scale to answer many ecological questions. For example, current hurricane models are typically created on a regional or state-wide scale ($\gg 10^5$ km^2), appropriate in the context of insurance and disaster relief planning (e.g., Dinan, 2017; Jagger and Elsner, 2006; Katz, 2002), but less relevant for specific ecosystems which are most strongly impacted by local conditions. Here, we show how to downscale hurricane models to a much smaller geographic scope (e.g., $<10^3$ km^2). Conveniently, recent global climate models (GCMs) have improved spatial resolution, providing better data for ecologists to develop ecological models with ECEs.

To our knowledge, extreme value theory has only been used in a handful of ecological modeling studies. In 2000, Meehl et al. argued that ecologists should use the Gumbel distribution (a form of the GEV) to model extreme weather and climate events, but their suggestion was not widely adopted. Similarly, Burgman et al. (2012) demonstrated the improved fit of modeling monsoon events using the GEV compared with a normal or lognormal distribution, a case study relevant for the population viability analysis (PVA) of the Sindh ibex. However, at the time of writing, Burgman et al. (2012) only has 10 citations on CrossRef, despite its strong case that the GEV offers the most realistic representation of extreme events compared to other distributions. We know of four other studies that employ the GEV in ecological models; these are models in the context of: minimum and maximum sea surface temperatures (Gaines and Denny, 1993), sedimentation rate in paleohydrology (Katz et al., 2005), wave dynamics (Denny et al., 2009), and thermal performance curves and heat tolerance (Kingsolver and Buckley, 2017). However, these models do not include population-level effects or climate-driven changes in ECE frequency and intensity on an ecologically relevant temporal or spatial scale; we must systematically include predicted shifts in ECE frequency and intensity to understand the potential consequences of continued climate change.

Here, we try to fill some of the gaps in the field of ECE ecology by providing a modeling framework that includes ECE frequency *and* intensity and describe how to downscale this ECE model to an ecologically relevant spatial scale. Our model is based on a single type of extreme event, and is built from data gathered at several nearby study sites which experienced different ECE severities. We also employ a long time series of the type of ECE we consider, and then focus our temporal sampling across events that are spatially localized. In terms of population effects, we use data collected mainly from several nearby spatial sites over a shorter time period to construct a population model. We use our model to systematically evaluate the response of the stochastic population growth rate (λ_S) to changes in ECE frequency vs intensity in the context of climate change and model ECE magnitude using a GEV distribution to realistically represent ECE intensity and its potential increase with continued climate change. We also estimate the sensitivity of the stochastic growth rate to shifts in ECE-caused ecosystem damage and ecosystem recovery rates.

2 The GEV value distribution

2.1 A brief history of the GEV

The GEV was popularized by Emil Gumbel in his 1958 book "Statistics of Extremes" (Gumbel, 1958). This book was the first comprehensive review of extreme value theory, detailing the three types of GEV distribution: Gumbel, Fréchet, and Weibull (also known as types I, II, and III, respectively). Gumbel showed applications of the GEV across different disciplines, including meteorology, engineering, and hydrology (Gumbel, 1958).

Initially, extreme values were conceptualized as very large values from the tails of normal distributions, not as values from a distribution in and of itself (Gumbel, 1958). In 1927, Fréchet (1927) was the first to propose a separate distribution for extreme values. In the following year, independent of Fréchet, Fisher and Tippett (1928) published three asymptotic forms of extreme value distributions, later formally proved by Boris Gnedenko in 1943 (Gnedenko, 1943). Gumbel (1958) extended this theoretical work and extensively applied extreme value theory to empirical problems, such as human lifespan, floods, droughts, and material fatigue.

Since then, the GEV has been applied extensively across many disciplines. For example, the GEV accurately estimates "value at risk" for financial institutions, providing predictions of catastrophic risk (Bali, 2007). Also applied in epidemiology, extreme value theory accurately models emergency room visits and pneumonia and influenza deaths (Thomas et al., 2016). The GEV also models air quality data well, guiding regulations for pollutant monitoring and limits (Roberts, 1979a,b).

2.2 Definitions

An individual extreme event is quantified by an observable measure, such as sustained wind speed during a hurricane or temperature during a heatwave.

The distribution of this measure (e.g., wind speeds) during an individual event follows a generalized Pareto distribution (Coles, 2001). The maximum (or minimum) of the observed measure, v (e.g., maximum sustained wind speed during a hurricane), across many independent or weakly correlated events follows a GEV distribution (Coles, 2001). Here, we use the "block maxima" approach, taking the maximum value v across events in a "block" (here, our block is one calendar year) (Coles, 2001). The "Peaks over Threshold" approach, using all v above a certain threshold determined based on the historical data, may alternatively be used to model ECEs (Coles, 2001).

The GEV has three parameters: location (μ), scale (σ), and shape (ξ). Location specifies the shift of the distribution relative to the standard GEV; scale is the spread of the distribution; and shape characterizes the heaviness of the distribution's tails (see Fig. 1 for examples of changing μ, σ, and ξ compared to the standard GEV). The standard GEV has parameters $\mu = 0$, $\sigma = 1$, and ξ varies depending on the type of GEV. The canonical GEV probability density (PDF) and cumulative distribution functions (CDF) are

$$\text{PDF} = \frac{1}{\sigma}\left[t(x)^{\xi+1}e^{t(x)}\right]$$

$$\text{CDF} = e^{-t(x)}$$

where

$$t(x) = \left[1 + \xi\left(\frac{x-\mu}{\sigma}\right)\right]^{-1/\xi}$$

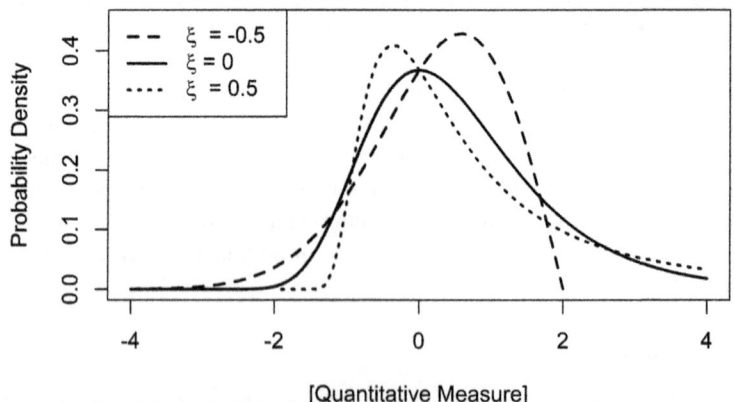

Probability Density Function for Standard GEV

FIG. 1 Probability density of the standard GEV with $\mu = 0$, $\sigma = 1$, and different shape parameter ξ values (indicated through line type).

There are three types of the GEV, differentiated based on the GEV's shape parameter, ξ:

$$
\begin{aligned}
\text{Gumbel (Type I)} &: \quad \xi = 0 \\
\text{Frechet (Type II)} &: \quad \xi > 0 \\
\text{Weibull (Type III)} &: \quad \xi < 0
\end{aligned}
$$

2.3 Parameterizing the GEV

Let z be the set of historical events in the sample with length m. We estimate the GEV parameters by maximizing the log-likelihood function (Coles, 2001):

$$
l(\mu, \sigma, \xi | z) = -m \log \sigma - \left(1 + \frac{1}{\xi}\right) \sum_{i=1}^{m} \log \left[1 + \xi\left(\frac{z_i - \mu}{\sigma}\right)\right]
$$
$$
- \sum_{i=1}^{m} \left[1 + \xi\left(\frac{z_i - \mu}{\sigma}\right)\right]^{-1/\xi} \tag{1}
$$

given

$$
1 + \xi\left(\frac{z_i - \mu}{\sigma}\right) > 0, \quad \text{for } i = 1, \ldots, m
$$

We use the extRemes package (Gilleland and Katz, 2016) in R (R Core Team, 2018) for log-likelihood parameter estimation.

2.4 Return level vs return period

We can model the value (return level, z_N) expected once every N years (the return period):

$$
z_N = \begin{cases}
\mu + \dfrac{\sigma}{\xi}\left(y_N{}^{\xi} - 1\right) & \text{if } \xi \neq 0 \\
\mu + \sigma \log y_N & \text{if } \xi = 0
\end{cases}
$$

where

$$
y_N = \log\left(\frac{N}{1 - N}\right)
$$

Using the historical data, we can compute the historical return period for events of a given magnitude z_N as a combination of the number of years in the historical event sample (N) and the rank of events from highest to lowest magnitude:

$$
z_N = \frac{N}{\text{rank}(z_N) - 0.5} + 0.5
$$

For comparison, we can plot event return levels vs return periods from the historical data and the GEV model.

2.5   Evaluating model fit

To assess the fit of the GEV compared to the historical data, we plot return period (N) vs return level (z_N) for the fitted GEV distribution and the historical data and compare (see Fig. 2 for an example). We also create quantile-quantile plots of the historical maximum observed measures v in the set of historical events z vs the GEV model estimates and compare the fitted GEV's CDF vs the historical data (see Fig. 3 for an example). Here, for the historical data,

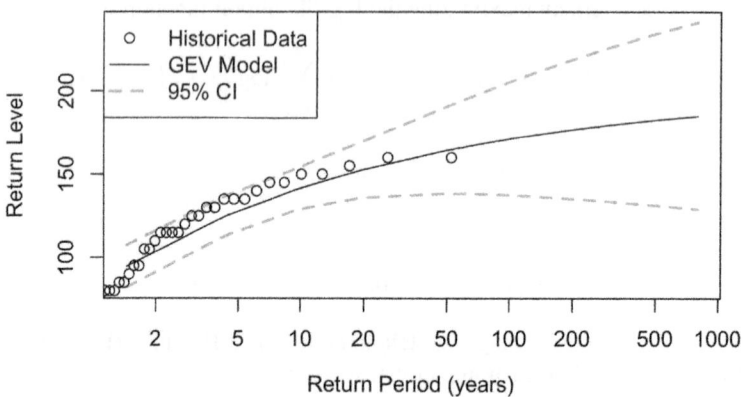

FIG. 2   Return level vs return period for southeast Florida study region. The *points* represent the historical data and the *solid line* represents the GEV distribution model. The *dotted lines* represent 95% confidence intervals.

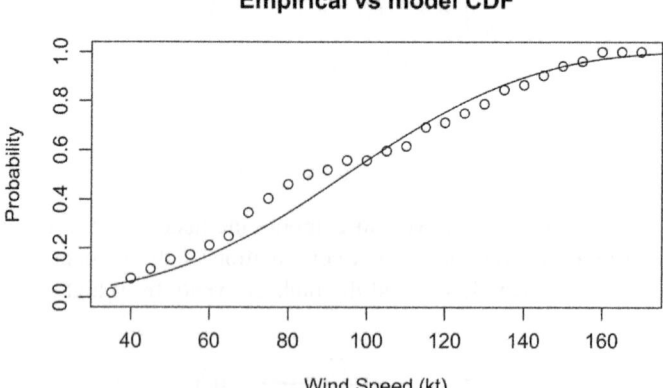

FIG. 3   Cumulative distribution of wind speeds for the GEV model and empirical data. The fitted GEV model is represented by the *solid line* and the historical data are represented by *hollow circles*.

we sum the number of storms for each maximum sustained wind speed and normalize by the total number of storms in the historical record (m).

2.6   Applying climate change to the GEV

We assess the effects of changing ECE frequency (p) and intensity. We can increase (or decrease) ECE frequency by increasing (or decreasing) the probability the event occurs in a given time period (increasing or decreasing p). For example, wildfires in the state of California, USA, are predicted to increase in frequency with continuing climate change (Westerling and Bryant, 2008). To increase (or decrease) event intensity, we shift the GEV's parameters: μ and σ. Here, we keep ξ constant while increasing μ and/or σ, assuming the shape of the GEV distribution remains the same (following the work of Lustenberger et al. (2014)). See Fig. 4 for an example of the influence of increasing the GEV's parameters.

3   Case study: Hurricanes and *Ardisia escallonioides*

We use the hurricane-disturbed subtropical shrub *A. escallonioides* as a case study. We use the model originally developed by Pascarella and Horvitz (1998) following Hurricane Andrew in 1992 for *A. escallonioides* populations in Southeast Florida, USA. *A. escallonioides* is an understory shrub whose dynamics are driven by light availability. When a hurricane occurs, canopy trees are removed, increasing the light available to *A. escallonioides*. The Pascarella and Horvitz (1998) model for this system explicitly incorporates hurricane frequency and intensity as drivers of environmental state, allowing us to quantitatively evaluate the impacts of increasing hurricane frequency and intensity on the population dynamics of *A. escallonioides*.

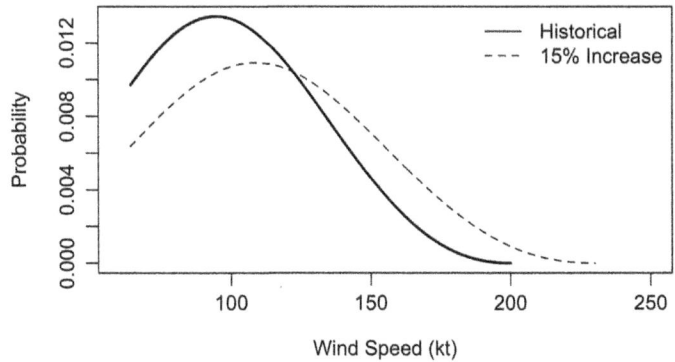

FIG. 4   The probability density function of hurricane wind speed fit to the historical data (*solid line*) and with a 15% increase in both μ and σ (*dashed line*).

3.1 Traditional hurricane modeling

Hurricanes are characterized using maximum sustained wind speed; wind speeds are much higher than the norm during a hurricane. The United States National Hurricane Center defines maximum sustained wind speed as the highest surface wind speed observed over a storm's lifetime lasting at least 1 min (Knapp et al., 2010). The Saffir–Simpson scale categorizes a hurricane's maximum sustained wind speed into one of five categories based on the storm's potential damage to humans (Saffir, 1973; Simpson, 1974). In our model, we discretize the continuous GEV wind speed distribution to these five categories to align with the Saffir–Simpson scale convention.

3.2 The data: HURDAT2

We use measurements from the World Meteorological Organization contained in the United States National Hurricane Center's updated HURDAT2 database v03r10 (Knapp et al., 2010). Wind speed measurements and hurricane positions are recorded every 6 h (Knapp et al., 2010). We only include observations from 1899 to 2016 because earlier wind speed measurements are usually less reliable (Jagger and Elsner, 2006). For storms from 1899 to 1930, 6-h storm wind speeds and positions were interpolated from once daily measurements; storm wind speeds and positions were interpolated from twice daily measurements for storms between 1931 and 1956; and from 1956 to present, storm wind speeds and positions were measured every 6 h. All maximum sustained wind speed measurements were rounded to the nearest multiple of 5. We assume storms are independent and make landfall at a random point in their lifetime (Emanuel and Jagger, 2010). The storm set we use to fit the GEV includes tropical storms (wind speeds ≥ 34 kt and < 64 kt) and hurricanes (≥ 64 kt) following Jagger and Elsner (2006) and we use the maximum sustained wind speed reached during a storm's lifetime, whether or not it occurred in our study area (Emanuel and Jagger, 2010).

3.3 Study region and storm selection criteria

To build the *A. escallonioides* population model, Pascarella and Horvitz (1998) collected data at three study sites hit by Hurricane Andrew in 1992: Matheson Hammock (25.68°, −80.27°), Castellow Hammock (25.56°, −80.45°), and Deering Estate (25.61°, −80.31°) (Fig. 5A). Following Pascarella and Horvitz (1998), we limited our study region to storms that crossed the southern Florida region coastal segments 23–25 and 28–29 (as described by Simpson and Lawrence, 1971) . We do not include segments 26–27 which represent the Florida Keys, islands off the coast of Florida. Limiting storm selection to these coastal segments downscales our storm set to an ecologically relevant spatial scale (Fig. 5B). When fitting the GEV to our historical storm set, we use the

A

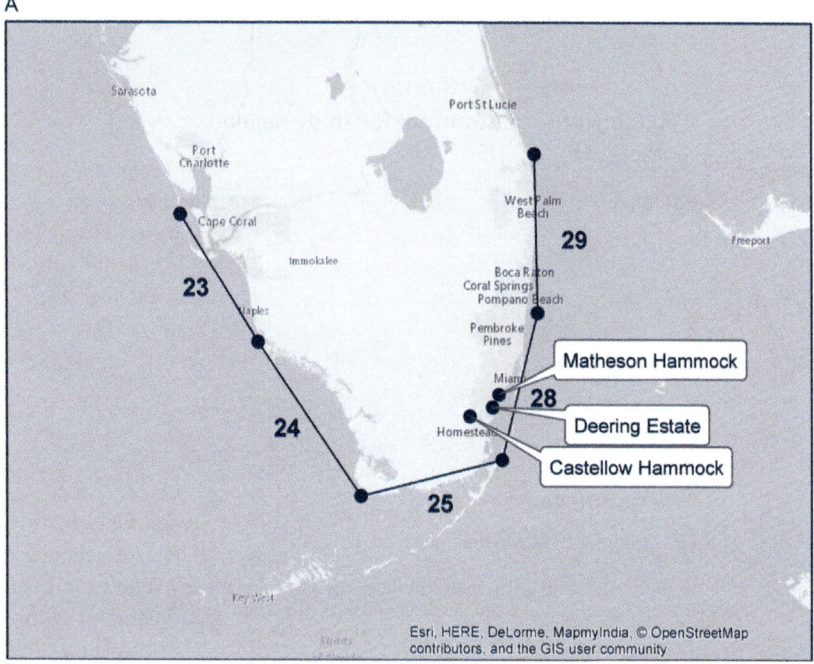

B

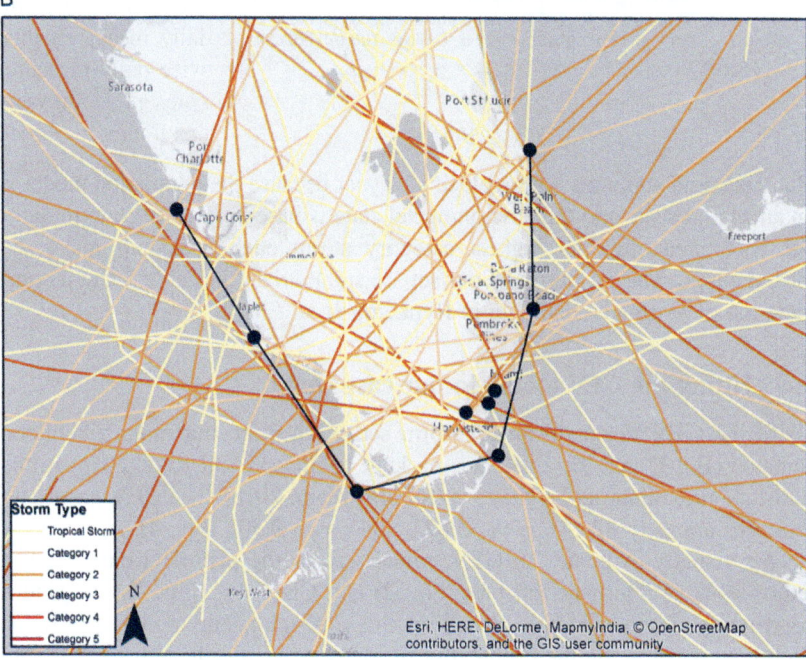

FIG. 5 Study region and historical storm tracks. (A) The three original study sites (Matheson Hammock, Castellow Hammock, and Deering Estate) are labeled. The *numbered lines* show the coastal segments from which historical storm tracks were sampled (segments from Simpson and Lawrence, 1971). (B) The study region with all historical storm tracks from 1899 to 2016. The *colors* of the *lines* represent the strength of the storm. *Lighter colors signify weaker storms and darker colors represent stronger storms. Originally published in Rypkema, D., Horvitz, C., Tuljapurkar, S., 2019. How climate affects extreme events and hence ecological population models. Ecology 100 (6), e02684.*

TABLE 2 Historical storm set for study region in southeast Florida, USA.

Category	Number
Tropical storm	11
1	13
2	5
3	3
4	12
5	8
Total number of storms	52
Total number of hurricanes	41

"Storms" refer to all events with maximum sustained wind speed ≥ 34 knots (1 knot = 1.852 km/h), including tropical storms and hurricanes. "Hurricanes" refer only to events with maximum sustained wind speed ≥ 64 knots.
Source: Originally published in Rypkema, D., Horvitz, C., Tuljapurkar, S., 2019. How climate affects extreme events and hence ecological population models. Ecology 100 (6), e02684.

block maxima approach with a block of 1 year because the population model is on an annual scale and the environmental state is based on the maximum damage to the canopy; the highest category storm each year will make the most damage. The 52 storms in our set are summarized in Table 2 and a complete list of storms is available in Rypkema et al. (2019).

3.4 Fitting the GEV model to our case study

To fit the GEV to our historical storm set, we use the maximum likelihood equation (Eq. 1) and the extRemes package (Gilleland and Katz, 2016) in the statistical programming language R (R Core Team, 2018). The GEV parameter estimates are: $\mu = 82.241$, $\sigma = 35.873$, and $\xi = -0.303$ (Fig. 4).

We visually evaluate the GEV model's fit to the historical data using a return level vs return period plot (Fig. 2) and CDF plot (Fig. 3), and find a good model fit to our data. Based on the CDF plot, our model appears to slightly underestimate the probability of lower wind speeds and overestimate the probability of higher wind speeds. In our context, this GEV model provides the best fit using maximum likelihood methods, but it may be possible to improve the fit by modifying the model in future work.

For more information on hurricane modeling using the GEV, see Jagger and Elsner (2006).

3.5   The demographic model

A. *escallonioides* is a subtropical understory shrub found in southeastern Florida (Pascarella and Horvitz, 1998; Tuljapurkar et al., 2003). As an understory plant, A. *escallonioides* is light-limited; hurricanes drive A. *escallonioides* population dynamics by removing canopy trees and increasing available light. We model light availability using the percent of canopy openness, divided into $i = 1, ..., 7$ environmental states (where state 1 is the most open canopy and state 7 is essentially closed canopy). The environmental states are $\geq 65\%$, $\geq 55\%$, $\geq 45\%$, $\geq 35\%$, $\geq 25\%$, $\geq 15\%$, and $< 5\%$ open, respectively. Each environmental state has a corresponding vital rate matrix (B_i; $i = 1, ..., 7$) with eight life history stages: seeds, seedlings, juveniles, prereproductives, and four size classes of reproductives. For complete model details, see Pascarella and Horvitz (1998).

3.6   Constructing the environmental transition matrix (P)

Each year, hurricanes occur with probability p (the event frequency). We calculate the frequency by dividing the number of hurricanes in our study area (41) by the number of years in our sample (117) to find the average number of hurricanes per year. We then divide by the total coastal segment length of our study region (5 coastal segments of 80.5 km each, for a total length of 402.5 km) and multiply by the median hurricane eye diameter (45 km, Weatherford and Gray, 1988). The result is a historical hurricane frequency of $p = 0.039$ per year.

Each year, a hurricane either does or does not occur. In a year without a hurricane, the canopy undergoes successional closure:

$$
S = \begin{bmatrix}
0.75 & 0 & 0 & 0 & 0 & 0 & 0 \\
0.25 & 0.75 & 0 & 0 & 0 & 0 & 0 \\
0 & 0.25 & 0.5 & 0 & 0 & 0 & 0 \\
0 & 0 & 0.5 & 0.25 & 0 & 0 & 0 \\
0 & 0 & 0 & 0.75 & 0.25 & 0 & 0 \\
0 & 0 & 0 & 0 & 0.75 & 0.25 & 0 \\
0 & 0 & 0 & 0 & 0 & 0.75 & 1
\end{bmatrix}
$$

We use the convention that transitions in **S** are from column to row: the probability that the environment moves to state i (row) at $t + 1$ given the environment is in state j (column) at time t. For example, $S[3, 2] = 0.25$ is the probability that the environment moves from state 2 (canopy openness $\geq 55\%$) to state 3 (canopy openness $\geq 45\%$ but $\geq 55\%$) by the next year.

In a year with a hurricane, the conditional probability that the strongest storm that year is a specific Saffir–Simpson category is w_j, $j = 1, \ldots, 5$, described by the vector w:

$$w = \begin{bmatrix} w_1 & w_2 & w_3 & w_4 & w_5 \end{bmatrix}$$

w is parameterized as:

$$w = \begin{bmatrix} 0.327 & 0.184 & 0.084 & 0.239 & 0.166 \end{bmatrix}$$

using the GEV model of the historical data.

The hurricane category then determines the damage to the canopy:

$$\mathbf{d} = \begin{bmatrix} 0.8 & 0.6 & 0.4 & 0.333 & 0.1 \\ 0.2 & 0.3 & 0.4 & 0.333 & 0.4 \\ 0 & 0.1 & 0.2 & 0.334 & 0.5 \end{bmatrix}$$

Columns in the damage $\mathbf{d}$ matrix correspond to Saffir–Simpson storm category and rows correspond to canopy damage levels: low, medium, and severe. For example, category 2 hurricanes have a 60% probability of causing low damage, a 30% probability of causing medium damage, and a 10% probability of causing severe damage to the canopy.

In this model, hurricane intensity is incorporated in two ways: the frequency of each storm category given a storm occurs (in w) and the damage resulting from a hurricane of a given category (in $\mathbf{d}$). In h, we combine $\mathbf{d}$ and w to find the probability of each canopy damage level given a hurricane occurs: low (p_l), medium (p_m), or severe (p_s).

$$h = \mathbf{d} * w' = \begin{bmatrix} p_l \\ p_m \\ p_s \end{bmatrix}$$

The apostrophe symbol indicates a transpose.

Using the probabilities in h, we create the transition matrix $\mathbf{D}$ to describe the probability of transitioning between canopy states given a hurricane occurs:

$$\mathbf{D} = \begin{bmatrix} 1 & 1 & 0.5*p_l + p_m + p_s & p_m + p_s & 0.5*p_m + p_s & p_s & 0.5*p_s \\ 0 & 0 & 0.5*p_l & 0.5*p_l & 0.5*p_m & 0.5*p_m & 0.5*p_s \\ 0 & 0 & 0 & 0.5*p_l & 0.5*p_l & 0.5*p_m & 0.5*p_m \\ 0 & 0 & 0 & 0 & 0.5*p_l & 0.5*p_l & 0.5*p_m \\ 0 & 0 & 0 & 0 & 0 & 0.5*p_l & 0.5*p_l \\ 0 & 0 & 0 & 0 & 0 & 0 & 0.5*p_l \\ 0 & 0 & 0 & 0 & 0 & 0 & 0 \end{bmatrix}$$

For example, if the canopy is $\geq 65\%$ or $\geq 55\%$ open (states 1 or 2), we expect that the canopy will be $\geq 65\%$ in the next time step after a hurricane occurs with probability 1. In contrast, if the canopy is in the most closed state ($<5\%$ open), we expect the canopy will open given a hurricane occurs (there is a probability of 0 that the canopy will remain in state 7 if there is a hurricane in that time step), and more severe storms will cause more damage, leading to the most open canopy states (e.g., $D[1, 7] = 0.5 * p_s$ and $D[2, 7] = 0.5 * p_s$ whereas $D[5, 7] = 0.5 * p_l$ and $D[6, 7] = 0.5 * p_l$).

We then combine the hurricane disturbance and successional dynamics transition matrices (D and S) in the overall environmental state transition matrix (P) weighted by the hurricane frequency p:

$$P = p * D + (1 - p) * S$$

For more information, see Pascarella and Horvitz (1998) and Rypkema et al. (2019).

3.7   Incorporating climate change in the environmental transition matrix (P)

Hurricanes are predicted to increase in magnitude between 2% and 11% by 2100 with continued climate change (Knutson et al., 2010). To incorporate the effects of climate change on hurricane intensity, we increase the GEV's μ and σ 0%–200% of the historical values in increments of 10% (maximum $\mu \approx 164.482$ and $\sigma \approx 71.746$) and keep ξ constant. We then compute the CDF for each GEV parameter combination using the extRemes package (Gilleland and Katz, 2016) in R (R Core Team, 2018) and bin these values into the discrete Saffir–Simpson categories (Table 3). We remove the bin for storms with maximum sustained wind speeds lower than the category 1 threshold (64 kts) and normalize the remaining probabilities to sum to 1 by dividing the probability of observing a storm from a specific hurricane category by the probability of observing a hurricane in any category. We use these normalized probabilities as our new category vector w, the probability of observing a storm of each category, given a hurricane occurs.

3.8   Population simulations

To compare the relative importance of shifts in hurricane frequency vs intensity, we run 1000 iterations of a 1000 year stochastic population simulation with a fully factorial combination of w and p values. The GEV's μ and σ parameters range from 0% to 200% of their historical values and hurricane frequency p ranges from 0.5 to 2 times the historical value (p ranges from 0.0195 to 0.078). We then compare the stochastic population growth rate $\log \lambda_S$ for *A. escallonioides* for each set of parameter combinations.

TABLE 3 Normalized probability of hurricane wind speed by category given a hurricane occurs.

Category	Minimum kt for category designation	Probability of observing this category	Normalized probability
Tropical storm or below	1	0.1534	Removed
1	64	0.2767	0.3269
2	83	0.1556	0.1838
3	96	0.0714	0.0843
4	113	0.2022	0.2389
5	137	0.1407	0.1662

Values are the normalized probability of observing wind speeds of a particular hurricane category given a hurricane occurs. The third column includes all wind speeds >0 kt. The fourth column is the normalized probability of observing a hurricane, given one occurs (the probability of observing a wind speed less than the hurricane threshold (<64 kt) is removed). Note: Both columns sum to 1.

In addition to *A. escallonioides'* $\log \lambda_S$, we compare the effects of shifting hurricane frequency and intensity on the mean canopy openness (the environmental state).

3.9   Sensitivity analysis

Climate change may alter hurricane damage levels and canopy recovery rates, in addition to hurricane frequency and intensity. For example, hurricanes of a given category based on wind speed may have greater storm surge in the future, causing more damage to the canopy and potentially slowing recovery (Dey and Das, 2015; Dinan, 2017). To further evaluate the effects of climate change, we calculate the sensitivity of $\log \lambda_S$ to small perturbations in four components of the environmental transition matrix **P**: (a) the frequency of storms, p; (b) the GEV's μ and σ parameters discretized into the w hurricane category vector; (c) the damage caused by a storm of a given category in **d**; and (d) the speed of recovery (canopy closure) after a storm contained in **S**. We increase or decrease p or μ and σ to represent changes in hurricane frequency and intensity, respectively. To increase hurricane damage in our model, we increase the probability of severe damage and decrease the probability of low damage for each storm category in the damage matrix **d** by 0.01 and 0.02, respectively. To alter canopy recovery dynamics, we increase or decrease the probability of remaining in the current canopy state in the canopy recovery matrix **S** by 0.01 and 0.02, respectively. We run 1000 simulations of all combinations of parameter shifts (e.g., intensity alone; intensity and frequency; intensity, frequency, and **d**; intensity, frequency, **d** and **S**). We find the sensitivity of the population growth rate using the slope of $\log \lambda_S$ vs small perturbations in the component of interest. Since the changes in the parameter(s) of interest are small, this method allows us to calculate the partial derivative of $\log \lambda_S$ with respect to each environmental component using a linear approximation.

3.10 Effects of changing hurricane frequency, intensity, damage levels, and/or canopy recovery rates

Increasing hurricane frequency has a much stronger effect on *A. escallonioides'* $\log \lambda_S$ than increasing hurricane intensity (Fig. 6). For example, if hurricane frequency is twice the historical ($p \approx 0.078$) and μ and σ remain at their historical values, $\log \lambda_S$ more than doubles to ~ 0.241 (compared to ~ 0.116 under historical conditions). In contrast, when hurricane frequency remains at the historical level ($p \approx 0.039$) and μ and σ are both increased 200%, $\log \lambda_S$ increases to ~ 0.169, a mere 0.053 increase relative to the historical population growth rate. Here, we show results for shifting both μ and σ to increase hurricane intensity. Effects on λ_S of shifting only μ or only σ are comparable and are presented in the supplementary material of Rypkema et al. (2019).

Similarly, we find that hurricane frequency has a stronger effect on mean canopy openness than increasing hurricane intensity (Fig. 7). At the historical hurricane frequency and intensity, the mean percent canopy openness is 12.2%. When frequency is doubled ($p \approx 0.078$) and intensity remains the same, mean canopy openness increases by 8.9%–21.1%. However, if hurricane frequency remains at the historical level and the GEV's μ and σ parameters double, the mean canopy openness increases by 4.5%–16.7%.

A. escallonioides is an understory plant and its population dynamics are strongly driven by light availability. Under historical conditions, the darkest state is the most common environmental state. When hurricanes are more intense but occur at the same frequency, the canopy enters a more open state after the storm, then returns to the darkest state and remains there until another storm happens. In contrast, when hurricanes are weaker but occur more frequently, the canopy will return to an open state more often, supplying more light to *A. escallonioides*.

Our sensitivity results further support these findings, with $\log \lambda_S$ most sensitive to slower canopy recovery rates, followed by increased canopy damage level, hurricane frequency, and hurricane intensity (Table 4). *A. escallonioides* population growth rate responds positively to all four components; in contrast, the canopy is negatively impacted by all four changes. With climate change, global hurricane frequency is likely to remain the same or decrease but hurricane intensity will likely increase (Kunkel et al., 2013). In light of this

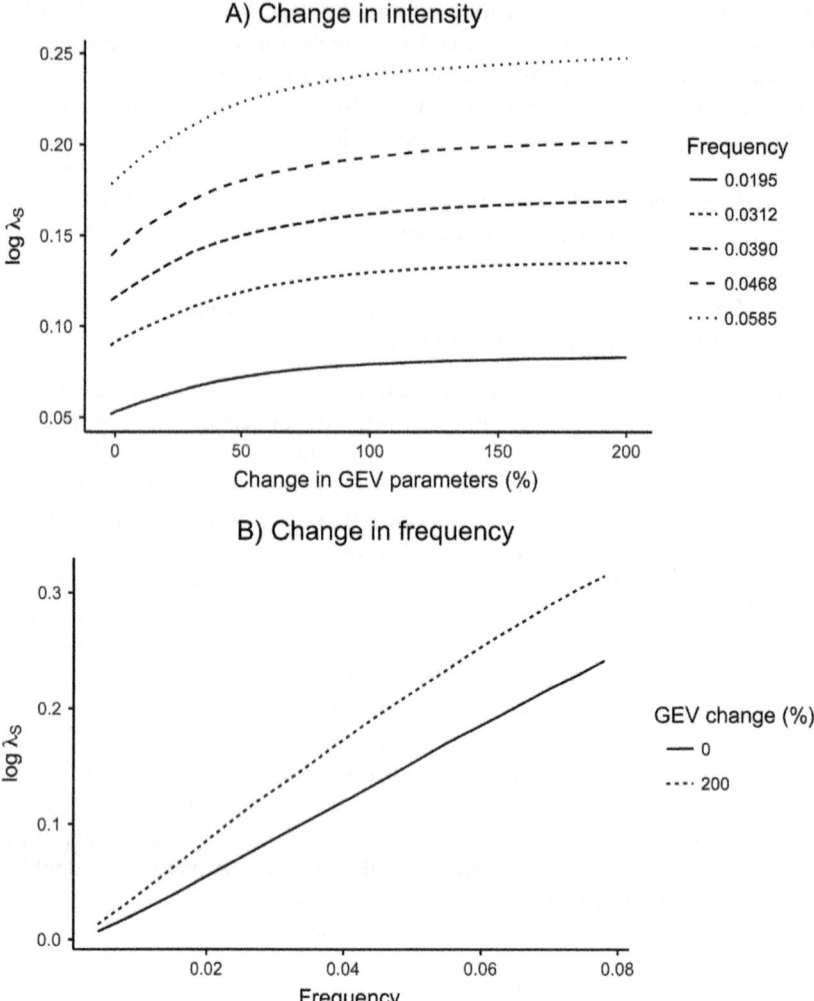

FIG. 6 Log of the stochastic growth rate ($\log \lambda_S$) for Ardisia escallonioides with changing hurricane magnitude and frequency. (A) Stochastic growth rate over different hurricane magnitudes (percent change in generalized extreme value (GEV) distribution parameters) for a given frequency. *Line* types represent hurricane frequencies (0.5x, 0.8x, 1x, 1.2x, and 1.5x historical frequency, respectively). (B) Stochastic growth rate over different hurricane frequencies, for 0% change in the GEV distribution parameters (*solid line*) and a 200% increase in the GEV distribution location and scale parameters (*dashed line*). Standard deviations for growth rate estimates are <0.005 and are not displayed on figures. *Originally published in Rypkema, D., Horvitz, C., Tuljapurkar, S., 2019. How climate affects extreme events and hence ecological population models. Ecology 100 (6), e02684.*

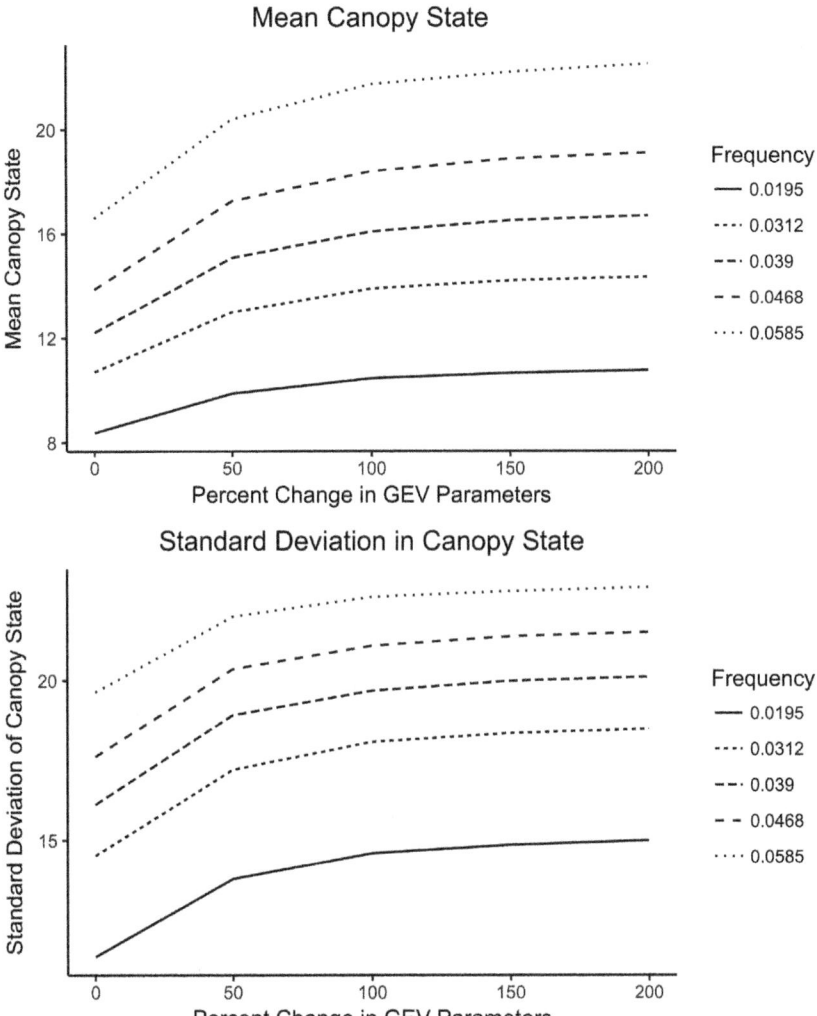

FIG. 7 (A) Mean percent canopy openness for a range of GEV and parameters and hurricane frequencies. (B) Standard deviation of percent canopy openness across a range of GEV and values and hurricane frequencies. *Line* types represent hurricane frequencies (0.5x, 0.8x, 1x, 1.2x, and 1.5x historical frequency, respectively).

prediction, our results suggest that climate change may not have strong effects on *A. escallonioides* populations. However, our model is single-species; the canopy may respond more strongly to higher magnitude hurricanes (e.g., through slowed canopy regrowth), changing the light environment and, in turn, *A. escallonioides* dynamics.

TABLE 4 Sensitivity of *Ardisia escallonioides* stochastic population growth rate ($\log \lambda_S$) to perturbations in storm frequency, intensity, canopy damage, and canopy recovery dynamics.

Factor	Slope
Frequency	0.1266
Intensity	0.0971
Canopy damage level	0.1990
Canopy recovery dynamics	0.2549

Source: Originally published in Rypkema, D., Horvitz, C., Tuljapurkar, S., 2019. How climate affects extreme events and hence ecological population models. Ecology 100 (6), e02684.

4   Challenges

4.1   Defining extreme events

The GEV is a useful model of ECE intensity and provides a way to systematically evaluate the impact of increasing ECE severity on ecosystems as the climate continues to change. Although the GEV is a strong model with vast empirical support, there are some challenges with its implementation in ecological settings.

The first challenge is deciding how to quantitatively characterize an extreme event. In certain cases, there are commonly accepted criteria used to define the event. For example, hurricanes have historically been represented by maximum sustained wind speeds and there is a pre-defined threshold of 64 knots; once a storm's maximum sustained wind speed exceeds 64 knots, it is deemed a hurricane (Shott et al., 2019). However, wind is not the only component of a hurricane; winds are accompanied by storm surge which may be just as (or more) destructive to property and ecosystems than strong winds (Dey and Das, 2015; Dinan, 2017). For a hurricane of a given category, storm surge has increased in recent years and is predicted to continue increasing in the future (Dey and Das, 2015; Dinan, 2017; IPCC, 2013). In other words, not only are hurricanes becoming more intense (more category 4 and 5 storms), they are causing more damage than previous storms of that category (Dey and Das, 2015; Dinan, 2017; IPCC, 2013). In the model presented here, we only model hurricanes and their effects using maximum sustained wind speed, however future modeling efforts for this system may need to include storm surge to realistically encompass the increasing destructiveness of hurricanes of a given wind speed.

In other cases, it is more difficult to select an appropriate metric for an extreme event. For example, wildfires are quantified by a variety of metrics, including total area burned, the duration of the fire, the severity of burn, and/or the proportion of high severity burn in the burned area (Tedim et al., 2018). The choice of one metric over another may influence which wildfires are designated as "extreme" in the model and can have downstream effects on results, so it is important to critically evaluate which option (or combination of options) is most appropriate for the situation.

Defining which events are extreme rather than high or low instances of a "typical" event can also be challenging. Some ECEs have commonly accepted criteria, such as hurricanes which are either above or below the maximum sustained wind speed threshold outlined by the Saffir–Simpson categorization. However, other ECEs are more difficult to define, especially temperature and/or precipitation extremes. For example, a heatwave can be defined based on the event's intensity and duration (Xu et al., 2016). To quantify intensity, a "temperature metric" needs to exceed a relative (e.g., percentile) or absolute threshold (Xu et al., 2016). Popular temperature metrics include maximum temperature, mean temperature, heat index, apparent temperature, or a combination of these (e.g., maximum and mean temperatures) (Xu et al., 2016). Xu et al. (2016) conducted a meta-analysis of the impacts of heatwave definition on human mortality estimation and found that criteria varied substantially. Studies with absolute temperature thresholds ranged from a minimum of 25–37°C while studies using relative temperatures used thresholds ranging from the 90th to 99.5th percentile of their chosen temperature metric (Xu et al., 2016). Minimum high temperature duration for these studies ranged from 2 to 6 days (Xu et al., 2016). All of these definitions aim to describe a heatwave as an abnormally hot time period, however the specifics differ.

These definitional differences can lead to alternate categorizations of heatwaves across studies and regions. The discrepancies in heatwave criteria led studies to define specific heatwaves differently. For example, the 2003 European heatwave lasted as short as 10 days (4–13 August 2003, Johnson et al., 2005) or as long as 92 days (1 June–31 August 2003, Grize et al., 2005) depending on the study's thresholds and metrics. Depending on heatwave thresholds, the 2004 Australian heatwave also has different definitions and, consequently, different conclusions about mortality associated with this heatwave. Analyzing the effects of the February 2004 heatwave in Brisbane on excess deaths, one study estimated the heatwave caused 49 excess deaths (Wang et al., 2012) and another estimated 75 excess deaths (Tong et al., 2010). The smaller estimate came from the study with a stricter definition of a "heatwave": a daily maximum temperature greater than or equal to 37°C for at least two consecutive days (Wang et al., 2012); implementing these criteria, the heatwave lasted from 21 to 22 February 2004. In contrast, Tong et al. (2010) defined the heatwave using daily mean, minimum, and maximum temperatures relative to historic warm seasons in the region and determined the heatwave lasted from

7 to 26 February 2004. In another study, Tong et al. (2014) apply a variety of relative heatwave definitions to daily mean temperatures for the warm season in Brisbane between 1996 and 2004, finding between 3 and 34 heatwave events in this time period depending on the percentile threshold ($\geq$99% to $\geq$90%).

The geographic scope of a study must be incorporated when choosing ECE definition criteria as well. For example, a study assessing the impacts of heatwaves across the United States would need to consider its specific research questions: if the study aims to analyze the effects of extremely high temperatures on human mortality, perhaps an absolute temperature threshold may be sufficient; if the study's goal is ecologically focused, a relative threshold may be more appropriate, providing for comparison of context-dependent ecosystem responses (e.g., an unusually high temperature in the Southwest United States is likely to be much higher than in the Pacific Northwest).

Another consideration is the time period used as the basis for "average" conditions. As the climate continues to change, conditions once considered "extreme" are becoming the new normal. If an ECE is defined relative to historical conditions (e.g., the minimum temperature must be $\leq$10th percentile of the region's historical temperature record to be considered a "coldwave"—an extreme cold event), the reference time period can shift which days are designated as extreme. Should our baseline shift to keep ECEs rare? Or, should we maintain our current baseline, increasing the frequency of ECEs, as ECEs continue to strongly impact ecosystems and society by providing extreme conditions the world has not yet adapted to consider "normal"?

Extreme event definitions do not have to be purely meteorological either. The traditional hurricane modeling framework, the Saffir–Simpson Scale, was developed to correspond maximum sustained wind speeds with damage to human lives and property (Saffir, 1973; Shott et al., 2019; Simpson, 1974) (hence why the category bins are unequal). The scale includes specific anticipated outcomes for each wind category, such as "Large branches of trees will snap and shallowly rooted trees may be toppled," for a Category 1 storm up to "A high percentage of framed homes will be destroyed, with total roof failure and wall collapse," for a Category 5 hurricane (Shott et al., 2019). Depending on the application, this framework may be helpful while defining "extreme"— the Saffir–Simpson Scale even includes information about expected power loss (e.g., "Near-total power loss is expected with outages that could last from several days to weeks," for Category 2 storms, Shott et al., 2019) and habitability of the area (e.g., "Most of the area will be uninhabitable for weeks or months," for Category 5 storms, Shott et al., 2019). However, in an ecological model (such as presented here), the damage relevant to the ecosystem may not correspond exactly to the damage relevant for insurance and disaster relief planning purposes.

4.2   Measuring the impacts of extreme events

Measuring ECE impacts brings up a second set of challenges. To accurately determine the effects of an ECE, a description of the system *before* the event is necessary for a meaningful comparison. However, ECEs are generally unpredictable (or predictable only on short time scales, such as hurricanes), so observations of "before" conditions need to be ongoing; this underscores the importance of long-term datasets, which are relatively rare in ecology.

Climatologically, collecting more spatially even physical data globally would be incredibly useful. There is a disproportionate amount of research on the "Western" world, neglecting other areas, such as most of the African continent (James et al., 2018). Without on-the-ground measurements of variables such as soil moisture and runoff, climate models and projections are poorly calibrated, creating high uncertainty when modeling these undersampled regions (James et al., 2018; Mengistu et al., 2019). Some of the poorly surveyed regions are challenging to model using methods developed to model and widely applied to "Western" systems and instead require physical, on-the-ground measurements. For example, tropical regions typically have frequent cloud cover, rendering the remote sensing datasets with frequent temporal resampling (and likely improved accuracy for climate modeling) less useful than for less cloudy regions (e.g., Sano et al., 2007).

Additionally, observation error can play a role in ECE determination and impact measurement. Over time, our ability to detect and measure ECEs has improved significantly, particularly during the latter half of the 20th century. For example, hurricanes were likely severely underestimated until satellite imagery was used to detect storms starting in 1960 (Landsea et al., 2004; Sheets, 1990). Before satellites, hurricanes were recorded based on sailors' reports when at sea and using observations when storms made landfall (prior to 1937), using radiosondes (starting in 1937), radar (starting during the World War II), and reconnaissance aircraft (starting in 1943) (Sheets, 1990). These methods missed many storms that never made landfall; Landsea et al. (2004) estimated that 0–6 tropical storms and hurricanes were "missed" per year from 1851 to 1885 and 0 to 4 were "missed" per year from 1886 to 1910. For storms that did make landfall, their intensity was sometimes underestimated because they reached their highest maximum sustained windspeed offshore (Landsea et al., 2004). Significant improvements in technology must be taken into account when analyzing the historical record, otherwise ECEs will appear to be increasing in frequency and/or intensity when, in reality, we are just better at detecting and measuring them.

ECEs may also appear to be more frequent and/or severe because the human development has increased over time. The population has grown substantially, so a higher number of people are affected by a given ECE and people now live in previously undeveloped locations, increasing the affected human-inhabited spatial extent (Dinan, 2017; IPCC, 2012). For example, Landsea et al. (2004) attribute the decrease in "missed" storms from 1886 to 1910 compared to 1851 to 1885 to increased coastal settlement, recorded meteorological observations, and shipping traffic. As climate change continues, more people will be exposed to a given ECE, increasing loss of life and property, decreasing disaster relief funding per individual, and potentially causing more conflict over resources and increasing the likelihood of climate refugees (Below et al., 2009; IPCC, 2012).

5   Open questions and future applications

The GEV is broadly applicable across systems, providing a flexible tool for ecological modelers. In addition to demographic studies, the GEV can be incorporated to answer a variety of ecological questions, modeling entire ecosystems, ecological communities, or species invasions (among others). In our specific case, *A. escallonioides* has a seed predator, *Periploca* sp. (*Cosmopterigidae*). This moth can remove 90% of *A. escallonioides'* fruits when the canopy is in its historically most common state, the most closed state (Horvitz et al., 2005). Although we implicitly incorporate historic moth predation in *A. escallonioides* state-specific vital rate matrices, the moth–*A. escallonioides* interactions can be explicitly modeled (see Horvitz et al., 2005). Previous work found that *A. escallonioides'* stochastic growth rate is most sensitive to changes in moth abundance in the moderately open habitat states (3 and 4) and found that hurricane frequency was a bigger driver of *A. escallonioides'* stochastic growth rate than moth abundance (Horvitz et al., 2005). However, previous work did not include changes in hurricane intensity (the most likely consequence of climate change), which may alter whether disturbance or species interactions have a greater effect on the stochastic growth rate.

Furthermore, our modeling framework can be adapted to include more nuanced models than the one presented here. For instance, feedbacks between ECEs and simultaneous events and drivers can be included. An example is the combination of heatwaves and droughts: the compound effects of incredibly dry and hot conditions can increase the probability and/or severity of a wildfire (Flannigan et al., 2009; IPCC, 2012). Conditions that would not be considered extreme when occurring alone can also combine to increase the likelihood of an extreme event, such as flooding as the result of heavy (but not extreme) precipitation coupled with fully saturated soil moisture (IPCC, 2012). Additionally, there may be an ecological feedback interacting with the physical environment, such as low precipitation coupled with high evapotranspiration leading to low soil moisture which can result in a drought (Miralles et al., 2019).

Human–environment impacts may also be important, with recent research suggesting that land use and land cover change may increase extreme precipitation (Pathirana et al., 2014).

6 Summary

Here, we outline how to fit the GEV and incorporate ECE frequency, intensity, and damage, and system recovery into ecological population models. We provide a framework that downscales GEV models to an ecologically relevant spatial scale and discuss challenges and considerations when applying GEVs to different event types and systems. The GEV is well-supported, flexible, and proves to be a useful tool, with a long history as the modeling standard for extreme events. Although the GEV is not as detailed as some climate simulations (e.g., the Coupled Model Intercomparison Project, Eyring et al., 2016), it is very straightforward to use and interpret—providing an opportunity for ecologists to explicitly model ECEs in their study system with a low barrier to entry. By explicitly including ECE frequency and intensity, we can estimate the strongest sources of ecosystem change, modeling damage and recovery during and after an ECE. Our case study underscores the importance of explicitly modeling ECE characteristics; by incorporating the effects of climate change on our system through the lenses of both hurricane frequency *and* intensity, we found (surprisingly) that the stronger hurricanes of the future may not significantly impact our focal species. Without explicitly modeling hurricane frequency and intensity, we would not have been able to parse out the effects of probable future climate change scenarios on *A. escallonioides* and may have overestimated the potential influence of future storms on our study species. Better understanding the impacts of ECEs on ecosystems can improve our plans for the future, supporting conservation prioritization and effective natural resource management. As climate change continues, it is imperative that we accurately and explicitly model extreme events so we can understand current ecosystem dynamics, project future conditions, and plan for the future.

7 Code availability

Code for model fitting and simulations is available on GitHub: https://doi.org/10.5281/zenodo.2575083.

Acknowledgments

We thank John Pascarella for original population data collection and original model design. We also thank Carol Horvitz for original model design and subsequent updates, and concept development. Figs. 5 and 6 and Tables 2 and 4 were originally published by Rypkema, Horvitz, and Tuljapurkar in *Ecology* as "How climate affects extreme events and hence ecological population models" (2019).

References

Alexander, L.V., Zhang, X., Peterson, T.C., Caesar, J., Gleason, B., Klein Tank, A.M.G., Haylock, M., Collins, D., Trewin, B., Rahimzadeh, F., Tagipour, A., Rupa Kumar, K., Revadekar, J., Griffiths, G., Vincent, L., Stephenson, D.B., Burn, J., Aguilar, E., Brunet, M., Taylor, M., New, M., Zhai, P., Rusticucci, M., Vazquez-Aguirre, J.L., 2006. Global observed changes in daily climate extremes of temperature and precipitation. J. Geophys. Res. Atmos. 111 (D5). https://doi.org/10.1029/2005JD006290.

Bailey, L.D., van de Pol, M., 2016. Tackling extremes: challenges for ecological and evolutionary research on extreme climatic events. J. Anim. Ecol. 85, 85–96.

Bali, T.G., 2007. A generalized extreme value approach to financial risk measurement. J. Money Credit Banking 39, 1613–1649.

Below, R., Wirtz, A., Guha-Sapir, D., 2009. Disaster category classification and peril terminology for operational purposes. Common Accord Centre for Research on the Epidemiology of Disasters (CRED) and Munich Reinsurance Company (Munich RE). cred.be/sites/default/files/DisCatClass_264.pdf.

Burgman, M., Franklin, J., Hayes, K., Hosack, G., Peters, G., Sisson, S., 2012. Modeling extreme risks in ecology. Risk Anal. 32 (11), 1956–1966.

Carstensen, J., Conley, D.J., 2019. Baltic sea hypoxia takes many shapes and sizes. Limnol. Oceanography Bull. 28 (4), 125–129. https://doi.org/10.1002/lob.10350.

Chernokulsky, A.V., Kurgansky, M.V., Mokhov, I.I., 2017. Analysis of changes in tornadogenesis conditions over Northern Eurasia based on a simple index of atmospheric convective instability. Doklady Earth Sci. 477, 1504–1509. https://doi.org/10.1134/S1028334X17120236.

Clarke, H., Lucas, C., Smith, P., 2013. Changes in Australian fire weather between 1973 and 2010. Int. J. Climatol. 33 (4), 931–944.

Coles, S., 2001. An Introduction to Statistical Modeling of Extreme Values. Springer, London.

Corvalan, C., Hales, S., McMichael, A., 2005. Millenium Ecosystem Assessment: Ecosystems and Human Well-being. Island Press, Washington, DC.

Daily, G.C., 1997. Nature's Services: Societal Dependence on Natural Ecosystems. Island Press, Washington.

Denny, M.W., Hunt, L.J.H., Miller, L.P., Harley, C.D.G., 2009. On the prediction of extreme ecological events. Ecol. Monogr. 79 (3), 397–421. https://doi.org/10.1890/08-0579.1.

Dey, A., Das, K., 2015. Modeling extreme hurricane damage using the generalized Pareto distribution. Am. J. Math. Manage. Sci. 35 (1), 55–66. https://doi.org/10.1080/01966324.2015.1075926.

Dinan, T., 2017. Projected increases in hurricane damage in the United States: the role of climate change and coastal development. Ecol. Econ. 138, 186–198. https://doi.org/10.1016/j.ecolecon.2017.03.034.

Dreesen, F.E., De Boeck, H.J., Janssens, I.A., Nijs, I., 2014. Do successive climate extremes weaken the resistance of plant communities? An experimental study using plant assemblages. Biogeosciences 11, 109–121.

Easterling, D.R., Meehl, G.A., Parmesan, C., Changnon, S.A., Karl, T.R., Mearns, L.O., 2000. Climate extremes: observations, modeling, and impacts. Science 289 (5487), 2068–2074. https://doi.org/10.1126/science.289.5487.2068.

Emanuel, K., Jagger, T., 2010. On estimating hurricane return periods. J. Appl. Meteorol. Climatol. 49 (5), 837–844. https://doi.org/10.1175/2009JAMC2236.1.

Eyring, V., Bony, S., Meehl, G.A., Senior, C.A., Stevens, B., Stouffer, R.J., Taylor, K.E., 2016. Overview of the coupled model intercomparison project phase 6 (CMIP6) experimental design and organization. Geosci. Model Dev. 9, 1937–1958.

Fisher, R., Tippett, L., 1928. Limiting forms of the frequency distribution of the largest or smallest member of a sample. Math. Proc. Cambridge Philosoph. Soc. 24, 180–190.

Flannigan, M., Krawchuk, M., Wotton, M., Johnston, L., 2009, 01. Implications of changing climate for global wildland fire. Int. J. Wildland Fire 18, 483–507. https://doi.org/10.1071/WF08187.

Fréchet, M., 1927. Sur la loi de probabilité de l'écart maximum. Annales de la societe Polonaise de Mathematique 6, 93–116.

Frölicher, T.L., Fischer, E.M., Gruber, N., 2018. Marine heatwaves under global warming. Nature 560, 360–364. https://doi-org.proxy.library.cornell.edu/10.1038/s41586-018-0383-9.

Gaines, S.D., Denny, M.W., 1993. The largest, smallest, highest, lowest, longest, and shortest: extremes in ecology. Ecology 74 (6), 1677–1692. https://doi.org/10.2307/1939926.

Gilleland, E., Katz, R.W., 2016. extRemes2.0: an extreme value analysis package in R. J. Stat. Softw. 72 (8). https://doi.org/10.18637/jss.v072.i08.

Gnedenko, B., 1943. Sur la distribution limite du terme maximum d'une serie aleatoire. Ann. Math. 44 (3), 423–453.

Grant, P.R., Grant, B.R., Huey, R.B., Johnson, M.T.J., Knoll, A.H., Schmitt, J., 2017. Evolution caused by extreme events. Philosoph. Trans. R. Soc. B Biol. Sci. 372 (1723), 20160146. https://doi.org/10.1098/rstb.2016.0146.

Greenfield, P., Lampcov, M., 2020. Beetles and fire kill dozens of 'indestructible' giant sequoia trees. The Guardian. https://www.theguardian.com/environment/2020/jan/18/beetles-and-fire-kill-dozens-of-california-indestructible-giant-sequoia-trees-aoe.

Gumbel, E., 1958. Statistics of Extremes. Columbia University Press.

Gutschick, V.P., BassiriRad, H., 2010. Biological extreme events: a research framework. Eos, Trans. Am. Geophys. Union 91, 85–86.

Harris, R.M.B., Beaumont, L.J., Vance, T.R., Tozer, C.R., Remenyi, T.A., Perkins-Kirkpatrick, S.E., Mitchell, P.J., Nicotra, A.B., McGregor, S., Andrew, N.R., Letnic, M., Kearney, M.R., Wernberg, T., Hutley, L.B., Chambers, L.E., Fletcher, M.S., Keatley, M.R., Woodward, C.A., Williamson, G., Duke, N.C., Bowman, D.M.J.S., 2018. Biological responses to the press and pulse of climate trends and extreme events. Nat. Clim. Change 8, 579–587.

Harvey, H.T., Shellhammer, H.S., Stecker, R.E., 1980. Giant Sequoia Ecology: Fire and Reproduction. vol. 12. U.S. National Park Service, Washington, DC.

Ho, S.S., Bond, N.R., Thompson, R.M., 2013. Does seasonal flooding give a native species an edge over a global invader? Freshwater Biol. 58, 159–170.

Horvitz, C.C., Tuljapurkar, S., Pascarella, J.B., 2005. Plant-animal interactions in random environments: habitat-stage elasticity, seed predators, and hurricanes. Ecology 86, 3312–3322.

IPCC, 2012. Managing the Risks of Extreme Events and Disasters to Advance Climate Change Adaptation. A Special Report of Working Groups I and II of the Intergovernmental Panel on Climate Change. Cambridge University Press, Cambridge, United Kingdom and New York, NY, USA.

IPCC, 2013. Climate Change 2013: The Physical Science Basis. Contribution of Working Group I to the Fifth Assessment Report of the Intergovernmental Panel on Climate Change. Cambridge University Press, Cambridge, United Kingdom and New York, NY, USA.

IPCC, 2019. Global Warming of 1.5°C. An IPCC Special Report on the Impacts of Global Warming of 1.5°C Above Pre-industrial Levels and Related Global Greenhouse Gas Emission Pathways, in the Context of Strengthening the Global Response to the Threat of Climate Change, Sustainable Development, and Efforts to Eradicate Poverty. Cambridge University Press, Cambridge, United Kingdom and New York, NY, USA.

Jagger, T.H., Elsner, J.B., 2006. Climatology models for extreme hurricane winds near the united states. J. Clim. 19 (13), 3220–3236. https://doi.org/10.1175/JCLI3913.1.

James, R., Washington, R., Abiodun, B., Kay, G., Mutemi, J., Pokam, W., Hart, N., Artan, G., Senior, C., 2018. Evaluating climate models with an African lens. Bull. Am. Meteorol. Soc. 99 (2), 313–336. https://doi.org/10.1175/BAMS-D-16-0090.1.

Jentsch, A., Kreyling, J., Beierkuhnlein, C., 2007. A new generation of climate-change experiments: events, not trends. Front. Ecol. Environ. 5 (7), 365–374. https://doi.org/10.1890/1540-9295(2007)5[365:ANGOCE]2.0.CO;2.

Johnson, H., Kovats, R.S., McGregor, G., Stedman, J., Gibbs, M., Walton, H., Cook, L., Black, E., 2005. The impact of the 2003 heat wave on mortality and hospital admissions in England. Health Stat. Q. 25, 6–11.

Katz, R.W., 2002. Techniques for estimating uncertainty in climate change scenarios and impact studies. Climate Res. 20, 167–185.

Katz, R.W., Brush, G.S., Parlange, M.B., 2005. Statistics of extremes: modeling ecological disturbances. Ecology 86 (5), 1124–1134. https://doi.org/10.1890/04-0606.

Kingsolver, J.G., Buckley, L.B., 2017. Quantifying thermal extremes and biological variation to predict evolutionary responses to changing climate. Philosop. Trans. R. Soc. B Biol. Sci. 372 (1723).

Knapp, K.R., Kruk, M.C., Levinson, D.H., Diamond, H.J., Neumann, C.J., 2010. The international best track archive for climate stewardship (IBTrACS). Bull. Am. Meteorol. Soc. 91 (3), 363–376. https://doi.org/10.1175/2009BAMS2755.1.

Knutson, T.R., McBride, J.L., Chan, J., Emanuel, K., Holland, G., Landsea, C., Held, I., Kossin, J.-P., Srivastava, A., Sugi, M., 2010. Tropical cyclones and climate change. Nat. Geosci. 3 (3), 157–163. https://doi.org/10.1038/ngeo779.

Kossin, J., Emanuel, K., Vecchi, G., 2014. The poleward migration of the location of tropical cyclone maximum intensity. Nature 509, 349–352. https://doi.org/10.1038/nature13278.

Kunkel, K.E., Karl, T.R., Brooks, H., Kossin, J., Lawrimore, J.H., Arndt, D., Bosart, L., Changnon, D., Cutter, S.L., Doesken, N., Emanuel, K., Groisman, P., Katz, R.W., Knutson, T., James, O., Paciorek, C.J., Peterson, T.C., Redmond, K., Robinson, D., Trapp, J., Vose, R., Weaver, S., Wehner, M., Wolter, K., Wuebbles, D., 2013. Monitoring and understanding trends in extreme storms: State of knowledge. Bull. Am. Meteorol. Soc. 94 (4), 499–514. https://doi.org/10.1175/BAMS-D-11-00262.1.

Grize, L., Huss, A., Thommen, O., Schindler, C., Braun-Fahrländer, C., 2005. Heat wave 2003 and mortality in Switzerland. Swiss Med. Weekly 135, 200–205.

Lande, R., 1993. Risks of population extinction from demographic and environmental stochasticity and random catastrophes. Am. Nat. 142 (6), 911–927. https://doi.org/10.1086/285580.

Landsea, C.W., Anderson, C., Charles, N., Clark, G., Dunion, J., Fernandez-Partagas, J., Hungerford, P., Neumann, C., Zimmer, M., 2004. The Atlantic hurricane database re-analysis project: Documentation for 1851-1910 alterations and additions to the HURDAT database. In: Hurricanes and Typhoons: Past, Present, and Future. Columbia University Press, New York, NY, pp. 178–221.

Li, X.X., 2020. Heat wave trends in Southeast Asia during 1979–2018: the impact of humidity. Sci. Total Environ. 721, 137664. https://doi.org/10.1016/j.scitotenv.2020.137664.

Lustenberger, A., Knutti, R., Fischer, E.M., 2014. Sensitivity of European extreme daily temperature return levels to projected changes in mean and variance. J. Geophys. Res. Atmos. 119 (6), 3032–3044. https://doi.org/10.1002/2012JD019347.

Martins, E.S., Stedinger, J.R., 2000. Generalized maximum-likelihood generalized extreme-value quantile estimators for hydrologic data. Water Resour. Res. 36 (3), 737–744. https://doi.org/10.1029/1999WR900330.

Meehl, G.A., Karl, T., Easterling, D.R., Changnon, S., Pielke Jr., R., Changnon, D., Evans, J., Groisman, P.Y., Knutson, T.R., Kunkel, K.E., Mearns, L.O., Parmesan, C., Pulwarty, R., Root, T., Sylves, R.T., Whetton, P., Zwiers, F., 2000. An introduction to trends in extreme weather and climate events: observations, socioeconomic impacts, terrestrial ecological impacts, and model projections. Bull. Am. Meteorol. Soc. 81 (3), 413–416.

Mengistu, A.G., van Rensburg, L.D., Woyessa, Y.E., 2019. Techniques for calibration and validation of SWAT model in data scarce arid and semi-arid catchments in South Africa. J. Hydrol. Reg. Stud. 25, 100621. https://doi.org/10.1016/j.ejrh.2019.100621.

Miralles, D.G., Gentine, P., Seneviratne, S.I., Teuling, A.J., 2019. Land-atmospheric feedbacks during droughts and heatwaves: state of the science and current challenges. Ann. N. Y. Acad. Sci. 1436 (1), 19–35.

Neria, Y., Nandi, A., Galea, S., 2008. Post-traumatic stress disorder following disasters: a systematic review. Psychol. Med. 38, 467–480.

Nydick, K., Brigham, C., Bradshaw, G., 2017. A climate-smart resource stewardship strategy for sequoia and kings canyon national parks. National Park Service, U.S. Department of the Interior.

Okoth, F.O., Odhiambo, B.D.O., 2018. Relationship between flooding and out break of infectious diseases in Kenya: a review of the literature. J. Environ. Public Health 2018. https://doi.org/10.1155/2018/5452938.

Pardo, D., Jenouvrier, S., Weimerskirch, H., Barbraud, C., 2017. Effect of extreme sea surface temperature events on the demography of an age-structured albatross population. Philos. Trans. R. Soc. B Biol. Sci. 372. https://doi.org/10.1098/rstb.2016.0143.

Parmesan, C., Root, T.L., Willig, M.R., 2000. Impacts of extreme weather and climate on terrestrial biota. Bull. Am. Meteorol. Soc. 81 (3), 443–450. https://doi.org/10.1175/1520-0477(2000)081¡0443:IOEWAC¿2.3.CO;2.

Pascarella, J.B., Horvitz, C.C., 1998. Hurricane disturbance and the population dynamics of a tropical understory shrub: megamatrix elasticity analysis. Ecology 79 (2), 547–563. https://doi.org/10.1890/0012-9658(1998)079[0547:HDATPD]2.0.CO;2.

Pathirana, A., Denekew, H.B., Veerbeek, W., Zevenbergen, C., Banda, A.T., 2014. Impact of urban growth-driven landuse change on microclimate and extreme precipitation–a sensitivity study. Atmos. Res. 138, 59–72.

Core Team, R., 2018. R: A Language and Environment for Statistical Computing. R Foundation for Statistical Computing, Vienna, Austria. https://www.R-project.org/.

Roberts, E.M., 1979a. Review of statistics of extreme values with applications to air quality data, Part I. J. Air Pollut. Control Assoc. 29 (6), 632–637.

Roberts, E.M., 1979b. Review of statistics of extreme values with applications to air quality data. J. Air Pollut. Control Assoc. 29 (7), 733–740.

Robinson, P.J., 2001. On the Definition of a Heat Wave. J. Appl. Meteorol. 40 (4), 762–775. https://doi.org/10.1175/1520-0450(2001)040¡0762:OTDOAH¿2.0.CO;2.

Rodrigues, J.A.M., Viola, M.R., Alvarenga, L.A., de Mello, C.R., Chou, S.C., de Oliveira, V.A., Uddameri, V., Morais, M.A.V., 2020. Climate change impacts under representative concentration pathway scenarios on streamflow and droughts of basins in the Brazilian Cerrado biome. Int. J. Climatol. 40 (5), 2511–2526. https://doi.org/10.1002/joc.6347.

Russo, S., Marchese, A.F., Sillmann, J., Immé, G., 2016. When will unusual heat waves become normal in a warming africa? Environ. Res. Lett. 11 (5), 054016.

Rypkema, D., Horvitz, C., Tuljapurkar, S., 2019. How climate affects extreme events and hence ecological population models. Ecology 100 (6), e02684.

Saffir, H.S., 1973. Hurricane wind and storm surge. Mil. Eng. 423, 4–5.

Sano, E.E., Ferreira, L.G., Asner, G.P., Steinke, E.T., 2007. Spatial and temporal probabilities of obtaining cloud-free Landsat images over the Brazilian tropical savanna. Int. J. Remote Sens. 28 (12), 2739–2752. https://doi.org/10.1080/01431160600981517.

Schwager, M., Johst, K., Jeltsch, F., 2006. Does red noise increase or decrease extinction risk? Single extreme events versus series of unfavorable conditions. Am. Nat. 167, 879–888.

Schwartz, M.W., Butt, N., Dolanc, C.R., Holguin, A., Moritz, M.A., North, M.P., Safford, H.D., Stephenson, N.L., Thorne, J.H., van Mantgem, P.J., 2015. Increasing elevation of fire in the Sierra Nevada and implications for forest change. Ecosphere 6 (7), art121. https://doi.org/10.1890/ES15-00003.1.

Sheets, R.C., 1990. The national hurricane center-past, present, and future. Weather Forecast. 5, 185–231.

Shott, T., Landsea, C., Hafele, G., Lorens, J., Taylor, A., Thurm, H., Ward, B., Willis, M., Zaleski, W., 2019. The Saffir-Simpson hurricane wind scale. United States National Oceanic and Atmospheric Administration National Hurricane Center. https://www.nhc.noaa.gov/pdf/sshws.pdf.

Simpson, R.H., 1974. The hurricane disaster potential scale. Weatherwise 27, 169–186.

Simpson, R., Lawrence, M., 1971. Department of Commerce, National Oceanic and Atmospheric Administration, Atlantic hurricane frequencies. NWS SR-58.

Smith, M.D., 2011. An ecological perspective on extreme climatic events: a synthetic definition and framework to guide future research. J. Ecol. 99 (3), 656–663. https://doi.org/10.1111/j.1365-2745.2011.01798.x.

Snow, R.W., Kibuchi, E., Karuri, S.W., Sang, G., Gitonga, C.W., Mwandawiro, C., Bejon, P., Noor, A.M., 2015. Changing malaria prevalence on the Kenyan Coast since 1974: climate, drugs and vector control. PLoS One 10 (6), e0128792. https://doi.org/10.1371/journal.pone.0128792.

Tabari, H., Willems, P., 2018. More prolonged droughts by the end of the century in the Middle East. Environ. Res. Lett. 13, 104005. https://doi.org/10.1088/1748-9326/aae09c.

Tedim, F., Leone, V., Amraoui, M., Bouillon, C., Coughlan, M.R., Delogu, G.M., Fernandes, P.M., Ferreira, C., McCaffrey, S., McGee, T.K., Parente, J., Paton, D., Pereira, M.G., Ribeiro, L.M., Viegas, D.X., Xanthopoulos, G., 2018. Defining extreme wildfire events: difficulties, challenges, and impacts. Fire 1 (1). https://doi.org/10.3390/fire1010009.

Thomas, M., Lemaitre, M., Wilson, M.L., Viboud, C., Yordanov, Y., Wackernagel, H., Carrat, F., 2016. Applications of extreme value theory in public health. PLoS One 11 (7), e0159312. https://doi.org/10.1371/journal.pone.0159312.

Tong, S., Ren, C., Becker, N., 2010. Excess deaths during the 2004 heatwave in Brisbane, Australia. Int. J. Biometeorol. 54 (4), 393–400.

Tong, S., Wang, X.Y., FitzGerald, G., McRae, D., Neville, G., Tippett, V., Aitken, P., Verrall, K., 2014. Development of health risk-based metrics for defining a heatwave: a time series study in Brisbane, Australia. BMC Public Health 14, 435.

Tuljapurkar, S., Horvitz, C.C., Pascarella, J.B., 2003. The many growth rates and elasticities of populations in random environments. Am. Nat. 162 (4), 489–502. https://doi.org/10.1086/378648.

USGCRP, 2017. Climate Science Special Report: Fourth National Climate Assessment. vol. I U.S. Global Change Research Program, Washington, DC.

van de Pol, M., Jenouvrier, S., Cornelissen, J.H.C., Visser, M.E., 2017. Behavioural, ecological and evolutionary responses to extreme climatic events: challenges and directions. Philos. Trans. R. Soc. B Biol. Sci. 372 (1723). https://doi.org/10.1098/rstb.2016.0134.

van der Geest, K., de Sherbinin, A., Kienberger, S., Zommers, Z., Sitati, A., Roberts, E., James, R., 2019. The impacts of climate change on ecosystem services and resulting losses and damages to people and society. In: Mechler, R., Bouwer, L.M., Schinko, T., Surminski, S., Linnerooth-Bayer, J. (Eds.), Loss and Damage from Climate Change: Concepts, Methods and Policy Options. Springer International Publishing, pp. 221–236.

Vasseur, D.A., DeLong, J.P., Gilbert, B., Greig, H.S., Harley, C.D.G., McCann, K.S., Savage, V., Tunney, T.D., O'Connor, M.I., 2014. Increased temperature variation poses a greater risk to species than climate warming. Proc. R. Soc. B 281, 20132612. https://doi.org/10.1098/rspb.2013.2612.

Wang, X.Y., Barnett, A.G., Yu, W., FitzGerald, G., Tippett, V., Aitken, P., Neville, G., McRae, D., Verrall, K., Tong, S., 2012. The impact of heatwaves on mortality and emergency hospital admissions from non-external causes in Brisbane, Australia. Occup. Environ. Med. 69 (3), 163–169.

Weatherford, C.L., Gray, W.M., 1988. Typhoon structure as revealed by aircraft reconnaissance. Part II: Structural variability. Mon. Weather Rev. 116 (5), 1044–1056. https://doi.org/10.1175/1520-0493(1988)116¡1044:TSARBA¿2.0.CO;2.

Westerling, A.L., Bryant, B.P., 2008. Climate change and wildfire in California. Clim. Change 87, 231–249. https://doi-org.proxy.library.cornell.edu/10.1007/s10584-007-9363-z.

Westerling, A.L., Hidalgo, H.G., Cayan, D.R., Swetnam, T.W., 2006. Warming and earlier spring increase Western U.S. forest wildfire activity. Science 313 (5789), 940–943. https://doi.org/10.1126/science.1128834.

White, P.S., Pickett, S.T.A., 1985. The Ecology of Natural Disturbances and Patch Dynamics. Academic Press, New York, NY, pp. 3–13.

Xu, Z., G., F., Guo, Y., Jalaludin, B., Tong, S., 2016. Impact of heatwave on mortality under different heatwave definitions: a systematic review and meta-analysis. Environ. Int. 89–90, 193–203.

Zampieri, M., Ceglar, A., Dentener, F., Toreti, A., 2017. Wheat yield loss attributable to heat waves, drought and water excess at the global, national and subnational scales. Environ. Res. Lett. 12 (6), 064008. https://doi.org/10.1088/1748-9326/aa723b.

Section II

Engineering sciences data

Chapter 3

Blockchain technology: Theory and practice

Srikanth Cherukupally*
Indian Institute of Science, Bengaluru, India
Corresponding author: e-mail: sricheru1214@gmail.com

Abstract

In the physical world, money exchange between two persons can happen without notice and intervention of a trusted third party. Such a money exchange transaction is called peer-to-peer transaction. When it comes to digital world, money exchange between two on-line entities is always monitored by a trusted third party like bank. The cryptographic research community constantly asked for over two decades the question: *"can the concept of peer-to-peer transactions be introduced into the digital world?"* The landmark proposal of Bitcoin crypto-currency (Nakamoto, 2009) provides a perfect answer to the question. However, Bitcoin suffers from criticism from many, given its uncontrollable nature. Despite this setback, what has caught more attention of the community is its underlying technology: *Blockchain.*

The blockchain technology is touted to be the next big wave in industry as it promises the features: transparency of data, accountability of transactions, transactions without intermediary, and data symmetry. The banking sector and supply chain systems are two fields that benefit most from this new technology and will witness a dramatic transformation. In addition to these two fields, blockchain has potential to show its implications to water management, health analytics, and several other areas. However, from a development point of view, the technology is still in rudimentary stage. It demands expertise in the subjects: distributed networking and cryptography. The features promised by the technology seek a new type of cryptographic algorithms, which are sort of open issues to be resolved. These open issues form the current active research of the cryptographic community.

The main focus of the present chapter is to teach the fundamentals of the blockchain technology and its use cases. The use cases are discussed through the blockchain platforms: Hyperledger, Ethereum, and other upcoming platforms. The chapter discusses the open research problems that are to be solved in having a full-fledged blockchain environment.

Keywords: Blockchain, Bitcoin, Hyperledger, Ethereum, Distributed consensus algorithm, Cryptographic secure hash, Public key encryption, Symmetric key encryption, Digital signature, Zero-knowledge proof

Handbook of Statistics, Vol. 44. https://doi.org/10.1016/bs.host.2020.10.001

1 Introduction

Consider the scenario: Alice is in debt to Bob a sum of 100 dollars. To pay back the indebted amount, Alice has two options: (i) pay in cash to Bob, and (ii) transfer the amount from her bank account to Bob's bank account. In the first option, the event of paying in cash is entirely private to them, meaning no third person intervenes in between them for the money transaction. To put differently, no other entity acts as an intermediate authority to settle the money exchange transaction. Such a transaction is called *"peer-to-peer transaction"*, which is formally defined as follows.

Definition: A peer-to-peer (P2P) transaction is a transaction that happens between two entities and does not require the presence of an intermediate party to settle (or validate) the transaction.

In the first option, Alice and Bob are happy about their P2P transaction being private. It is indeed a wish of any person to maintain privacy of their transactions. Money exchanges through physical cash achieve privacy of transactions. In the second option, Alice transfers the amount from her bank account to Bob's bank account. In this case, the money transfer transaction needs approvals from both banks. Unlike the first option, this digital transaction is not a P2P transaction, since third party is present for the money settlement between Alice and Bob. It had been the constant question in the research community that if the concept of P2P transactions could be introduced into the digital world. The question remained unanswered until the landmark proposal of Bitcoin network, an electronic payment system that supports P2P transactions. For historical reasons, there was an electronic cash payment system (David, 1981), but the system was failure as its core idea was way ahead of its time that there was no communication and computing platform to host it.

1.1 Bitcoin network

Digital transactions are always monitored by a trusted third party such as bank. The bitcoin platform completely eliminates the presence of a trusted third party. If no authority is present for verifying the transactions, then the question is: *who does validate the transactions?* The answer is provided in the paper (Nakamoto, 2009). The core idea of the paper is as follows: Imagine a network of N computing machines (or nodes) that are connected with a secure communication mechanism. The machines combinely take the responsibility of validating a transaction. The validated transaction will be recorded in the storage of each computing machine. A transaction that is newly posted into the network is broadcast to all machines of the network. The new transaction will be verified against the storage of all previously validated transactions at each machine. Only transactions, which do not violate the storage of any machine, will be validated and stored at each machine eventually. The logic of Bitcoin platform is such that at least $\frac{N}{2}$ machines are sufficient to come to an agreement in validating a transaction and remaining machines accept eventually the validated transaction and record in their storage. The process of validating a transaction can be done by any $\frac{N}{2}$ machines of the network. So, this validation logic does not bring immediate consensus across the total network. Any given time each machine will have its own history of validated transactions and machines verify the transactions based on their local storage. With this logic, it appears the total network does not arrive at a consensus at all on the validated transactions. Thanks to the underlying technology of Bitcoin: Blockchain, by which total network will eventually evolves with the same history of validated transactions at each machine (Fig. 1).

1.2 Blockchain

In the Bitcoin network, each machine has its own history of previously validated transactions. So, a new transaction posted in the network is validated independently at each machine based on its local database. The history of validated transactions is stored in the form of blocks. Each block is created at a particular point of time in the time history of the network. Each block

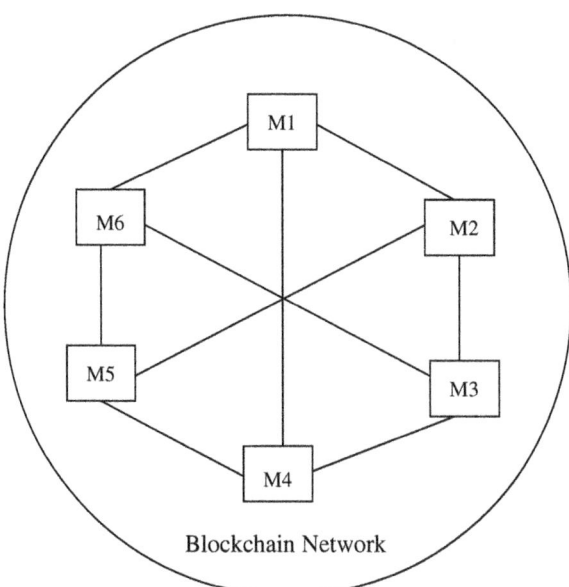

FIG. 1 A blockchain network of $N = 6$ machines which together validate transactions correctly, with the main assumption that $\frac{N}{2} = 3$ machines are honest. For the Bitcoin network, the value of N is several thousands.

comprises of some number of transactions. These blocks are chained one after another to form a chain of blocks, giving rise to the name blockchain. The new transactions that have been validated will form a new block which will be appended to the existing blockchain. Blockchain length grows as new transactions are getting validated. Whenever a new block is added, cryptographic hash of complete chain is calculated, and the hashed value is made public. With this simple idea of hashing, blockchain becomes immutable, meaning it is computationally infeasible to modify a transaction in a block. A malicious entity trying to modify a single transaction in a block (for its own benefit) has to rebuild entirely a new chain of blocks after the modified so that the hash output of new blockchain is same as that of the existing blockchain. The one-way computability nature of cryptographically secure hash functions (Section 2) makes the blockchain immutable.

As new transactions are posted into the network, some active machines of the network try to create news blocks. The network also contains dishonest machines (i.e., controlled by malicious users), which also actively try to add new blocks to the blockchain. A transaction is validated by at least N/2 nodes. To successfully attack the network, an adversary should be having computing power more than combined power of N/2 honest machines of the network to construct its own blockchain. As the length of blockchain increases, it becomes more and more difficult for an adversary to mount an attack on the network. In security sense, Bitcoin network is a Byzantine Fault Tolerant (BFT) system (Cachin et al., 2016; Castro and Liskov, 2002), with the underlying assumption that the number of dishonest nodes does not exceed half the total number of nodes.

1.3 How is blockchain different from present database systems

Existing database system maintains different copies of the same data by replication. There will be a main server which constantly gets data from transactions posted by applications. After storing data in its database, the main server replicates the same data in secondary servers. This setup of a main server and multiple secondary servers form a crash-tolerant system. In the event of a server crash, data is still intact as other servers have a backup of the same data. If main server crashes, one of secondary servers take the role of main sever. But the system is not fault-tolerant. Once the main server is compromised and is controlled by a malicious entity, its corrupt data can spread across other servers without notice.

Unlike the above database systems, blockchain system (network) does not replicate the data just by copying the data from one machine to other machine. The network follows a completely different mechanism which ensures that all the machines of the network eventually stores the same data. To understand this mechanism, imagine a collection of programs, called "Blockchain

Program Module (BPM)."[a] Suppose the module resides in all the machines of the network. Each machine will maintain its own database, which is independent of other machines'. Our objective is to make database modifications, which should be reflected at all machines. A modification to database is suggested through a transaction proposal. When a transaction is submitted into the network, each machine receives it. The received transaction become input to BPM. The output of BPM at a machine is a function of the received transaction and the local database of the machine. Based on their output of BPM, each machine makes changes to its local database. Thus, modifications to database of each machine is independently carried out with the same BPM. With this mechanism, if one machine gets compromised, it will become out of synchronization with other machines of the network.

In the case of present crash-tolerant database systems, a function module (like BPM) resides only at a central server. Any modifications to the database happen at the central server first, and the same will be replicated to secondary servers. Let us discuss this core functionality difference between present database system and a blockchain system with an example. Consider a BPM, which consists only of one function "CreateAccount(NAME,NUMBER, bal_amount)". The function creates an account with a given name, account number, with a given bal_amount as balance of the created account. Suppose there is an account with balance 100$. If some adversary modifies the balance as 200$ at the central server, it will be replicated in all secondary severs. The tampering of account balance might go unnoticed as all accounts show the same account balance. The tampering may be noticed in a careful audit process later point of time but damage has already happened and the tampering cannot be reversed. In case of blockchain network, the modification of account balance at one machine will be immediately caught as other machine still show up balance as 100$. The adversary's task of modifying the balance is much heavier here as he has to modify in all the machines. The main assumption of the Bitcoin network is that an adversary has to modify the balance at more than half the number of machines (called 51% attack) to achieve his objective.

1.3.1   Distributed consensus algorithm

Based on the output of BPM, each machine of blockchain network makes changes to its local database. By looking at these modifications, one might wonder if the network would ever have the same database (blockchain) at all the machines. Any given time, different machine will have a different view of blockchain. However, by a distributed consensus algorithm, the network will eventually have the same view of blockchain. The consensus algorithm

---

[a]In case of Ethereum network, BPM is called "smart contract." In case of Hyperledger network, it is called "chain code."

essentially tells the machines when to commit a transaction. Based on consensus algorithm, the blockchain network are classified into two different types of network: (i) public blockchain network and (ii) private blockchain network. In a public blockchain, any machine can validate transactions, whereas in a private blockchain, only a certain set of machines are allowed to validate transactions.

1.3.2 Organization of the chapter

The objective of the chapter is to teach fundamentals of cryptography that find relevance in the context of blockchain security. The chapter also presents blockchain types with examples.

Section 2 presents cryptographic foundations; Section 3 discusses the core idea of blockchain and what are commonality between public and private blockchains; Section 4 discusses the concepts of public blockchain; and Section 5 discusses private blockchains.

2 Cryptographic foundations

This section presents cryptographic algorithms that form the crux of the security of blockchain.

2.1 Secure hash function

The hash function is applied on the content of the newly added block and the hash of the existing chain (see Fig. 2). In mathematical terms, when a new block B is added to the chain, the hash function H is applied on B and h, where h is the hash output of the existing chain.

$$h' = H(B|h).$$

Here, $|$ is concatenation operation. The function H takes a message of arbitrary length as input and produces a random output of fixed length.

Definition: A function H is called *cryptographically secure hash function* if satisfies the following three properties:

1. given y, it should be computationally infeasible to compute x such that

$$y = H(x).$$

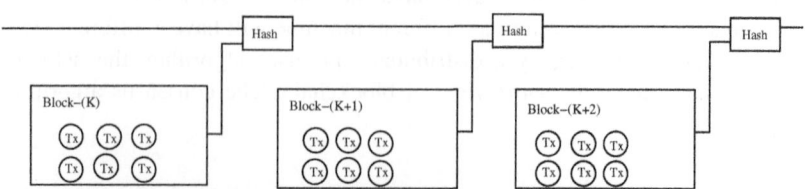

FIG. 2 Each block is formed with some number of transactions. Blocks are chained to form *Blockchain*. With addition of a new block, hash of the complete chain is computed.

2. given x, it should be computationally infeasible to compute $x_1(\neq x)$ such that

$$H(x_1) = H(x).$$

3. it should be infeasible to compute x_1 and x_2 such that $H(x_1) = H(x_2)$.

The property (a) implies that the hash function H is preimage resistant, meaning given any element y in the range of H it is difficult compute its preimage $x = H^{-1}(y)$. The property (b) implies that H is second-preimage resistant. The property (c) implies that H is collision resistant. However, given x, it is easy to compute $H(x)$. Informally speaking, the function H is one-way computable. The output of H is random: a small change in input produces a completely different output.

Examples of hash function are MD family and SHA family.

The output of the hash function is called *digest*. For example, the well-known SHA-256 function produces an output (digest) of length 256 binary bits. Since hash output is a finite number, there always exist two different messages that map to the same hash output. For example, there exist two different strings S1 and S2, from any set of $2^{256} + 1$ randomly chosen messages, that will have the same SHA digest, which means a hash collision. From the definition, it is difficult to find a collision. The brute-force method requires $2^{256} + 1$ messages to find a collision. However, one requires only 2^{128} different messages to find a collision with very high probability, due to Birthday paradox problem.

Birthday paradox problem: *What is the minimum size of a group of people of which there are at least two people in the group whose birthday fall on the same day with high probability?*

Considering a normal year of 365 days, if there are 366 people in the group, there will definitely be two people in the group with same day as their birthday. This is the trivial size of the group. The argument is deterministic and the collision is certain. However, with probability arguments, we could bring down the size of the group drastically in finding a collision.

Suppose G is a randomly chosen group of $n(<365)$ people. Let $P(n)$ be the probability that no two people of G share the same birthday. In order for G to get all distinct birthdays, the first person of G has 365 choices to choose his/her birthday. Then, second one has 364 choices, so on, and finally the last will have $365 - n$ choices. So,

$$p(n) = \frac{364}{365} \times \frac{363}{365} \times \frac{362}{365} \times \frac{365 - n}{365}. \tag{1}$$

Then,

$$\overline{p(n)} = 1 - \left(1 - \frac{1}{365}\right)\left(1 - \frac{2}{365}\right)\dots\left(1 - \frac{n}{365}\right)$$

is the probability that G has at least two people with the same birthday.

Consider the Taylor series $e^y = 1 + y + \frac{y^2}{2!} + \frac{y^3}{3!} + \dots$. When $y < 1$, $e^y \approx 1 + y$. With this approximation, we have

$$
\begin{aligned}
\overline{p(n)} &= 1 - (1 - \frac{1}{365})(1 - \frac{2}{365})\dots(1 - \frac{n}{365}) \\
&\approx 1 - e^{-\frac{1}{365}} \cdot e^{-\frac{1}{365}} \dots e^{-\frac{n}{365}} \\
&= 1 - e^{\frac{-(1+2+3+\dots+n)}{365}} \\
&= 1 - e^{\frac{-n(n+1)}{2 \times 365}}.
\end{aligned}
\tag{2}
$$

If $\overline{p(n)} = 1/2$, then n is about 23. This means that a random group of 23 people will have two people with same birthday with probability 1/2. From (2), it is clear that as group size increases, the success probability of birthday collision will increase.

The birthday problem can be applied to hash function for finding collisions. Suppose a function $f : A \rightarrow B$ on a given input x takes one of d different values (the size of B is d). Let $p(n, d)$ be the probability that a randomly chosen subset of A of n points will have all points that map to distinct elements of B. Using arguments similar to that of (1), the probability is

$$
p(n, d) = \frac{d-1}{d} \times \frac{d-2}{d} \times \frac{d-n}{d}.
$$

The probability that a randomly chosen subset of A of n points will have two points map to the same element of B is given by

$$
\overline{p(n, d)} \approx 1 - e^{\frac{-n(n+1)}{2d}}
$$

If $\overline{p(n, d)} = 1/2$, then $n^2 + n \approx 4d\log_e(2)$. So, n is approximately $\sqrt{d}$.

With the above result, a collision for a given function with k-bit output can be found by trying out $2^{\frac{k}{2}}$ input messages. Notice that the above probability arguments do not assume any distribution on the range space of the function. So, the collision argument can be applied to cryptographic hash functions, whose digest is uniformly distributed over its all possible values. In conclusion, if one tryout out with different 2^{128} input messages, one can a collision in SHA-256 with probability 1/2. The following section shows how difficult the process of finding a collision is by existing computing facilities of different capabilities.

2.1.1 A note on computational hardness

Let us consider a computing machine with processing speed of 2.2 GHz. Such a machine generates 2^{31} cycles per second. In a year, the machine generates about 2^{55} cycles. With considerable RAM size and pipelining facility for parallel computations, suppose one SHA-256 computation on a single input message takes 1/100 cycle. To find a collision for SHA-256, one needs to try out 2^{128} input messages. So, the entire collision-finding process consumes about $\frac{2^{128}}{100} \approx 2^{122}$ cycles. The machine has to run continuously for $\frac{2^{122}}{2^{55}} = 2^{67}$ years, a gigantic time period. So, the problem is computationally infeasible. The infeasibility of the problem is discussed in supercomputer environment and quantum era as follows.

- As of 2018, the world's fastest supercomputer is IBM Summit, which has capability of carrying out 122 Peta floating point operations per second (Pflops). One peta flop means $10^{15} \approx 2^{49}$ floating point operation per second. Such a supercomputer can execute 2^{73} cycles per year. With this machine, SHA-256 collision takes 2^{49} years, still a gigantic time period.
- Due to birthday paradox, the complexity of the collision-finding problem becomes the square root of total message space. In advent of a large scale quantum computer, the Grover's search algorithm (Grover, 1996) finds a collision in time $d^{1/4}$ where d is the total message space. For SHA-256, we have $d = 2^{256}$. Still, 2^{64} is a huge number. However, to counter the quantum attack, the length of hash digest can be increased.

In the above discussion, we have seen that finding collision for SHA-256 is computationally infeasible in computing environments of different capabilities. In general, any mathematical problem that consumes 2^{80} operations for its solutions is considered to be computationally infeasible. At the time of writing this chapter, any cryptographic algorithm in practice has their security parameters large enough such that the underlying hard problem require 2^{80} operations to be solved. For security reasons, if underlying hard problem of a cryptographic algorithm is search related, the sizes of security parameters need to be doubled keeping in mind the Grover's algorithm.

2.2 Open issue in theoretical computer science

Blockchain platforms use SHA-256 algorithm. The reason for the adoption of SHA-256 is that the cryptographic community, after a decade of streamlined research work, has come to a conclusion that there is no easy way out for finding collisions in SHA-256. The infeasibility of finding collisions means that SHA-256 is *one-way computable*. Like SHA-256, there exist problems which are one-way computable based on which encryption methods are designed.

Definition: A function $f : X \rightarrow Y$ is called "one-way computable", if it satisfies the following two conditions:

- It is easy to compute image of f for any $x \in X$
- It is computationally infeasible to compute preimage $f^{-1}(y)$ for any given $y \in Y$

The interesting question is "Do one-way functions exist?" The question is related to the well-known open issue in Theoretical Computer Science:

P vs NP problem. The class **P** is the collection of problems which can be solved in time polynomial in input size of the problem. The class **NP** is the collection of problems whose solving time is exponential in the size of input but their solution (if known) can be verified in polynomial time. The P vs NP problem states that **P≠NP**, which means there exist some computational problems which cannot be solved in polynomial time.

Three well-known problems of the class **NP** are presented below.

1. Integer factoring problem (IFP) is the problem of computing the prime factors of a given number N. No efficient algorithm exists for solving hard instances of the problem. The best algorithm called Number Field Sieve (Buhler et al., 1993; Lenstra and Lenstra, 1993) takes time exponential in $\log_e N$ for finding the prime factors of N. An hard instance of the problem is a large number N which is the product of two prime numbers p, q of same size. Given the prime factors p,q, it is easy to verify whether $N = p * q$.

2. Discrete Logarithm problem (DLP) is to compute x given $y = g^x$ (mod p). The problem is hard for a large prime p. The current best algorithm for solving the problem is Number Field Sieve (NFS) (Gordon, 1993) whose running time is exponential in $\log_e p$. However, given the solution x it is easy to verify if $y \equiv g^x$ (mod p).

3. Graph Isomorphism Problem (GIP) is to verify whether given graphs G1, G2 are copy of each other. The current best algorithm for solving GIP has running time which is exponential in the size of the input graphs. However, once the solution (mapping between vertex set of G1 and G2) is given, it can be verified quickly if the graphs are isomorphic.

The well-known encryption methods are proposed based on IFP and DLP. The security of the encryption methods assumes computational hardness of these two problems. The two problems, belonging to class **NP**, will fall into class **P** in the advent of a large scale quantum computer. The two problems will be solved in polynomial time using Shor's algorithm (Shor, 1997) on a quantum computer.

The Graph Isomorphic Problem, though out of context of the present chapter, is an interesting open problem in Graph theory. The recent result (Babai, 2016) shows that the problem can be solved in quasi-polynomial time [b] in k where k is the number of vertices. The result (Babai, 2016) is considered to be a major breakthrough in the field of computational complexity.

From the above discussion, it is important to note the following. When a problem is said to be in **NP**, it means the current state-of-art technique for solving the problem takes exponential time in the size of the input. The absolute one-way ness of a problem is subject to solving the **P ≠ NP** problem.

---

[b]Quasi-polynomial time in k means $e^{\log k^{\alpha}}$; polynomial time in k means k^{α}; exponential time in k means $e^{\alpha k}$ for some constant α. It can be verified that $k^{\alpha} < e^{\log k^{\alpha}} < e^{\alpha k}$, means quasi-polynomial time is bigger than polynomial time and is less than exponential time.

2.3 Encryption methods

The usual practice for a secure communication between entities/devices (of a system) is to encrypt messages. Being a decentralized network, blockchain system witnesses a lot of communication among its machines. It is thus important for the machines of the network to use encryption for keeping data secure. There are two types of encryption:

(i) public key encryption
(ii) symmetric key encryption

The main difference between the two types is the number of keys that entities of the system use for encryption and decryption. In public key encryption, each entity owns a pair of public key and private key. The public keys of entities are meant for encryption, and corresponding private keys are meant for decryption. In symmetric key encryption, entities (sender and recipient) of a communication agree upon on a common secret key. Both encryption and decryption of messages are performed using the common secret key.

2.3.1 Public key encryption system

Consider the scenario of two entities: Alice and Bob. Suppose Alice holds a pair of keys: one is public key p_a, and other is private/secret key s_a. Similarly, Bob's key pair is (p_b, s_b) where p_b is his public key and s_b is corresponding private key. Let $E(,)$ be the encryption function, and $D(,)$ be the decryption function. The functions E and D are publicly known. The functions are designed such that

$$x = D(E(x, p), s)$$

for any public key private key pair (p, s) in the system. Such x form the set of messages that are admissible in the system. Suppose Alice wants to send a plain message m to Bob, she computes cipher message c using Bob's public key, where $c = E(m, p_b)$. Bob, upon receiving cipher message c, recovers the message m by performing $D(c, s_b)$. Since s_b is private to Bob, it is only him who can recover m from c. Like Alice, using p_b any one can send Bob encrypted messages, which can only be decrypted by Bob.

Similarly, if Bob wants to send message m' to Alice, he computes $c' = E(m', p_a)$, and Alice performs $m' = D(c', s_a)$.

2.3.2 Authenticity of public key and private key pairs

An entity (say U) of the public key system will a pair of keys: (i) public key pk_u, and (ii) secret key sk_u. A sender will obtain pk_u for encrypting messages that are meant for U. The main question is how will the sender know the public key pk_u is valid. This problem is resolved by issuing cryptographic certificate to U by an well-known, authorized certificate authority (CA).

The certificate issued to U by CA comprises of essential details such as pk_u, encryption algorithm to be used, hashing algorithm to be used, certificate validity period. It is important to note that the issued certificate will be digitally signed by CA. Now the sender can verify the certificate (see Section 2.4) first to verify the veracity of pk_u. After successful verification of the certificate, the sender can now use pk_u for encrypting the messages that are meant for U.

2.3.3 Symmetric key encryption system

Here, the main ingredient is a pseudo-random function $F(x)$, which generates a random looking key stream on input number x. Suppose Alice and Bob agree upon a common secret key k. Let m be a messages that Alice wants to send Bob securely. She generates a key stream KS by $F(k)$. Then, she takes a bitwise XOR of the generated key stream and message stream [c] from m to produce cipher stream. Bob, on receiving end, generates the same key stream by $F(k)$. He then takes bitwise XOR of cipher stream and the generated key stream to recover the message stream.

2.3.4 Comparison

Compared to public key methods, symmetric key methods are faster, so will be used for bulk encryption. The sizes of secret parameters used in symmetric key methods are much smaller compared to sizes of secret parameters used in public key methods. It is known that, for using a symmetric key method, sender and receiver have to agree upon a common secret key. Public key encryption methods will be used for sharing the common secret key. In practice, a hybrid system (consisting of both public key and symmetric key methods) is used in a network of machines for secure transmission of data.

2.3.5 Usage in blockchain network

Being a decentralized network, a plenty of message transmissions happen between the machines. Suppose a machine is down for some time. When the same machine is up, it has to fetch updated ledger from other machines. The transmission of these messages happen securely. The size of the message to be obtained is small, a public key method can be used. If a machine joins the network recently, it will have to fetch complete ledger from other machines. Suppose a single machine is only reachable at that time. In this case the newly joined machine may have to request that single machine for obtaining the complete ledger securely. The entire ledger is bulk, so it has be transmitted by means of symmetric key encryption. The veracity of the obtained ledger can be verified against ledgers of other machines later point of time.

---

[c]Message stream is nothing but the binary representation of the message.

2.4   Digital signature mechanism

Message integrity is an important feature of messages that should be ensured for communication between entities/devices of a system. The feature ensures that message is not tampered in the middle before it reaches receiving entity. In other words, the feature ensures the receiving device that received message is same as the message that the sending entity had actually sent. Digital signature mechanism ensures message integrity.

Digital signature mechanism is complementary of public key encryption. The main difference is in the usage of public key and private key pairs of entities: private key is used to sign messages, and public key is used for verifying the signed messages. Signing on a message is done by only one entity with its private key. Since corresponding public key is known to all, verification of the signed messages can be done by any entity.

To illustrate further, consider the scenario of Alice and Bob, with their respective public key, private key pairs: (p_a, s_a), (p_b, s_b). Suppose $\sigma()$ is signature function, and $V()$ is the corresponding signature verifying function. Suppose H is the a cryptographically secure hash function. Let m be the message Alice wants to send Bob. Like before she sends $c = E(m, p_b)$. In addition to cipher text, she sends also her signature on $H(m)$ using her private key s_a, i.e., $q = \sigma(H(m), s_a)$. Upon receiving the cipher text and signature, Bob first recovers m by performing $D(c, s_b)$. Then, using Alice's public key p_a, he checks for message integrity of recovered message m by evaluating $V(q, p_a)$. The function V will be designed such a way that it provides a proof that the private key s_a was used for signing m.

2.5   Pseudo random number generators

It is important to note that communication between machines is secure using encryption algorithms but messages being exchanged are always transmitted over a insecure channel. An adversary can intercept the channel to obtain all messages that are exchanged between machines. The adversary can store all exchanged messages and try to extract information from them to his advantage at a later time. Since messages are encrypted, he may not be able to decrypt but try to read a pattern[d] in the underlying plain messages. He can achieve his objective if encrypted messages are not random strings. Finding patterns in underlying plain messages is as equivalent as breaking the encryption method used to encrypt the plain messages. The main idea of encryption algorithm is to suppress the patterns of the plain messages with use of a random key to produce random encrypted messages. Hence, it is of utmost

---

[d]Plain messages are nothing but sentences from a language human beings speak. Thus there is definitely a pattern in the messages.

importance to use random numbers for encryption algorithms. We discuss below two examples showing the importance of random numbers in crypto-graphic algorithms.

In Section 2.3, we have seen that sender and recipient have to agree upon a common secret integers for using symmetric key encryption method. Suppose the shared secret key is generated by a pseudo-random function whose output is of length 128 binary bits. If the output is random, meaning output value is uniformly distributed over all 2^{128} possible values. In this case, adversary has to try out with all 2^{128} values. From the note on computational hardness (Section 2.1), it will be infeasible for the adversary to guess correctly the key. If there is a clear bias in the key, security of the method will be compromised. For example, for the generated 128-bit secret key, suppose bits at 40 specific positions are not random. If adversary know those specific positions, then he will have to guess remaining 78 bits only. This leads to 128-bit security level coming down to 78-bit security level only. The adversary will store all cipher texts produced from the algorithm, and can mount a passive attack on recovering plain messages, perhaps by guessing remaining 78 bits of the secret key.

2.5.1   Appending random numbers to messages

For encryption of bulk messages, symmetric key methods are used. For small messages, a public key method can be used directly. However, encrypting a small message as it is will lead to fully recover the message. For example, m is the plain message being encrypted, which is of length less than 32 bits. From Section 2.3, Alice wants to send m to Bob. Using Bob's public key p_b, she computes cipher $c = E(m, p_b)$. A man in the middle (adversary) will come to know both c and p_b. Since message is a binary string of length 32 bits. He can try out with every $x \in \{0, 1, 2, \ldots 2^{32}\}$, and verify whether $c = E(x, p_b)$. Suppose computing the output of encryption function $E(x, p_b)$ takes one cycle. So, knowing message m requires only 2^{32} cycles. From the Note in Section 2.1, the entire process is doable in 1 s. The reader is requested to do little calculation to see that even a message of length 50 bits is recoverable with machine with substantial computing power.

Though the ability to recover small messages may not imply complete break-ing of the encryption method used as the trial and error method will not work for very big messages. But, this a serious issue as messages being encrypted are small more often. This issue can be avoided by appending a big random number r of a fixed length to m. With this modification, Alice now computes $c = E(m|r, p_b)$. The adversary will not succeed with the trial and error method.

2.6   Zero-knowledge protocols

Zero-knowledge game is a question-and-answer game that is played between two entities: (1) challenger and (2) prover. The prover wants to prove to the challenger that he has a secret. In the game, the challenger asks some questions one after another for which the prover has to provide answers. Each question is independent of other questions. If the prover provides answers correctly for finite number of rounds the challenger will get convinced that prover indeed has a secret. In the process the challenger will learn nothing other than prover's answers to the questions posed, thus the game is zero-knowledge. To achieve the objective of the game, it should be played infinite number of rounds. Here, the game is played only for a finite number of rounds. there is a probability associated with with success rate of the prover. So, the game is probabilistic.

The game is illustrated with an example that involve coin tossing. Suppose Tom and Harry are two friends. Tom claims that he can guess the outcome of a coin tossing correctly. Like anyone, Harry is skeptical of Tom's claim and becomes curious about verifying the claim. Tom is the prover and Harry is the challenger. He challenges Tom with the question: "what is your guess?" before he tosses the coin. Tom's answer is either Heads or Tails. If Tom answers incorrectly, the game ends and Tom's claim is proved wrong. If Tom answers correctly, Harry asks again what Tom's guess is before he tosses the coin again. As Tom answers correctly for each coin tossing, Harry gets convinced more and more about the claim. Since the game cannot be played for infinite number of rounds, Harry will stop at one point and will get convinced that Tom's claim is right. In the course of the game, Harry learns only Tom's guesses and will not learn anything about how to guess the coin tossing correctly. So, Harry cannot own the Tom's claim and he cannot play the same game with other person. Hence the game is zero-knowledge.

Let us discuss the above game mathematically. Assuming the uniform distribution on the outcome of coin tossing, two outcomes {H, T} occur with equal probability, i.e., 1/2 each. Further, we assume that the outcome of one coin tossing is independent of the outcome of other coin tossing. Suppose Tom and Harry play the game for n coin tossing. Let Tom's guessing string be $g_1 g_2 ... g_{n-1} g_n$ for n coin tossings when $g_i \in \{H, T\}$. The total number of binary strings of length n is 2^n. By the uniform distribution on the outcome of coin tossing, any n-length string occurs with probability $\frac{1}{2^n}$. Thus, if Tom does not know any magical method of guessing the outcome of a coin tossing correctly, the probability that Tom succeeds in guessing correctly for all n coin tossings is $\frac{1}{2^n}$. As n tends to ∞, $\frac{1}{2^n}$ tends to 0. For $n = 100$, the success probability for Tom is very small. Thus, it is reasonable for Harry to play the game for 100 coin tossings to get convinced that the Tom's claim is correct.

If we observe the above zero-knowledge game based on coin tossing, there is no clear quantitative measure for the secret procedure that Tom is claiming to have. This defect makes it usable in cryptographic world. The zero-knowledge protocols used in practice are based on number theoretic problems. In the landmark paper (Goldwasser et al., 1989), the notion of zero-knowledge protocol is introduced.

2.6.1 Formal notion

Zero-knowledge protocol is a method, by which a party (Prover P) can prove to another party (Verifier V or Challenger) about a statement S, which satisfy the following three properties.

1. *Completeness:* If the statement S is true, an honest prover can prove S to an honest verifier.
2. *Soundness:* If the statement S is false, no dishonest prover can prove to an honest verifier that S is true.
3. *Zero-knowledge:* If the statement S is true, the verifier does not learn anything other than the fact that S is indeed true.

The protocol expects both prover and verifier to be honest, meaning they follow the rules of the protocol correctly. Note that there is a small probability, the soundness error, that a cheating prover can convince verifier that a false statement is true. So, zero-knowledge protocols are probabilistic. As the protocol is about proving the prover's claim, zero-knowledge protocols are also called zero-knowledge proofs which is not actually so in mathematical sense. A mathematical proof is always deterministic which either proves or disproves a statement deterministically.

There are two types of zero-knowledge protocols:

- Interactive protocol
- Noninteractive protocol

In an interactive protocol, prover and verifier interact for more than one round for having to verify the prover's claim. For example, the coin tossing game between Tom and Harry discussed earlier is interactive. In a noninteractive protocol, prover produces a complete proof about his claim at once to the verifier.

a Zero-knowledge protocol is defined based on the computational hardness of two problems: (1) Integer Factoring problem and (2) Discrete Logarithm problem, discussed in Section 2.2.

2.6.2 Interactive proof (example 1)

Recall that Integer factoring problem (IFP) is the problem of computing prime factors of a given number N. There exists no efficient algorithm exists for solving hard instances of the problem. The best algorithm called Number Field Sieve takes time exponential in $\log_e N$ for finding the prime factors of N. An hard instance of the problem is a large number N which is the product of two prime numbers p, q of same size. A number r is called a **quadratic residue** modulo N if there exist a number x such that $x^2 \equiv r \pmod{N}$. The problem of computing square roots of a quadratic residue (mod N) is equivalent to finding the prime factors of N. For example, suppose $x^2 \equiv y^2 \equiv a \pmod{N}$. If $y \equiv \pm x \pmod{N}$, then the greatest common divisor (gcd) of

$y \pm x$ and N will yield a prime factor of N. $4^2 \equiv 1^2 \equiv 1 \pmod{15}$, and the square roots of 1 are 1,14,4,11. Thus, gcd of 4 ± 1 and 15 results in the prime factors of 15. Based on the equivalence between computing square roots and IFP, an interactive protocol is as follows.

Tom has the number N. He claims that $N = p \times q$ and (p, q) are prime numbers which are his secret integers. In other words, Tom claims that he knows the prime factors of N. Harry is the verifier (challenger). Tom picks a random $r < N$ and computes $R \equiv r^2 \pmod{N}$. He makes R public. Harry picks a random $r_1 < N$ and computes $R_1 \equiv r_1^2 \pmod{N}$. He sends R_1 to Tom. Harry now asks for a square root of one of R and $R \times R_1$ based on coin tossing. Tom can provide square root of any one of two, since he has the knowledge of p and q. If he does not know the factors, he cannot provide a square root of $R \times R_1$. Like in the coin toss game discussed earlier, the success rate for Tom winning the game in all k rounds is $\frac{1}{2^k}$, which is negligible for a large enough k. After k rounds, Harry accepts that Tom knows the prime factors (p, q).

2.6.3 Interactive proof (example 2)

Discrete Logarithm problem is to compute x given $g^x \pmod{p}$. The problem is hard for a large prime p. The current best algorithm for solving the problem is Number Field Sieve (NFS) whose running time is exponential in $\log_e p$. Based on this hardness assumption, an interactive protocol is as follows.

Tom wants to prove to Harry that he knows secret x such that $y = g^x \pmod{p}$ where (y, g, p) are known. He picks a random r such that $R = g^r \pmod{p}$. Harry will ask for discrete logarithm of either y or yR. In other words, he will ask for one of r or $x + r$. If Tom provides r, Harry verifies that $R = g^r \pmod{p}$. If Tom provides $x + r$, Harry verifies that $yR = g^{r+x} \pmod{p}$. Harry asks for one of the two Discrete logarithms following uniform distribution. Harry does coin tossing to choose of one of r and $x + r$. if coin tossing is Heads, he asks for r. If it is Tails, he will ask for $x + r$.

2.6.4 Noninteractive proof (example)

In public key encryption system, the authenticity of public key and private key pair of an entity is ensured by a certificate authority. Harry (verifier) wants to verify that Tom has a private key (i.e., a legitimate user of the system), he can directly check the authenticity of the certificate issued to Tom. The certificate is served as a noninteractive proof.

In a blockchain network, verification of a node's identity uses noninteractive proofs explained as above. Further, zero-knowledge proofs can be used to validate a transaction without revealing identity of sender and recipient. Zcash (Ben-Sasson et al., 2014) is a crypto-currency that provides anonymity of entities involved in its transactions.

3   Blockchain platforms

Blockchain network is a live network of computing machines[e] Each machine acts an access point for the network, meaning transactions can be submitted into the network via any one of the machines. Based on a consensus algorithm, the machines together validate submitted transactions. The validated transactions will be pushed into blockchain. Once a transaction is stored into blockchain ledger, it is copy will resides will reside in ledger at each machine for ever.

The participating machines of the network are not controlled by a single authority. The machine can be located in different geographic locations. The assumption is that a machine may join the network or leave the network any time.[f] So, the machines that are running take the responsibility of retaining the functionality of the network. We have the basic assumption that machine may join the network or leave the network any time. This means that the identification of machines will not be known to other machines, and thus the network offers a trust-less environment. However, one can create a blockchain network which offers a trustful environment. This is achieved by creating a trust mechanism for identifying the machines. Based on the level of trust that is expected between the machines, we have two paradigms (or categories) of a blockchain network:

1. Public blockchain
2. Private blockchain

The public blockchain is also a "Permissioned-less blockchain," since any machine can join the network any time just by importing the blockchain software and downloading the ledger from the network. The synonymous name for Private blockchain is "Permissioned blockchain" as participating machines have a clear permission to be part of the network. The permissions are identified and recognized in form of the cryptographic certificates issued from certificate authority (CA).

Public blockchain networks: Bitcoin (Bitcoin, n.d.) and Ethereum (Ethereum, n.d.) *Private blockchain networks*: Hyperledger Fabric (Hyperledger, n.d.)

3.1   Commonalities of public and private blockchains

Based on the trust assumption we have two categories of blockchain network. The trust assumption is a deciding factor how two networks operate completely in a different manner. In a private blockchain network, we have

---

[e]Machine, we mean by, is simply a computer which has core blockchain programs (software) installed on it. At the time of writing the chapter there exist no hardware devices designed specially for blockchain.

[f]This assumption is crucial as a machine may be shutdown suddenly any given point of time. The same machine will be up after later point of time, and try to join the network.

participating machines identified by cryptographic certificates issued from a trusted party, so there is a clear structure defined for the network. Whereas, a public network is completely a trust-less environment, which may accommodate vulnerable machines any given time. So, adversarial model for the two networks will be entirely different. Also, agreement protocols used for achieving a consensus among the machines will be different for the two networks. We study these points in detail on the two networks in Sections 4 and 5. Before the detailed discussions, it is of interest to ask the following question from a development point of view.

"What are the common features shared by public and private blockchain networks?"

To answer the question pin pointedly, the both networks obey the very definition of blockchain: once the data is pushed into blockchain successfully, it persists untampered forever. In addition to this immutable nature, some basic functionalities of machines of both networks are same. Meaning, a participating machine, either it is belonging to a public blockchain or to a private blockchain network, consists of the following two key components.

(i) State
(ii) Ledger (Blockchain)

We can think of the Blockchain State (or simply called State) as a database storing up-to-date values of created data elements such as accounts, assets, etc. Any modification to State is suggested via transactions. When a transaction is posted into the network, it will be validated by each machine of the network with respect to their State. Once the transaction is validated, the modifications suggested in the transaction will apply on State, and the transaction will be registered into Ledger (blockchain). So, State contains the actual data, and blockchain registers a list of the validated transactions which have successfully done modifications to State so far. To understand this concept, let us consider a set of transactions:

Create-tx(Alice,1729,0): *"create an account named Alice, with account number: 1729 with 0$ balance"*

As the transaction Create-tx(Alice,1729,0) is posted into the network, each machine will receive it. The transaction will go through validation phase at each machine. Note that the validation of the transaction at each machine is independent as it is done with respect to machine's State. Remember that State is a database comprising all created accounts. Each machine will look into its State if there exist already any account with the account number 1729. It there exist no such account, a new account will be created on name Alice will account number 1729. The balance of the account will be 0$.

Let us consider a new transaction:

Create-tx(Bob,1730,5): *"create an account named Bob, with account number: 1730 with balance 5$"*

Similar to the above transaction, Bob's account will be created if it does not exist already. So, each machine will have created Bob's account in its State, and the transaction: Create-tx(Bob,1730,5) will get registered in the corresponding Ledger.

3.2 Consistency of Ledger across the network

From the above section, we have two transactions: (i) Create-tx(Alice,1729,0), and (ii) Create-tx(Bob,1730,5). After execution of the transactions, each machine will have two accounts created in its State, and the two transactions will be registered in its Ledger. Now the interesting question is about the ordering of the two transactions. Since blockchain network is a distributed network with inherent communication latencies, machines may receive the two transactions in different ordering. Suppose some machines might have received Create-tx(Bob,1730,5) first, then Create-tx(Alice,1729,0). If it is so, the Ledgers of the machines get out of sync. At this point of time, it seems that two different orderings of the transactions do not create any problem for the network. Now let us consider the following transaction.

> **Transfer-tx(Bob,Alice,3)**: *"Transfer 3\$ from Bob's account to Alice's account"*

Suppose a machine gets Transfer-tx(Bob,Alice,3) before Create-tx(Bob, 1730,5). Then, the transfer transaction will become invalid with the ordering: Create-tx(Alice,1729,0)||Transfer-tx(Bob,Alice,3)||Create-tx(Bob,1730,5). It is clear that ordering of the transactions matter. It is important for the network to maintain not only correct execution of transactions, but also have same ordering of the transactions at each machine. Then only, Ledger will be consistent across the network.

3.2.1 Consensus algorithm

Maintaining the consistency of Ledger across the machines is subjected to a consensus algorithm employed by the network. The trust assumption we have for blockchain network will play a major role in choosing a consensus algorithm. For private blockchain networks, such as Bitcoin and Ethereum, Proof-of-Work (PoW) consensus algorithm is employed. Any machine of the network is eligible for running PoW algorithm. In case of a public network like Hyperledger, as participating machines are clearly identified, one can choose a set of machines which will endorse (or sign) the submitted transactions, and remaining machines can just do modifications to their State after successful verification of all endorsements.

4 Public blockchains

In this section, we discuss two public blockchain networks: (i) Bitcoin network, and (ii) Ethereum network.

4.1 Bitcoin network

The Bitcoin network is the first ever decentralized network. The network runs in a trust-less environment, meaning there is no trust assumption among the machines of the network. The underlying assumption is that some machines can act in a malicious way, some of them machines can be compromised and some machines do tweaking in the network for their own advantage. In presence of all these adverse conditions, the paper proves that the network assures the correct execution of peer-to-peer (P2P) transactions, under the main assumption that at least half of the total machines of the network are honest. The network utilizes the well-known cryptographic algorithms.

4.1.1 Proof-of-Work (PoW) algorithm

As transactions are submitted into the network, each machine gathers them and validate against their State. Then, each machine tries to create a block locally out of the transactions it has just validated. Now, the step for machines is to run Proof-of-Work algorithm on newly created block until a "desired" output appears. Getting a desired output is equivalent to solving a hash puzzle. Essentially, hash algorithm (SHA-256) applied to the new block twice should result in an output with a certain pattern, in specific, k Most Significant Bits (MSB) in the hash output should be zero. To achieve the target hash, one has to try appending a number only once (*nonce*) with the new block.

$$\text{SHA} - 256(\text{SHA} - 256(\text{new block}\|\text{nonce})) < 2^{256-k} \qquad (3)$$

where "$\|$" representation the concatenation operation. The hash difficulty is 2^k, meaning one has try out with 2^k different nonce to get a hash output with k MSB zeros.

Whenever a machine gets a desired hash output with a particular nonce, it will announce its result immediately to the network. The result comprises of three parts: (i) new block, (ii) nonce, and (iii) hash output. The other machines will receive the result and verify if the desired result is correct by verifying Eq. (3). Upon successful verification of the equations, the machines then validate the transactions that are contained in the new block against their own State. For each validated transactions, modifications suggested in it will be effected in the State.

To summarize the above procedure, the steps followed by nodes of the network in adding a new block to existing chain are as follows.
1. New transactions are broadcast to all nodes.
2. Each node validates new transactions, and then collects them into a block.
3. Each node works on PoW algorithm on the newly formed block.
4. When a node find a proof-of-work, meaning it finds a desired hash output mentioned in Eq. (3), it broadcast the result in the network.
5. Each node accepts the received block only if all transactions are valid and not already spent (the validation of transactions with respect to its State).
6. Nodes accept finally the new block by appending it to the existing chain.

It is important to notice that there is no requirement as such for a node to get involved in running the PoW algorithm. A node can just accept blocks from other nodes without running the PoW algorithm.

4.1.2   Miners

A node that run the PoW algorithm is called "miners." There will not be any fixed set of miners defined for the network. Any given point of time, a node wanting to be a miner can become so. The network's transaction history is always protected by a set of honest miners whose combined decision prevents an adversary from pushing fake transactions (for his benefit) into the history. When a miner announces a new block with produced hash value satisfying (3), other nodes will verify that if every transaction is the new block is valid and the new block hashes to the announced hash. When a node verifies the new block successfully against its local database, it will accept the new block and add it to its blockchain. The miner who announces the new block will be incentivized with some Bitcoins.

4.1.3   Cost of successful mining

We have sha-256(sha-256(block$\|$nonce)) $< 2^{256-k}$ only if k most significant bits are zero. If the MSB part is nonzero, then sha-256(sha-256(block$\|$nonce)) $> 2^{256-k}$. So the probability that, for all z chosen nonces, sha-256(sha-256 (block$\|$nonce)) $> 2^{256-k}$ is $\left(1 - \frac{1}{2^k}\right)^z$. The probability that, sha-256(sha-256 (block$\|$nonce)) $< 2^{256-k}$ for at least one of z nonce is

$$1 - \left(1 - \frac{1}{2^k}\right)^z \approx 1 - e^{-\frac{z}{2^k}}$$

$$= 1 - e^{-1} \quad \text{when } z = 2^k$$

So, a miner has to try out with at least 2^k different nonce, which means 2^{k+1} hash computations need to be carried out on an average for getting the desired hash output. The cost of mining increases exponentially with increasing k. Suppose $k = 40$, on an average 2^{41} hash computations are required.

4.2   Ethereum network

Ethereum (n.d.), like the Bitcoin network, is a public decentralized network. The Ethereum network adds more functionalities to the Bitcoin network while retaining its core properties. Like Bitcoin, the transactions of the network are validated by a set of honest nodes. A major difference is that the network is not just an electronic payment system (like Bitcoin), it has capability to accommodate new systems (models) which seek decentralization. The network hosts crypto-currencies including its own currency called *Ether*.

Ethereum network has a instruction set which is Turing-complete, meaning the network can simulate any script. The instruction set of Bitcoin network is not Turing-complete as it is restricted to executing only payment transfer script.

To deploy a model M, the script of the model (called smart contract) will be placed at every node of the network. When a transaction of M is posted into the network, smart contracts of M get executed. Execution of transaction will be charged. There will be a transaction fee in terms of "gas" used for transaction execution. This pricing mechanism prevents spam and ensures that transaction execution is finite. Thus the result of a transaction is notified when: (1) transaction is validated, (ii) gas is finished.

4.2.1   Proof-of-work

As in Bitcoin network, Ethereum miners try to find a block of latest transactions whose hash gives desired output. A successful miner will get some Ether like a Bitcoin miner is rewarded with some number of bitcoins. The problem with this mechanism is transaction time increases as more stringent conditions are put on the hash output. Thereby, applications posting transactions into the network need to wait for longer time for their transaction to finish. The extended validation time is unsolicited. A new consensus algorithm, called proof-of-stake, is adopted in reducing validation time.

4.2.2   Proof-of-stake

The network achieves distributed consensus based on the stakes held by nodes wanting to participate in decision making. In this mechanism, a node with more coins gets a chance to create a new block. The negative side of the mechanism is that a single rich node with a huge number of coins could acquire control over the consensus algorithm and become equivalent of a central authority. To counter this, there are several different methods have been proposed. One method considers coin-age, the number of days coins are with participating node. A member with more coins with age t will have more weight-age to creating a new block. Once the owner node participates in creating a block, all its coins start with zero coin-age so that the same node will not participate in validating transactions.

4.2.3   Smart contracts

The Blockchain Program Module (BPM) for Ethereum is called a *smart contract*. It is a collection of programs designed for a model M, which will be deployed at each machine of the network. When a web-browser or Mobile application A posts a transaction T belonging to M into the network, the smart contracts of M will be executed at all machines which receive T. The charges for an execution of T will be based on its runtime and resources it consumes. Once T is validated, it will be notified to the application A.

5   Private blockchains

A private blockchain network runs in a trustful environment, in which participating nodes are clearly identified and authenticated. The network has roles clearly defined for each node of the network. There are special nodes in the network which are dedicated to validating transactions. The remaining nodes simply accept validated transactions and store in their local database. When a client application (web/mobile) wants to fetch the information about a validated transaction, it will approach some "special" nodes of the network. If the information fetched from all special nodes is same, then the client application considers the transaction to be genuine based on an endorsement policy defined a priori.

5.1   Hyperledger fabric

Hyperledger (Hyperledger, n.d.) is an example for a private blockchain network. The network executes the transactions in a very controlled environment. The Blockchain Program Module (BPM) of the network is called *chain code*. Based on functionality, nodes of the network are classified into different categories.

1. Peer node
2. Ordering node
3. Endorsing node
4. Client node

A participating machine of the network will accommodate at least one of the above nodes.

Peer node is vital component for storing the data as it maintains both State and Ledger. Each node of the network is essentially a peer node. As discussed in Section 3.1, the node verifies the received transactions with respect to its State. If a transaction meets endorsement policy (discussed below) and do not violate its State, then changes suggested by the transaction will be written into State. The validated transaction will be registered into its Ledger. Some peer nodes will be selected a priori as endorsing nodes (discussed below).

Endorsing node is a peer node which is selected for signing on incoming transactions. The endorsement policy for a given chain code clearly mentions a set of endorsing nodes out of the complete set of peer nodes. From a chosen set of endorsing nodes, the policy further mentions at least how many endorsing nodes have to sign on a transaction. For example, m-out-n policy imposes the condition that m out of total n endorsers have to agree on the result of the transaction.

Client node provides an access point for the blockchain network. Transactions will be posted into the network using this node. The network has a notion of Admin cards, which will be possessed by client nodes, through which the network is accessed.

Ordering node is a node which takes the responsibility of receiving transactions from client nodes, and broadcasting the received transactions to all peer nodes of the network. It neither validates transactions nor stores the data of transactions. It just uses a broadcasting mechanism for transactions to flow across the network. So, every transaction has to go through an ordering node toward its final validation. The collection of all orderer nodes is called an Ordering service (such as data center), which can be hosted by an independent server. An example for the ordering service is Kafka message system, which is crash tolerant.

5.1.1 Transaction cycle

When a transaction is posted into the network, it passes through different phases before reaching the commit phase. The life cycle of a transaction is as follows.

- A client application posts a transaction T, which will be broadcast to all endorsing nodes of the network.
- As a response to T, each endorsing node comes up with a pair (R,W) based on its local database. Here, R is the read set which comprises of the present database values of variables of T, and W is the write set which consists of modified values of the same variables that are supposed to be written to the database when T is committed. The endorsing nodes send their (R,W) to the client application without committing W to their local database. The endorsing nodes sign on (R,W).
- The client application receives (R,W) sets from endorsing nodes. Suppose the endorsing policy is that m out of n endorsing nodes have to agree on the response (R,W).
- If the endorsement policy is satisfied, the client application sends T,(R,W) to orderer node which will broadcast to all peer nodes of the network. Once received by the orderer, all peer nodes (including endorsing nodes) now verify (R,W) against their local database and commit W to their local database.

5.1.2 Endorsement policy

Each endorser node puts signatures on (R,W) set of T and sends to the client application. Having got endorsed (R,W) sets, it is the time for the client application to decide on the validity of T, based on *m-out-of-n endorsement policy*. The policy is such that out of total n endorsers, at least m endorsers have to agree on (R,W) set. If the policy is satisfied, T,(R,W) is valid and thus will be broadcast to all nodes. If the policy is not satisfied, T is invalid. One can see that the endorsement policy is application dependent. The question of who endorses transactions is answered based on the problem statement.

5.1.3 Modularity

The Hyperledger design is modular. The main functionalities of ordering service and endorsing are independent of each other: one functionality can be modified without affecting the other functionality. Similarly, the functionality of peer node and client node can be modified without affecting the functionalities of other node types.

5.2 Applications

The Bitcoin network is the first ever electronic payment system that offers P2P transactions. Ethereum network is a modification over the Bitcoin network which brings in more flexibility of deploying one's own model other than the payment system. But, given their public nature, these two networks will not be suitable for a closed system such as Banking sector and Supply chains. Hyperledger, being private and more controlled network, has plenty of use cases in Banking sector and Supply chain systems, which can benefit from the features: immutability of transactions, transparency in transaction execution and data integrity.

We briefly discuss below how Hyperledger can add new features to Banking sector and Supply chains and how it can change forever the way the two systems operate now.

5.2.1 Banking sector

Current international money transfers use intermediaries. The use of intermediaries adds significant additional cost to a money transfer and also its increased processing time as the money transfer involves financial entities belonging to different geographic locations.

Consider the scenario: a consortium of financial entities (e.g., Banks) form a closed Hyperledger network. Each entity of the group owns a computing machine which will be a node of the network. The network has a privacy protocol which ensures the privacy of transactions. Here, a transaction is a money transfer from an account of a financial entity (Bank A) to an account of other financial entity (Bank B). The endorsement policy dictates which financial entity should sign on a transaction for authentication. For example, a transaction $T_{A,B}$ that involve Bank A and Bank B will only be signed by Bank A and Bank B and can be verified by other financial entities, and privacy protocol ensures that complete details of $T_{A,B}$ should only be seen by Bank A and Bank B. Other endorsing nodes can authenticate the transaction by verifying digital signatures of Bank A and Bank B made on the Hash of $T_{A,B}$. The very feature of $T_{A,B}$ is that it does not involve intermediaries unlike a traditional financial transaction does. By this, $T_{A,B}$ will witness a significant reduction in both transaction time and cost of its execution to completion.

Money transfers without intermediaries is already achieved in the Bitcoin network. However, it is important to note that money involved in the Bitcoin network is not state controlled, i.e., no financial institution has control over Bitcoins. In the above use case, Hyperledger network involves money transfers that are of state controlled nature and are subject to well-defined regulations by different states.

5.2.2   Supply chains

A product of a manufacturing unit passes through different distributing units before reaching an end customer. Each unit (either manufacturer or distributor) in the supply chain maintains its local database of products (received/sent). There will be considerable data delays between these database systems. As a result, the logistics details of a product in transit will be different at different units for considerable amount of time. Thus, any give time the manufacturing unit and distributing units will not be having the same view of products in transit.

Consider the scenario: a group of manufacturing units and distributing units form a Hyperledger network. Each unit owns a computing machine which will be part of the network. The network defines a privacy protocol which ensures the privacy of transactions, with rules: (i) manufactures can see all transactions that involve only its own products, and (ii) a distributing unit can see only transactions that involve products which pass through it. The endorsement policy has the rule: a transaction T that involves a product P should be signed by the manufacturing unit of P and a distributing unit which P passes through, and other units will accept T as they verify digital signatures made (by relevant unit) on Hash of T. With this setup, transaction execution will result into the same database at machines of the network. Hence, the entities of the system will have the same view of the product P at any given point of time.

5.3   Discussion and conclusions

The current supply chain systems are entirely governed by Enterprise Resource Planning (ERP) software. Blockchain will surely disrupt supply chain systems and has potential to eliminate ERP system completely. However, the industry is not yet ready for adopting blockchain solution as the present ERP solution is well deployed across the industry and contributes to multi billion Dollar revenues annually. Furthermore, key players of supply chain systems have to come together to form a consortium to build a blockchain network and define relevant data regulation policies.

Hyperledger is in development phase, whose versions (hlf 1.2, hlf 1.3) (Hyperledger, n.d.) offer a good platform for Proof-of-Concept and pilot demonstrations of blockchain use cases. Hyperledger community has recently released a new version: hlf 2.0. For production grade applications, the industry has to wait for some more time. Further, each blockchain use case requires customization in Hyperledger application layer, and poses challenges in scaling up the solution. The scalability of Hyperledger with respect to Proof-of-Work (endorsement policy) and Byzantine fault tolerance is presented in Vukolić (2015).

In conclusion, the chapter presents cryptographic foundations in the important context of the security of a blockchain network and its two paradigms (public & private network). A public blockchain network, such as the Bitcoin network and Ethereum, has applications to P2P cash payment system. A private network, such as Hyperledger, has applications to closed systems (Banking and Supply chains). However, the applications require customization in transaction flow, endorsement and privacy policy and scalability aspect of the network. The reader is encouraged to give his/her own insights into these applications to overcome the present challenges in executing them to a production grade. The reader is requested to make a note that the blockchain technology has both theoretical and application flavor, as discussed throughout the chapter.

References

Babai, L., 2016. Graph isomorphism in quasipolynomial time. In: Proceedings of the Forty-Eight Annual ACM Symposium on Theory of Computing (STOC'16), pp. 684–697.

Ben-Sasson, E., Chiesa, A., et al., 2014. Zerocash: decentralized anonymous payments from Bitcoin. In: Proceedings of the IEEE Symposium on Security & Privacy Oakland, vol. 2014, pp. 459–474.

Bitcoin, n.d., http://bitcoin.org.

Buhler, J.P., Lenstra Jr, H.W., Pomerance, C., 1993. Factoring integers with the number field sieve. In: The Development of the Number Field Sieve, Springer, pp. 50–94.

Cachin, C., Schubert, S., Vukolić, M., 2016. Non-determinism in Byzantine fault-tolerant replication. In: 20th International Conference on Principles of Distributed Systems (OPODIS).

Castro, M., Liskov, B., 2002. Practical Byzantine fault tolerance and proactive recovery. ACM Trans. Comput. Syst. 20 (4), 398–461.

Ethereum, n.d., http://ethereum.org.

David, C., 1981. Untraceable electronic mail, return addresses, and digital pseudonyms. Commun. ACM 24 (2), 84–90. https://doi.org/10.1145/358549.358563.

Goldwasser, S., Micali, S., Rackoff, C., 1989. The knowledge complexity of interactive proof systems. SIAM J. Comput. 18 (1), 186–208.

Gordon, D.M., 1993. Discrete logarithms in GF(p) using the number field sieve. SIAM J. Discret. Math. 6 (1), 124–138.

Grover, L.K., 1996. A fast quantum mechanical algorithm for database search. In: Proceedings, 28th Annual ACM Symposium on the Theory of Computing, pp. 212–219.

Hyperledger, n.d., http://www.hyperledger.org.

Lenstra, A.K., Lenstra Jr., H.W., 1993. The Development of the Number Field Sieve. Lecture Notes in Math, Springer.

Nakamoto, S., 2009. Bitcoin: A Peer-to-Peer Electronic Cash System. http://www.bitcoin.org/bitcoin.pdf.

Shor, P., 1997. Polynomial-time algorithms for prime factorization and discrete logarithms on a quantum computer. SIAM J. Comput. 26 (5), 1484–1509.

Vukolić, M., 2015. The quest for scalable blockchain fabric: proof-of-work vs. BFT replication. In: International Workshop on Open Problems in Network Security (iNetSec), pp. 112–125.

Further reading

Goldwasser, S., Micali, S., 1984. Probabilistic encryption. J. Comput. Syst. Sci. 28, 270–299.

Rabin, M.O., 1978. Digitalized signatures. In: Foundations of Security Computation, pp. 155–168.

Rivest, R.L., Shamir, A., Adleman, L., 1978. A method for obtaining digital signatures and public key cryptosystems. Commun. ACM 21, 120–126.

Chapter 4

Application of data handling techniques to predict pavement performance

Sireesh Saride*, Pranav R.T. Peddinti, and B. Munwar Basha
Department of Civil Engineering, Indian Institute of Technology Hyderabad, Kandi, Sangareddy, India
*Corresponding author: e-mail: sireesh@ce.iith.ac.in

Abstract

The present study discusses the design of pavements and the importance of big data handling in improving their performance. A comprehensive framework based on a simple natural language processing technique is presented to reduce the computational time and error in data handling for pavement applications. The application of the proposed method to automate a graphical user interface (UI) adopted in pavement design is demonstrated. The proposed method was found to reduce the run-time by about 83% as compared to the conventional procedures. The proposed framework is highly flexible and can be adapted to extract data from various file formats and automate UIs at ease. To present the potential of this framework, about 0.2 million data sets representing pavement geometry and material properties were generated using language processing algorithms. Further, robust non-linear regression equations for calculating pavement damage in terms of fatigue and rutting strains were developed by using automated data processing through the pavement design interface.

Keywords: Data handling, Automation, User interface (UI), Regression, Fatigue, Rutting, Pavement

1 Introduction

A well-developed road network is one of the key indices of overall development in any country. The socio-economic factors such as trade, goods transport, public transit, defense mechanisms, economic growth, accessibility, and many more depend on the extent of the road network. Fig. 1 shows the top 10 largest road networks in the world, with the United States being the first followed by India and China. It is not only the initial cost of construction

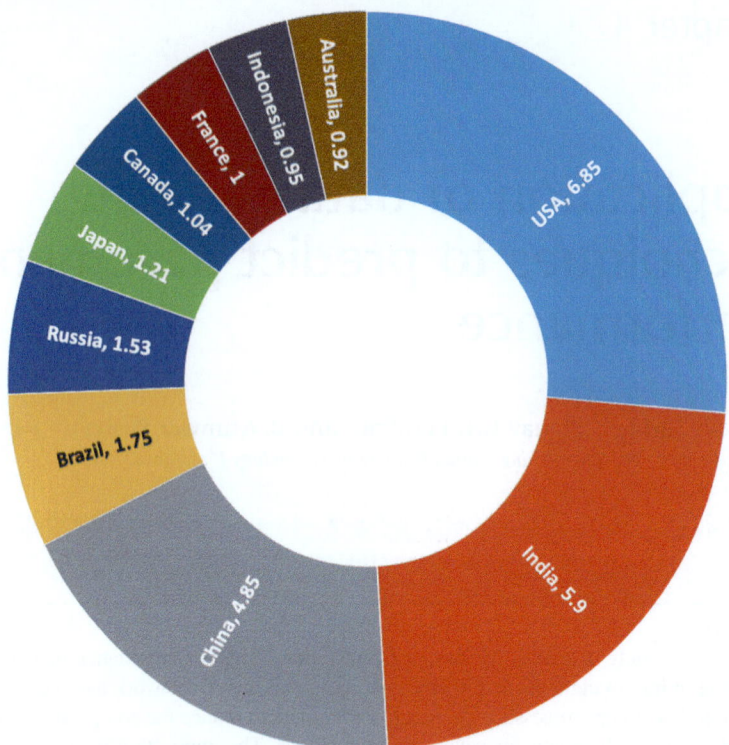

FIG. 1 World's top 10 largest road networks (in million km).

but also substantial annual repair and maintenance costs that define the quality of infrastructure. Most of the developed countries like the United States, United Kingdom, Japan, etc. are investing heavily in the aging infrastructure. Therefore, the investments in the construction and rehabilitation of roads are increasing globally. For instance, the World Bank lends more funds for road construction and rehabilitation as compared to education, health, and social services together (Berg, 2015). The bank has invested 1.8 billion USD between 2004 and 18 for highway development in India alone (The World Bank, 2018). Given the enormous investments and prominence of pavement networks, researchers, designers, and engineers in this field are expected to produce well-performing and reliable pavements to cater to the needs of increasing traffic.

The history of using structured pavements dates back to Romans in 200 AD, where stones, boulders, and soil are laid in a pattern to allow the movement of carts and army. By the mid-19th century, chemically bound materials such as bitumen and concrete were used in pavements. Therefore the thickness of pavement layers got reduced due to the high stiffness of these materials. The layer thicknesses adopted were based on the field experience, judgment, and purpose of construction, which led to premature failures in pavements due to variation in material quality, increased traffic, and climatic conditions. Therefore, a need for developing a scientific procedure for the design of pavements has become essential.

With the advent of mechanistic methods to determine the interface stresses and strains in homogenous and layered pavement systems (Ahlvin and Ulery, 1962; Boussinesq, 1885; Burmister, 1945; Foster and Ahlvin, 1954; Huang, 1968; Jones, 1962; Peattie, 1962; Timoshenko and Goodier, 1951), material behavior in terms of elastic modulus and Poisson ratio are also considered while designing the pavements. Generally, pavements are classified as a flexible and rigid type based on the materials and construction procedures adopted. The present discussion focuses on flexible pavements (Fig. 2) as they are widely adopted across the world. A typical flexible pavement is constructed on a prepared natural subgrade and by placing, in sequence, a granular subbase (GSB)—a drainage layer, a base course (BC) layer, dense bitumen macadam (DBM), and finally a surface course. The performance of the flexible pavement is generally assessed based on strains mobilized at specific critical locations, which are mainly categorized as fatigue and rutting modes. Fatigue represents the surface cracking of a bituminous layer and is characterized by the magnitude of tensile strains (ε_t) observed at the interface of the bituminous and base layer (Saal and Pell, 1960). On the other hand, rutting represents the permanent deformation of the pavement due to the compressive

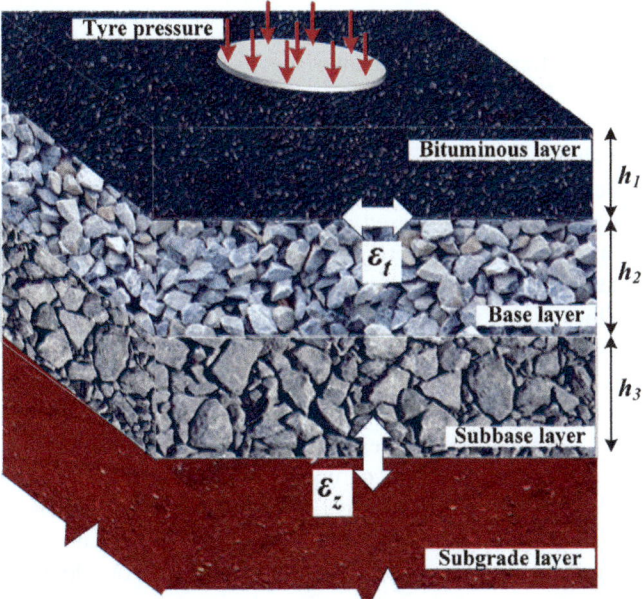

FIG. 2 Schematic of a typical flexible pavement.

FIG. 3   Typical failures of flexible pavements.

vertical rutting strain (ε_z) at the interface of subbase and subgrade layers (Kerkhoven and Dormon, 1953). In standard terms, the alligator/skin cracking observed on the road surface is known as fatigue cracking, and the depressions found along the wheel path are referred to as rutting, as shown in Fig. 3.

The basic principle of designing a pavement is to determine the thicknesses of different pavement layers based on the existing condition of the subgrade, intended traffic, and the quality of materials being used. The adequately designed pavement can withstand the desired number of traffic cycles without failing in either of the failure modes for the desired period. Several damage analyses-based-empirical models were also developed by researchers to determine the design life (limiting number of traffic cycles) of a pavement, which can withstand before failing in fatigue or rutting (Asphalt Institute, 1982; Shook et al., 1982). Although these design methods are robust and consider material behavior, they are based on unrealistic assumptions such as homogeneity and isotropy, which are not valid in the case of pavement materials. Besides, huge variability is associated with the pavement materials, which generally originates from the quality of the parent rock, or adopting recycled materials in pavement construction. Therefore, pavements constructed by these methods often showed premature failures.

2 Importance of data in pavement designs

To overcome the drawbacks mentioned above and develop an accurate design method, collection of pavement performance data along with mechanistic behavior of pavement materials and climatic conditions from the existing pavements and using this experience for new pavement construction was felt essential. Consequently, the American Association of State Highway Officials (AASHO) attempted the first-ever data collection program by monitoring the pavement performance of in-service pavements. AASHO road test was performed during the late 1950s in Ottawa, Illinois, to collect pavement performance data from the interstate highway (IH 80). The test was performed on several different pavement geometries and materials with varied traffic loads. The stresses and deformations at critical locations were monitored periodically. Based on the ride comfort ratings taken from the experience of few experts, a new concept called pavement serviceability index (PSI) was introduced by the American Association of State Highway Transportation Officials (AASHTO) to represent the overall condition of the pavement (AASHTO, 1993). However, the serviceability assessment is based on the judgment of few experts on a limited number of local pavement sections and hence cannot be generalized for all traffic and climatic conditions. Therefore, several other countries adopted extensive performance data collection programs to develop performance models based on the local climate, materials, and traffic. The results obtained from data collection programs are used to introduce mechanistic behavior into the existing design procedures.

Consequently, the mechanistic-empirical pavement design guide (MEPDG, 2004) replaced the erstwhile empirical methods in many countries. The mechanistic behavior from the existing pavement sections is also considered along with the empirical transfer functions. Reliability concepts were also introduced into these designs to account for the material variability to improve accuracy. However, pavement design, construction, maintenance, rehabilitation, and risk assessment practices have experienced tremendous advancements in recent years. Therefore, variability considerations and empirical transfer functions used in the M-E methods were proven inappropriate by recent research (Retherford and McDonald, 2010; Saride et al., 2019).

To be more sustainable, recent research aims at developing completely mechanistic design methods to replace the erstwhile empirical and mechanistic-empirical (M-E) procedures. Importantly, these mechanistic models are completely data-driven, and the number of collected data sets is important for developing well predicting mechanistic models. A wide variety of data from multiple sources is observed in pavement engineering (Fig. 4). To handle such enormous data, transportation agencies have adopted several strategies for the collection, handling, storage, and management of data. Long-term pavement performance (LTPP) program by federal highway authority (FHWA), the United States has been collecting and maintaining records and

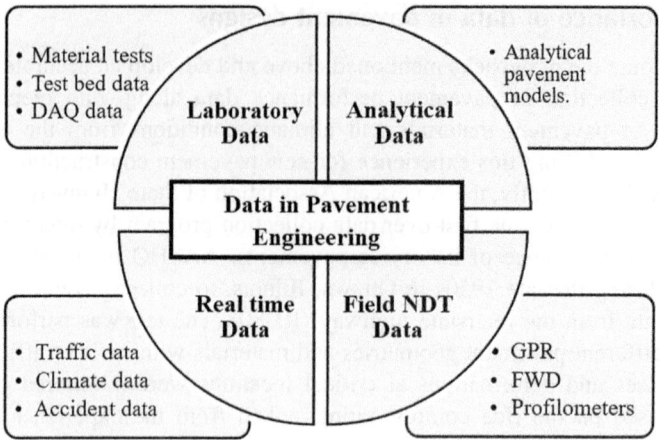

FIG. 4 Sources of data in pavement engineering.

data about 2500 North American test sections for the last three decades. The information collected in seven different modules includes pavement inventory, maintenance, monitoring, rehabilitation, material testing, traffic, and climate. Millions of data sets are associated with each of these modules and are expected to increase exponentially with time. To exemplify, the falling weight deflectometer (FWD) test data from 2577 LTPP sections currently contains 9,753,062 records. The LTPP FWD protocol (Schmalzer, 2006), primarily collects load and deflection data. In addition, distance, temperature, pavement crack width measurements made manually in several countries.

Similarly, ground-penetrating radar (GPR) profiles, profilometers, and video/imaging-based traffic analyzers deliver a variety of data in the form of text, datasheets, figures, or graphs. It is therefore understood that the pavement data constitutes an extensive range of aspects. Hence, efficient handling of these datasets is essential to take advantage of data collection in mechanistic pavement designs.

When numerous and complex datasets, generally referred to as "big data," are to be handled, technologies such as machine learning and artificial intelligence are being used (Moreno-Sandoval and Moro, 2015). Big data is characterized by one or more characteristics viz., volume, variety, velocity, and veracity (accuracy) and value (IBM, 2017). As discussed earlier, a huge volume and variety of data are generated in the pavement industry. With the demand for expanding the pavement network and advanced pavement testing methods, the volume of data generated is perhaps expected to gain velocity with time. Besides, databases such as LTPP are highly valuable in terms of accuracy (veracity), economy incurred in pavement construction, and time-saving. Effective large data handling is a prerequisite to bringing out an advantage of mechanistic models.

3 Challenges with pavement data handling

The big data in pavements is obtained from different sources, and therefore, retrieving the required data from different file formats to perform further analysis is the primary challenge, mainly when multiple data sets are to be extracted. Manual collection and processing of such highly unorganized discrete data are tedious. In addition, this data is often required to be fed to a computer interface to perform a given design process. Therefore, data handling techniques for data mining, retrieval, processing, and analysis are essential to handle this huge spectrum of data for developing sustainable pavement practices.

The mechanistic pavement design methods are more advantageous when they are calibrated to suit the local climate, traffic, and distress levels in the pavements. The basic challenges for local calibration of designs are data collection, handling, and incorporation of real-time data into the analysis. Lack of data manipulation, data fusion, and data entry techniques are the drawbacks that hinder the success of these methods (Guo and Timm, 2016). The success of the M-E pavement design methods is being withheld because of the poor data handling and data quality caused by increased human interaction and maneuvering time (Kang and Adams, 2007). Besides, for pavement design or rehabilitation, the collected field data must be processed through a text or graphical user interface (UI) adopted by the transportation agencies. Several pavement analysis programs such as CIRCLY (Wardle, 1977), MICHPAVE (Harichandran et al., 1989), MEPDG (ARA Inc., 2004), KENPAVE (Huang, 2004), IITPAVE (IRC, 2012) are some of the UIs in practice for flexible pavements. Numerous iterations of UI execution are required for local calibration, and consequently, considerable time and personal investments are necessary. A calibration study performed by Velasquez et al. (2009) required about 202,664 runs of the MEPDG program, which is highly tedious to carry out manually. The UIs available for pavement design is predominantly research-based and are helpful only when a smaller number of pavement combinations are to be investigated. A study by Schwartz et al. (2011) emphasizes that the MEPDG is a research-grade software that requires manual input of all data, and a sensitivity analysis is tedious to perform with such UI. In this regard, methods to reduce human intervention in UI operations are necessary to achieve accuracy and reduce computational time.

4 Data handling techniques in pavement analysis

Among the data handling methods, natural language processing (NLP) is vividly used because of its broad applicability to different types of data. NLP trains the computers to automatically recognize, modify, and generate natural text by using artificial intelligence. A computer-based approach for analyzing, representing, and manipulating natural language text to perform automatic summarization of text is demonstrated by recent research

(Chowdhury, 2003). Computer sciences and linguistics are combined to achieve proper data handling (Kumar, 2011). The pavement industry continuously poses a variety of data handling problems with the developments made in various sectors. Hence, there is a definite and continuous scope for the application of these techniques in pavement analysis and design.

Researchers have developed a few techniques for automating the UIs used in pavement engineering. Automation using macros has been suggested by Schram and Abdelrahman (2010) for transferring extensive data from Excel spreadsheets to MEPDG software. Other similar studies emphasized the advantage of using macro automation techniques for successful evaluation of sensitivity analysis and local calibration (Guo and Timm, 2015; Jadoun and Kim, 2012). Based on the type of application, the run time of the macro automation execution varied between 3 and 20 min/run. Macro recorder works with the principle of reading the data from spreadsheets, identifying the appropriate input window in the UI, entering the required values, and finally execute the UI automatically. The sensitivity analysis performed by Schwartz et al. (2011) required about 622 days on a single operating computer with a run time of about 20 min/run.

Advanced techniques such as automatic scripting, text identification, and image identification are used for developing several computer applications for automated form filling, browser operations, and website automation. Among these methods, image identification-based automation techniques deliver a quick response with UIs. Image identification-based automation involves pixel identification, pixel matching, and automated mouse/keyboard operations in a user-defined sequence. The program sequence could automate and iterate the UI operations, any number of times. Hence, the image recognition-based techniques, coupled with appropriate data handling algorithms, would reduce the run time and increase the efficiency of data handling techniques in pavements.

This procedure would enable the field engineers to compare and decide upon various pavement sections and material combinations at the planning stage. It can be used to evaluate the pavement distress at different locations and perform damage analysis with minimum computational time. The techniques can be extended for easy and quick data generation for applications like simulation studies, regression analysis, optimization studies, and reliability-based pavement design methods. Desired data sets can be transferred from real-time to the computer environment and vice versa. Since many countries are transforming from conventional pavement design methods to the mechanistic designs, such data handling framework is expected to aid in making the transition smooth and adaptable.

5 Data handling and automation framework for pavement design

To develop the framework mentioned above, data handling and automation of user interfaces used in pavement engineering may require all or some of the

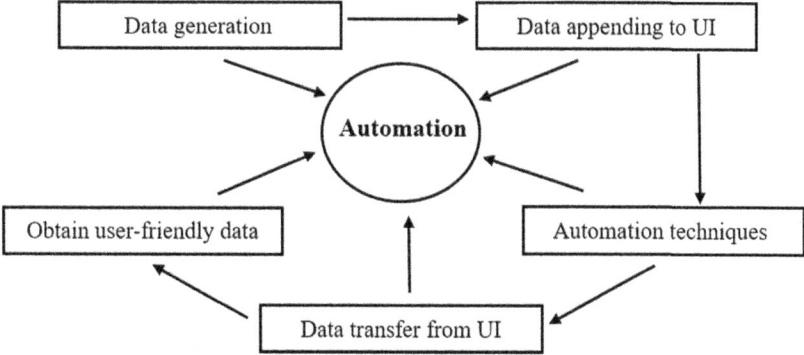

FIG. 5 Data handling techniques useful for UI automation.

techniques shown in Fig. 5. A comprehensive four-stage algorithm is used to perform data handling by these methods (Fig. 6). The proposed framework is demonstrated by automating the KENPAVE program (Huang, 2004), which is a widely accepted UI for performing design and damage analysis of pavements. The stages involved in this procedure are illustrated below.

5.1 Data generation

It is well known that the accuracy of mechanistic pavement models depends on the extent of data available. In some studies, it is required that a considerable amount of input data is to be generated by the user. In this study, an algorithm is developed in MATLAB$^{®}$ (2015) for generating the input data. To represent a typical four-layer flexible pavement section, the practical range of layer thicknesses (h) and resilient moduli (M_r) of materials are considered (Table 1). By considering suitable intervals within the range of each parameter, input data clusters are generated. A data cluster comprises of a set containing $\{h_1, h_2, h_3, M_{r1}, M_{r2}, M_{r3}, M_{r4}\}$ values. The subscript represents the layer number. A total of 172,800 such data sets were generated, and the data clusters are stacked in .xls files for further processing.

5.2 Data appending

Data appending procedures convert the input data sets to UI readable formats. A detailed representation of the data appending algorithm used in the study is presented in Fig. 6. Based on the type of UI, the readable formats include ".DAT" files ".TXT" files and automatic form filling. KENPAVE allows inputs in the form of predefined ".DAT" files. Each line in the file is represented by the number in the parenthesis, which consists of specific pavement input data (Fig. 7A). For example, line numbers 6 and 7 contain layer thickness and Poisson's ratio values, respectively. A file with all inputs is initially generated as a parent file. The parent file is then duplicated, and the input

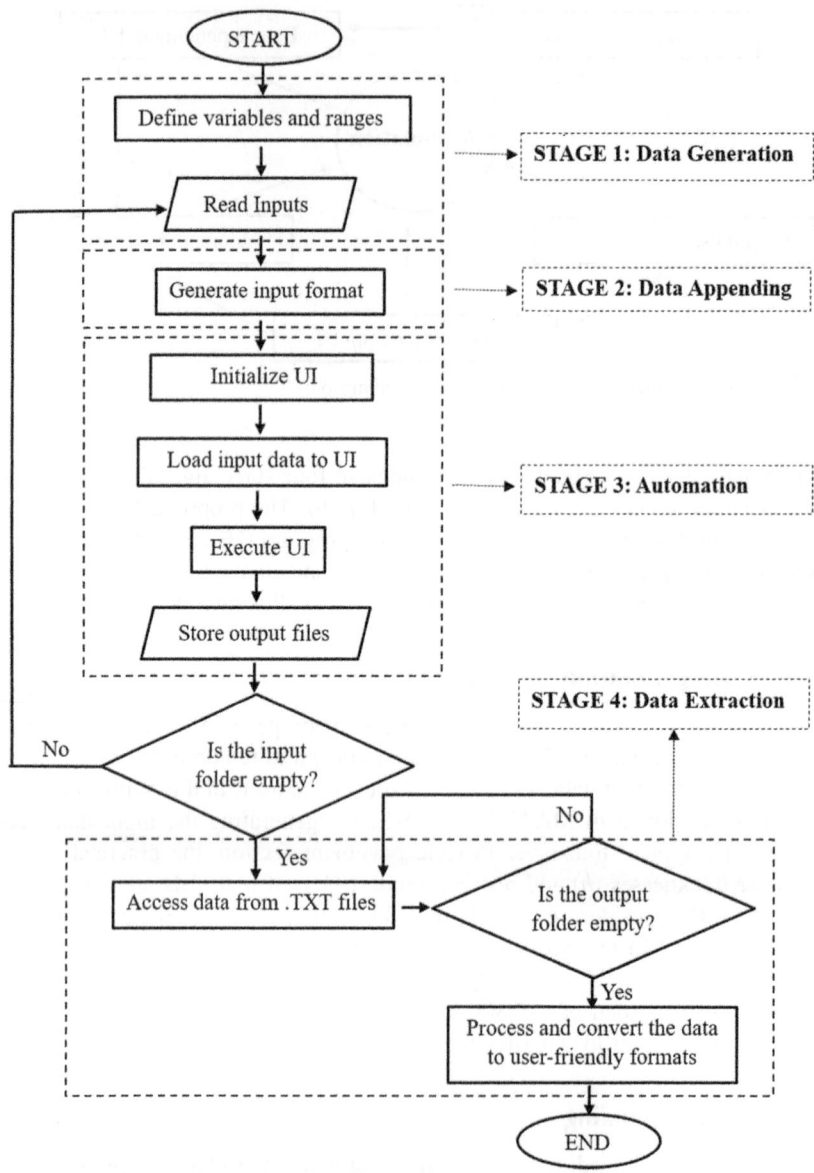

FIG. 6 Algorithm used for data handling and automation.

values corresponding to the 172,800 pavement combinations are appended to the specific locations in the file as string values. The sample file is loaded as a cell array in the MATLAB®, as shown in Fig. 7B. A cell array in MATLAB® contains a series of indexed cells in rows and columns. Each of the cells can contain any specific type of data or a combination of data types such as

TABLE 1 Input parameters and ranges.

Variable	Range
Surface layer thickness, h_1 (cm)	5–22.5
Base layer thickness, h_2 (cm)	10–30
Subbase layer thickness, h_3 (cm)	10–40
Resilient modulus of surface layer, M_{r1} (kPa)	1.5×10^6–3.5×10^6
Resilient modulus of base layer, M_{r2} (kPa)	1.0×10^5–6.0×10^5
Resilient modulus of subbase layer, M_{r3} (kPa)	1.0×10^5–6.0×10^5
Resilient modulus of subgrade layer, M_{r4} (kPa)	1.0×10^4–1.0×10^5

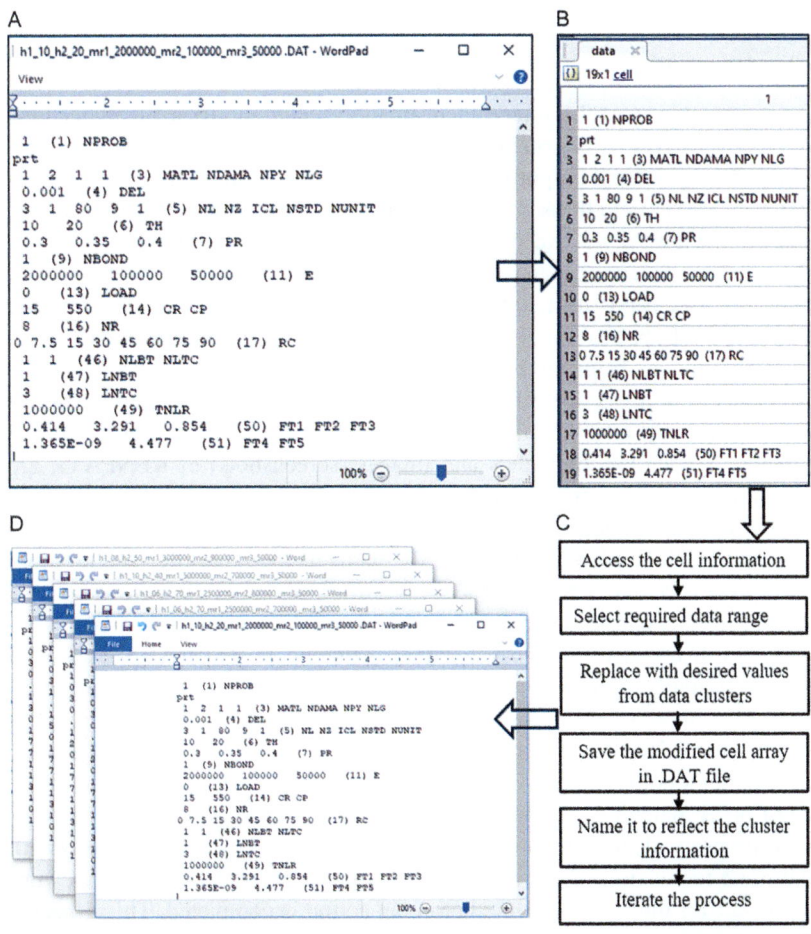

FIG. 7 (A) Input .DAT file for KENPAVE; (B) Cell array of the data file in MATLAB®; (C) Algorithm for data appending; (D) Generated input files.

strings, characters, and numbers of different sizes (Fig. 7B). Using the cell index and unit range of each character, the specific information in each cell can be accessed. The present study adopts the same technique to load each line of the ".DAT" file is shown in Fig. 7A as a cell, and the required position range within the cell is accessed using its cell index and unit identity range. The desired input values then replace the characters in the selected range. To avoid the file identity problem, the file name is also altered in every iteration. The file name is given as the set of inputs for easy identification of the users. The file depicted in Fig. 7C shows the file name as "h1_10_h2_ 20_mr1_2000000_mr2_100000_mr3_50000.DAT." Iterating the same process generates the required number of files (172,800) with the input values appended to specific locations in the input file. The generation of all input files has taken about 3 h by using an octa-core 2.26 GHz processor. Hence, the proposed algorithm saves a huge amount of computational time and manual intervention besides being accurate.

5.3 Automation

The automation uses the image identification technique to sequentially load the input files into UI to generate output files based on the targeted input parameters. The generated output files are identified and stored for further processing. The present study uses *Sikuli* (Yeh et al., 2009) for automation of the process in view of its exceptional image identification capability, whereas other programs recognize an object by its properties and nature. In the present study, the user needs to maintain a repository of images of all the windows used in the UI application. The program offers a variety of commands using which the images from the repository are used in a sequence to execute the program. Using image identification and mouse/keyboard keys, any sequence can be executed, and the same can be iterated using conventional looping methods. Sequential commands and images (screenshots) of KENPAVE are incorporated to load each input file and execute the program. The procedure adopted for automation is summarized in Fig. 8. The output is automatically saved as a ".TXT" file with the same file name as the input file. The program could process about 125 files/h (about 30 s/run), which is deemed to be very fast and saves a considerable amount of human intervention. The image comparison is also exact with pixel to pixel matching at the BitMap level. The system efficiency requirements to compile the automation program are minimum. The automation run-time was found to be the same when compiled in an octa-core 2.26 GHz, dual-core 2.6 GHz, and dual-core 2.9 GHz processors.

The methodology is reliable in terms of process execution because of the robust pixel matching capability. Based on the processing speed of the computer, property identification techniques may stop at any step when the system takes time to generate a particular window. In contrast, the present framework could wait for the specified time before an action is completed. This aspect is advantageous when the UI is poorly built.

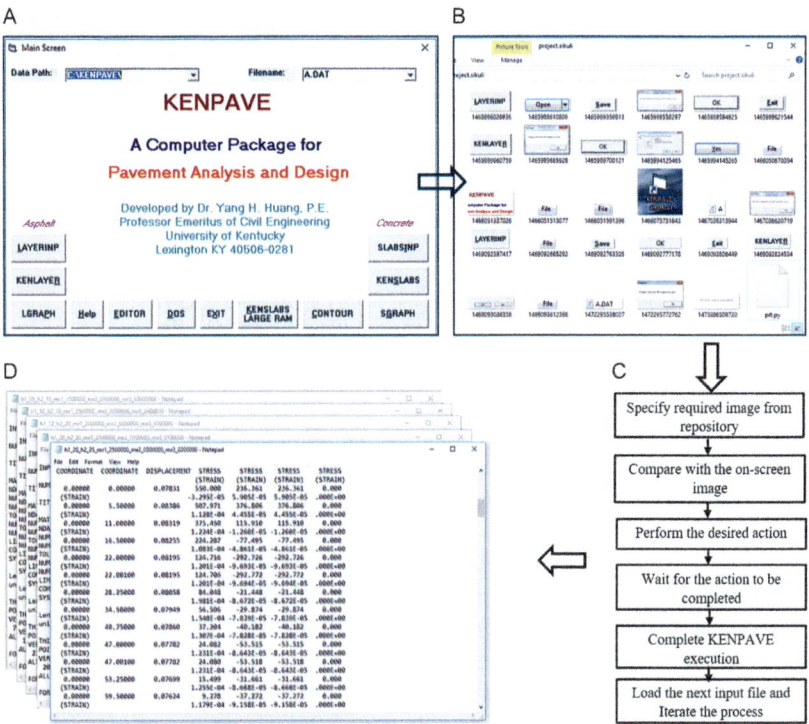

FIG. 8 (A) KENPAVE home screen; (B) Repository of images for execution; (C) Automation algorithm; (D) Generated KENPAVE output files.

5.4 Data extraction

Data pertaining to pavement design and damage analysis is present in the output files. An algorithm was used to access the desired data from files and save them as arrays or columns of a matrix to avoid confusion. Similar to the data appending, each line in the .txt output file is converted to a cell, and a specific location in the cell is accessed using number—string conversions. For example, the values of fatigue and rutting strains corresponding to all input combinations are accessed and stored in a matrix (Fig. 9). The columns of the matrices are then appended to user-friendly Excel files, as shown in Fig. 9 using MATLAB®. Based on the type of application, the data can be stacked in multiple ways to ease further processing.

6 Efficiency of the automation-based data handling framework

To quantify the performance of the proposed methodology, the time taken to execute each stage is compared with that of the conventional methods. Using the current methodology, the run time for UI automation was found to be 30 s/run. The run times for generating, appending, executing, and extracting

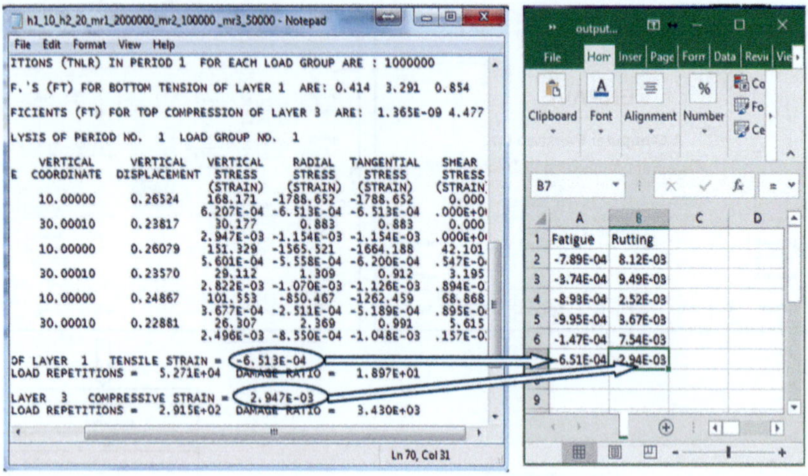

FIG. 9 Data extraction from UI output to user-friendly formats.

the data pertaining to 100 files are compared, as shown in Table 2. The time given for each methodology includes the time taken by both human maneuver and computer efforts, whichever is conventionally used. Human maneuver time is calculated as an average of time taken by three individuals to perform the task using a computer with an octa-core 2.26 GHz processor.

The time savings vary between 65% and 96% for various stages. The overall 83% saving in operational time depicts the usefulness and potential of the proposed methodology. Also, the percentage of time-saving increases exponentially when a huge number of data sets are processed. As human intervention is reduced and constant supervision is not necessary, the semi-supervised methodology can be adopted for the generation of large data sets.

7 Regression analysis using the generated data

To demonstrate the usefulness of the proposed data handling framework, an attempt has been made to develop damage prediction models for fatigue and rutting strains using the data generated from the framework mentioned above. The input thicknesses, resilient moduli values, and their corresponding fatigue and rutting strains are extracted into .xls format for performing regression analysis, as described in stage 4 of the data handling procedure. Initially, the extracted 172,800 data sets are split into 80% training and 20% test data using MATLAB®. As a first trial, conventional linear regression analysis is performed on the training data for predicting fatigue strain. The linear regression model (Eq. 1) showed a very poor coefficient of determination, R^2, and adjusted coefficient of determination, R^2_{adj} values, as shown in Model 1 of Table 3. The reason may be due to the huge variation in the magnitudes of

TABLE 2 Comparison of run-time savings.

Stage of execution	Conventional procedure	Time taken (min)	Proposed methodology	Time taken (min)	Saving in time (min)	Percentage savings in time
Data generation	Manual and Excel	2.5	MATLAB®	1	1.75	70.00%
Data Appending	Manual	75	MATLAB®	2.5	72.5	96.60%
KENPAVE execution	Manual	150	Sikuli	52	98	65.33%
Data extraction	Manual and Excel	145	MATLAB®	7	138	95.17%
Total		372.5		62.5	310	83.22%

Note: The values mentioned are for generation, execution and data extraction from 100 files.

TABLE 3 Coefficients of determination (R^2) of regression models for fatigue prediction.

S. No	Model	R^2	R^2_{adj}
1	Linear regression (LR)	0.67	0.67
2	LR with logarithmic values of independent variables	0.87	0.87
3	Regression model with logarithmic values of independent variables and second-order terms of M_{r2}, h_1	0.93	0.93
4	Regression model with logarithmic values of independent variables and second-order terms of all variables	0.94	0.94
5	Non-linear regression with logarithmic values of independent variables and the contributing second-order combinations of all variables	0.98	0.98

thicknesses, resilient moduli, and fatigue strain. For instance, the thicknesses are in the range of 5–40 cm, whereas resilient moduli can be as high as 3.5×10^6 kPa. To overcome this issue, logarithmic values of the thicknesses and resilient moduli of all layers are considered as $x_1 = \log_{10} h_1$, $x_2 = \log_{10} h_2$, $x_3 = \log_{10} h_3$, $x_4 = \log_{10} M_{r1}$, $x_5 = \log_{10} M_{r2}$, $x_6 = \log_{10} M_{r3}$, $x_7 = \log_{10} M_{r4}$ before performing the linear regression. The resulting linear regression model shown in Eq. (2), showed an improvement in the R^2 and R^2_{adj} (Model 2, Table 3). However, the accuracy of the model is expected to be much higher for pavements, particularly when probabilistic methods are used in designs.

To improve the model performance, the possible solution is to adopt higher-order terms in the model. However, when more independent variables (seven in this case) are available, many higher-order combinations are possible. To decide upon the addition of higher-order terms, one has to understand the level of influence of each independent variable on the performance functions like fatigue and rutting responses in the case of pavements. For instance, the correlation strength between each variable and fatigue strain is calculated from the entire data sets. The correlation coefficients for $\{h_1, h_2, h_3, M_{r1}, M_{r2}, M_{r3}, M_{r4}\}$ are observed to be $\{0.37, 0.03, 0.05, 0.16, 0.70, 0.09, 2 \times 10^{-6}\}$. A high correlation is observed with M_{r2}, h_1, and M_{r1}. To note, the influence is also reasonable as per the pavement mechanics. As fatigue is measured at the interface of bituminous and base layers as demonstrated in Fig. 2 with a horizontal arrow, the thickness of the bituminous surface layer (h_1) and the stiffness of both the layers (M_{r1} and M_{r2}) are expected to influence the strain mobilized at their interface. Adopting the higher-order terms, which contain the most influencing parameters, would increase the performance of the regression model. Therefore, the second-order terms $(\log_{10} h_1)^2$ and $(\log_{10} M_{r2})^2$, i.e., $(x_1)^2$ and $(x_5)^2$ are initially added to the existing logarithmic model (Eq. 3). The R^2 and R^2_{adj} are observed to be 0.93, depicting a well-performing model (Model 3, Table 3). Further, the model considering the second-order terms of all independent variables, i.e., $(x_1)^2$, $(x_2)^2$, $(x_3)^2$, $(x_4)^2$, $(x_5)^2$, $(x_6)^2$ and $(x_7)^2$ is considered for analysis. The model obtained from this analysis (Eq. 4) is observed to be performing similar to the previous model with most influencing terms (Model 4, Table 3). The addition of these terms only increases the complexity of the model.

The desired accuracy of the model further depends upon the type of post-analysis that the equations are intended for. To exemplify, reliability-based optimization studies require highly accurate models for damage analysis (Saride et al., 2019). In such cases, improving the performance of the present regression model can be achieved by adding a greater number of higher-order terms that contribute to the accuracy of the model. Hence, in this study, the remaining second-order combinations, such as x_1x_2 (i.e., $\log_{10} h_1 \times \log_{10} h_2$), x_1x_4 (i.e., $\log_{10} h_1 \times \log_{10} M_{r1}$) and x_4x_5 (i.e., $\log_{10} M_{r1} \times \log_{10} M_{r2}$) are further considered for the model. As the present analysis contains seven independent variables, 21 such combinations are possible. However, as observed in the case of Model 4 of Table 3, the simple addition of all the terms may not improve the accuracy and may further increase the model complexity. Un-supervised addition of higher-order terms may also lead to over-fitting models, which means that the model is highly convoluted (i.e., very high non-linearity) to suit the training data (Grant, 2019). Although the R^2 is high for such a case, the resultant model will exhibit high errors when it is used with the data other than the training dataset. Therefore, considering, truly contributing higher-order terms, is vital for developing accurate non-linear models. The influence of each higher-order term needs to be properly assessed before considering it in the models. This is why non-linear regression is often termed as supervised machine learning.

In the present study, the suitability of second-order products of individual variables to the model is supervised by using interaction plots before considering them in the model. Interaction plots are drawn for all possible second-order terms, some of which are as shown in Fig. 10. These plots depict the variation in fatigue strain for constant values of the first variable by varying the mean of the second variable within its range. The remaining values are kept constant at their midrange values. The intersection of the curves at one or more places depict a strong contribution to the model. As observed, the h_1h_2 combination exhibits more interaction (Fig. 10A) and can, therefore, be considered for the model, whereas $M_{r1}M_{r3}$ combination need not be considered (Fig. 10B). Based on the interaction analysis for all second-order terms, the non-linear regression model (Eq. 5) with an R^2 of 0.98 was developed. Using a similar analysis, the performance model for predicting rutting strain ($R^2 = 0.98$ and $R^2_{adj} = 0.98$) is developed (Eq. 6).

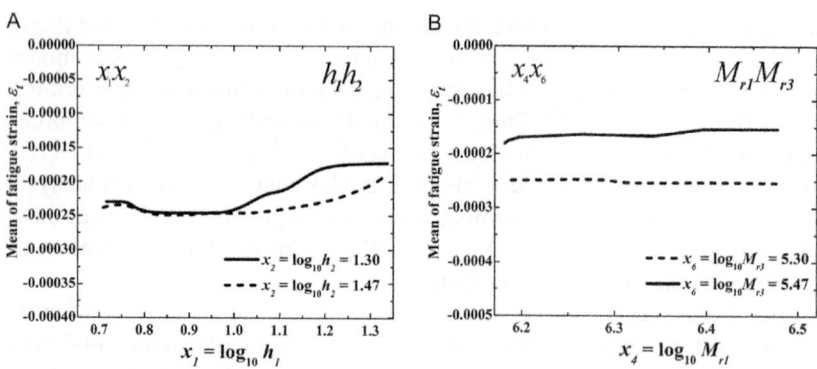

FIG. 10 Typical interaction diagrams to assess the influence of 2nd order terms. (A) h_1h_2; (B) $M_{r1}M_{r3}$.

$$\varepsilon_t = \left[1.68h_1 - 3.91h_2 - 2.30h_3 - 3.83 \times 10^{-5}M_{r1} + 2.89 \times 10^{-4}M_{r2} - 9.28\right.$$
$$\left. \times 10^{-5}M_{r3} - 6.49 \times 10^{-4}M_{r4}\right] \times 10^{-6} \tag{1}$$

$$\varepsilon_t = \left[147.73x_1 - 12.45x_2 + 3.87x_3 - 200.68x_4 + 229.35x_5\right.$$
$$\left. -31.05x_6 - 32.81x_7\right] \times 10^{-6} \tag{2}$$

$$\varepsilon_t = \begin{bmatrix} -974.76x_1 + 17.69x_2 + 26.62x_3 + 123.07x_4 - 628.10x_5 + 37.01x_6 - 0.92x_7 \\ +552.86x_1{}^2 + 85.88x_5{}^2 \end{bmatrix}$$
$$\times 10^{-6}$$
$$\tag{3}$$

$$\varepsilon_t = \begin{bmatrix} -948.31x_1 + 107.16x_2 + 99.89x_3 - 1420.80x_4 + 1084.30x_5 + 112.61x_6 - 21.26x_7 \\ +540.32x_1{}^2 - 36.46x_2{}^2 - 28.38x_3{}^2 + 122.00x_4{}^2 - 72.87x_5{}^2 - 6.87x_6{}^2 + 2.35x_7{}^2 \end{bmatrix}$$
$$\times 10^{-6}$$
$$\tag{4}$$

$$\varepsilon_t = \begin{bmatrix} 116.80x_1 - 1610x_2 - 1012x_3 - 407.50x_4 + 741.80x_5 - 119.10x_6 - 188.60x_7 \\ +548.80x_1{}^2 + 33.59x_5{}^2 + 65.97x_1x_2 + 34.99x_1x_3 + 428.90x_1x_4 - 791.20x_1x_5 \\ +6.78x_1x_6 + 74.31x_1x_7 - 82.69x_2x_3 + 213.10x_2x_4 + 180.60x_2x_5 - 124.0x_2x_6 \\ +153.60x_3x_4 - 18.59x_3x_5 + 41.46x_3x_6 - 77.30x_4x_5 + 46.08x_5x_6 + 19.52x_6x_7 \end{bmatrix}$$
$$\times 10^{-6}$$
$$\tag{5}$$

$$\varepsilon_z = \begin{bmatrix} -5860x_1 - 6299x_2 - 5977x_3 + 11710x_4 - 1399x_5 - 2388x_6 - 5234x_7 - 116.80x_1{}^2 \\ -278.40x_2{}^2 - 972.40x_4{}^2 - 27.44x_7{}^2 + 1132x_1x_2 + 1339x_1x_3 + 841.80x_2x_3 \\ +273.90x_2x_4 + 513.40x_2x_7 + 132.10x_3x_6 + 529.80x_3x_7 + 55.01x_4x_7 \\ +112.10x_5x_6 + 141.80x_5x_7 + 294.70x_6x_7 - 33.09x_1x_4 + 526x_1x_7 - 38.61x_3x_4 \end{bmatrix}$$
$$\times 10^{-6}$$
$$\tag{6}$$

8 Validation of proposed damage models

To validate the proposed damage models, 50 pavement sections from the split test data are selected. The fatigue and rutting strains calculated from KEN-PAVE and the models developed from the present study are compared for both fatigue and rutting modes, as shown in Fig. 11A and B, respectively. The models are found to be accurate in determining the fatigue and rutting strains. Further, a comparison of the present framework with a linear response surface equation developed by Dilip et al. (2013) for predicting strains in a three-layered flexible pavement is presented. The current framework is adopted to develop suitable models for three-layered pavements. Four different pavement sections are considered for comparison with the strains obtained from the KENPAVE program (Table 4). It is observed that the fatigue and rutting predictions from the present study are more closer (with a maximum error

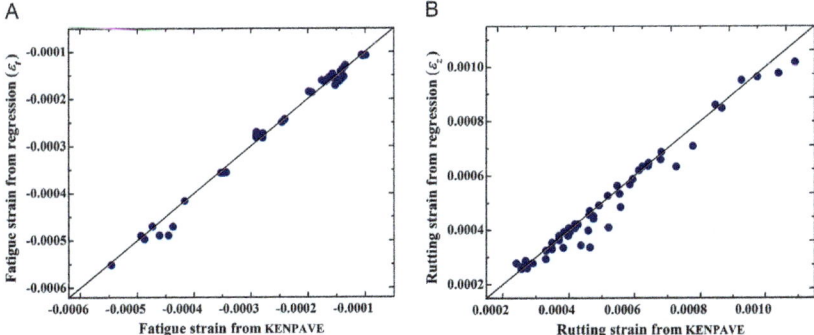

FIG. 11 Validation for regression models (A) Fatigue model (B) Rutting model.

TABLE 4 Comparison of the proposed models with Dilip et al. (2013).

Pavement section (h in cm; M_r in $\times 10^3$ kPa)	KENPAVE	Dilip et al. (2013)	Error (%)	Present study	Error (%)
Fatigue strain ($\times 10^{-6}$)					
$h_1 = 8$; $h_2 = 40$; $M_{r1} = 3500$; $M_{r2} = 400$; $M_{r3} = 30$	−190	−168	11.5	−194	2.1
$h_1 = 10$; $h_2 = 50$; $M_{r1} = 3500$; $M_{r2} = 100$; $M_{r3} = 100$	−159	−164	3.1	−155	2.5
$h_1 = 12$; $h_2 = 25$; $M_{r1} = 3000$; $M_{r2} = 300$; $M_{r3} = 30$	−152	−177	16.4	−146	3.9
$h_1 = 15$; $h_2 = 40$; $M_{r1} = 3500$; $M_{r2} = 300$; $M_{r3} = 50$	−188	−132	29.7	−188	0.01

Continued

TABLE 4 Comparison of the proposed models with Dilip et al. (2013).—Cont'd

Pavement section (h in cm; M_r in $\times 10^3$ kPa)	KENPAVE	Dilip et al. (2013)	Error (%)	Present study	Error (%)
Rutting strain ($\times 10^{-6}$)					
$h_1=8$; $h_2=40$; $M_{r1}=3500$; $M_{r2}=400$; $M_{r3}=30$	343	350	2.0	349	1.7
$h_1=10$; $h_2=50$; $M_{r1}=3500$; $M_{r2}=100$; $M_{r3}=100$	212	198	6.6	228	7.5
$h_1=12$; $h_2=25$; $M_{r1}=3000$; $M_{r2}=300$; $M_{r3}=30$	256	334	30.4	261	1.9
$h_1=15$; $h_2=40$; $M_{r1}=3500$; $M_{r2}=300$; $M_{r3}=50$	208	233	12	199	4.3

of 7.5% against 30%) to the actual values for all pavement combinations analyzed. Better performance of the proposed models is attributed to the consideration of the huge amount of data and consideration of non-linear terms for model development in the present study. With the huge amount of data sets made available by the data handling framework, the sensitivity of the influence of each variable and the decision on the addition of higher-order terms could be precisely understood within the practical ranges. Therefore, the models developed are robust and globally applicable within the practical ranges of thicknesses and resilient moduli of the pavement layers considered in the study.

9 Summary and recommendations

A framework based on the natural language processing technique is proposed to reduce the computational time and error in data handling for pavement applications. The application of the proposed method to automate a graphical user interface (GUI) adopted in pavement design is demonstrated. The present study demonstrated the form of data handling and automation to UIs by using the KENPAVE program. However, the framework proposed in the study (Fig. 6) can be easily extended to other UIs. The automation procedure presented in the study can be used with different user interfaces by suitably collecting the repository of images corresponding to the UI and executing an algorithm to perform the intended operations. Further, the data generation and data collection algorithms presented in the study can be modified accordingly to suit the different file formats used in various user interfaces. In the present case, the proposed method was found to reduce the run-time by about 83% as compared to the conventional procedures.

Damage prediction models for fatigue and rutting strains were proposed with a high coefficient of determination using the extensive data generated from the proposed framework. The proposed fatigue and rutting models were validated with an accuracy of at least four times the existing models. Adapting to multi-disciplinary research would help in developing mechanistic based models for design, risk assessment, maintenance, and rehabilitation of pavements. The developed strain equations are useful for the optimum design of flexible pavements considering the variability associated with thicknesses and resilient moduli.

References

AASHTO, 1993. Guide for Design of Pavement Structures. American Association of State Highway and Transportation Officials.

Ahlvin, R.G., Ulery, H.H., 1962. Tabulated values for determining the complete pattern of stresses, strains, and deflections beneath a uniform circular load on a homogeneous half space. Highw. Res. Board Bull. 342.

ARA Inc., 2004. ERES Consultants Division, Guide for Mechanistic-Empirical Design of New and Rehabilitated Pavement Structures, Final Report, NCHRP Project 1-37A.

Asphalt Institute, 1982. Research and Development of the Asphalt Institute's Thickness Design Manual (MS-1), ninth ed. Asphalt Institute Research Report 82-2.

Berg, C, 2015, How Far Do Roads Contribute to Development?, World Bank Forum, https://www.weforum.org/agenda/2015/12/how-far-do-roads-contribute-to-development/. accessed on March 15, 2019.

Big Data and Analytics Hub. IBM, 2017. https://www.ibm.com/analytics/hadoop/big-data-analytics. Accessed on August 20, 2017.

Boussinesq, J., 1885. Application des potentiels à l'étude de l'équilibre et du mouvement des solides élastiques: principalement au calcul des déformations et des pressions que produisent, dans ces solides, des efforts quelconques exercés sur une petite partie de leur surface ou de leur intérieur: mémoire suivi de notes étendues sur divers points de physique, mathematique et d'analyse. vol. 4Gauthier-Villars.

Burmister, D.M., 1945. The general theory of stresses and displacements in layered soil systems. III. J. Appl. Phys. 16 (5), 296–302.

Chowdhury, G.G., 2003. Natural language processing. Annu. Rev. Inf. Sci. Technol. 37 (1), 51–89.

Dilip, D.M., Ravi, P., Babu, G.S., 2013. System reliability analysis of flexible pavements. J. Transp. Eng. 139 (10), 1001–1009.

Foster, C.R., Ahlvin, R.G., 1954. Stresses and deflections induced by a uniform circular load. In: Highway Research Board Proceedings, vol. 33. HRB.

Grant, P., 2019. A Primer on Model Fitting. https://towardsdatascience.com/a-primer-on-model-fitting-e09e757fe6be. Accessed on June 6, 2020.

Guo, X., Timm, D.H., 2015. Local Calibration of MEPDG Using National Center for Asphalt Technology Test Track Data (No. 15-1032).

Guo, X., Timm, D.H., 2016. Mechanistic-Empirical Pavement Design Software Automation (No. 16-2079). In: Proc. 95th Annual Transp. Res. Board Meeting, Washington, D.C.

Harichandran, R.S., Baladi, G.Y., Yeh, M., 1989. Development of a Computer Program for Design of Pavement Systems Consisting of Bound and Unbound Materials. Department of Civil and Environmental Engineering, Michigan State University.

Huang, Y.H., 1968. Stresses and displacements in nonlinear soil media. J. Soil Mech. Found. Div. ASCE 94 (SM1), 1–19.

Huang, Y., 2004. Pavement Analysis and Design, second ed. Prentice Hall, Upper Saddle River, NJ.

Indian Road Congress (IRC), IRC: 37, Tentative Guidelines for the Design of Flexible Pavements, 2012, New Delhi; India.

Jadoun, F.M., Kim, Y.R., 2012. Calibrating mechanistic-empirical pavement design guide for North Carolina: genetic algorithm and generalized reduced gradient optimization methods. Transp. Res. Rec. 2305 (1), 131–140.

Jones, A., 1962. Tables of stresses in three-layer elastic systems. Highw. Res. Board Bull. 342, 176–214.

Kang, M., Adams, T.M., 2007. Local calibration for fatigue cracking models used in the mechanistic-empirical pavement design guide. In: Proceedings of the 2007 Mid-Continent Transportation Research Symposium.

Kerkhoven, R.E., Dormon, G.M., 1953. Some Considerations on the California Bearing Ratio Method for the Design of Flexible Pavements. Shell Petroleum Company.

Kumar, E., 2011. Natural Language Processing. IK International Pvt Ltd.

MATLAB, 2015. MATLAB Version 8.6. 0 (R2015b). The MathWorks Inc, Natick, MA.

Moreno-Sandoval, A., Moro, E., 2015. Big data versus small data: the case of 'gripe'(flu) in Spanish. Procedia Soc. Behav. Sci. 198, 339–343.

Peattie, K.R., 1962. Stress and strain factors for three-layer elastic systems. Highw. Res. Board Bull. 342.

Retherford, J.Q., McDonald, M., 2010. Reliability methods applicable to mechanistic-empirical pavement design method. Transp. Res. Rec. 2154 (1), 130–137.

Saal, R.N.J., Pell, P.S., 1960. Kolloid-Zeitschrift MI. Heft 1, 61–71.

Saride, S., Peddinti, P.R.T., Basha, M.B., 2019. Reliability perspective on optimum design of flexible pavements for fatigue and rutting performance. J. Transp. Eng. Part B Pave. 145 (2), 04019008.

Schmalzer, P.N., 2006. Long Term Pavement Performance Program Manual for Falling Weight Deflectometer Measurements. Washington, Department of Transportation, Federal Highway Administration. DTFH61-02-C-00007. FHWA-HRT-06-132. https:/www.fhwa.dot. gov/publications/research/infrastructure/pavements/ltpp/06132/06132.pdf. Accessed on August 18, 2017.

Schram, S.A., Abdelrahman, M., 2010. Integration of mechanistic-empirical pavement design guide distresses with local performance indices. Transp. Res. Rec. 2153 (1), 13–23.

Schwartz, C.W., Li, R., Kim, S., Ceylan, H., Gopalakrishnan, K., 2011. Sensitivity Evaluation of MEPDG Performance Prediction, NCHRP 1-47. Transportation Research Board of the National Academies, Washington, DC.

Shook, J.F., Finn, F.N., Witczak, M.W., Monismith, C.L., 1982. Thickness design of asphalt pavements-the asphalt institute method. In: Proceedings, 5th International Conference on Structural Design of Asphalt Pavements, vol. 1. Delft University of Technology, The Netherlands, pp. 17–44.

The World Bank, 2018. Press release https://www.worldbank.org/en/news/press-release/2018/05/31/government-india-world-bank-sign-usd500-million-additional-financing-project-for-all-weather-rural-roads-using-green-technologies. (accessed on 2nd December, 2019).

Timoshenko, S., Goodier, I.N., 1951. Theory of Elasticity. McGraw-Hill, New York.

Velasquez, R., Hoegh, K., Yut, I., Funk, N., Cochran, G., Marasteanu, M., Khazanovich, L., 2009. Implementation of the MEPDG for New and Rehabilitated Pavement Structures for Design of Concrete and Asphalt Pavements in Minnesota. Publication No. NM/RC 2009-06.

Wardle, L.J., 1977. Program CIRCLY: User's Manual. Commonwealth Scientific and Industrial Research Organization, Division of Applied Geomechanics.

Yeh, T., Chang, T.H., Miller, R.C., 2009. Sikuli: using GUI screenshots for search and automation. In: Proceedings of the 22nd annual ACM symposium on User interface software and technology. ACM, pp. 183–192.

Section III

Statistical estimation designs: Fractional fields, biostatistics and non-parametrics

Chapter 5

On the usefulness of lattice approximations for fractional Gaussian fields

Somak Dutta[a,*] and Debashis Mondal[b]
[a]Iowa State University, Ames, IA, United States
[b]Oregon State University, Corvallis, OR, United States
*Corresponding author: e-mail: somakd@iastate.edu

Abstract

Fractional Gaussian fields provide a rich class of spatial models and have a long history of applications in multiple branches of science. However, estimation and inference for fractional Gaussian fields present significant challenges. This book chapter investigates the use of the fractional Laplacian differencing on regular lattices to approximate to continuum fractional Gaussian fields. Emphasis is given on model based geostatistics and likelihood based computations. For a certain range of the fractional parameter, we demonstrate that there is considerable agreement between the continuum models and their lattice approximations. For that range, the parameter estimates and inferences about the continuum fractional Gaussian fields can be derived from the lattice approximations. Interestingly, regular lattice approximations facilitate fast matrix-free computations and enable anisotropic representations. We illustrate the usefulness of lattice approximations via simulation studies and by analyzing sea-surface temperature on the Indian Ocean.

Keywords: Argo floats, Discrete cosine transformation, Fractional Laplacian differencing, Geometric anisotropy, H-likelihood, Long range dependence, MLE, Power-law variogram, Regular lattice

1 Introduction

Fractional Gaussian fields have inspired extensive research in spatial statistics. Fractional fields generalize the notion of fractional noise in two or higher dimensions and are particularly important for studying power laws and modeling long-range dependencies. The early mathematical development of fractional fields can be traced to the works of Yaglom (1957), Whittle (1962),

McKean (1963), Gangolli (1968), Mandelbrot and Van Ness (1968), and others. Also notable are the works by Dobrushin (1979), Yaglom (1987), Granger and Joyeux (1980), Hosking (1981), Gay and Heyde (1990), Beran (1994), Ma (2003) and Kelbert et al. (2005). Recent surveys on the topic are provided in Chiles and Delfiner (2009), Cohen and Istas (2013) and Lodhia et al. (2016). Fractional Gaussian fields cover, as special cases, the de Wijs process or the Gaussian free fields (Matheron, 1970; Mondal, 2015; Sheffield, 2007), the thin plate spline (Gu and Wahba, 1993), higher-order intrinsic random fields (Matheron, 1970, 1973) and power variogram models. Fractional Gaussian fields can also be seen as limiting cases of the Matérn models. Their applications range from agriculture, hydrology and environmental science to cosmology, statistical physics, and quantum mechanics.

Advances in fractional Gaussian fields have been accompanied by the development of their discrete-space approximations. In one dimension, the discrete-space approximations emerged in the influential works of Granger and Joyeux (1980) and Hosking (1981) on fractional differencing and have received extensive treatments in time series analysis. Furthermore, there is an impressive array of works on intrinsic autoregressions that can be understood as discrete-space approximations of various intrinsic random fields; see, e.g., Künsch (1987), Besag and Kooperberg (1995), Besag and Mondal (2005), Rue and Held (2005), Lindgren et al. (2011), Cressie (2015), and Mondal (2018). In a recent paper, Dutta and Mondal (2015, 2016b) consider fraction Laplacian differencing as ways to approximate fractional Gaussian fields in two dimensions. These discrete-space approximations do not conflict, but rather establish a deeper connection with limiting, continuum fractional Gaussian fields, and help advance statistical computation.

The intent of this book chapter is to provide a basic introduction to fractional Gaussian fields with an emphasis on their interpretation, their statistical properties and on exploring their discrete-space approximations. We start with a basic definition of fractional Gaussian fields in Section 2.1, which arise when a fractional order of the Laplacian is applied to the Gaussian white noise on the two dimensional Euclidean space. We then present their spectral densities, variograms and, in Section 2.2, consider their discrete-space approximations. These discrete-space approximations are obtained by restricting the random fields on regular lattices and by replacing the Laplacian operator on the two dimensional Euclidean space with discrete Laplacians on regular grids. We primarily focus on Gaussian fields with geometric anisotropies which occur when variogram contours are formed by concentric ellipses, and standard statistical analysis presents further challenges.

In Section 3.1 we focus on a certain range of the fractional parameter and discuss maximum likelihood estimation for spatial models based on fractional Gaussian fields or their discrete-space approximations. We judge the effectiveness of the discrete-space approximations in terms of efficiency in approximating the continuum limits. For this, we consider simulation studies.

Using computer experiments, we demonstrate that discrete-space approximations provide as good estimates as the limiting, continuum model based on
the fractional Gaussian fields. We further demonstrate statistical scalability.
In Section 5, we present an analysis of the Indian Ocean surface temperature
obtained from the Argo floats devices, further establishing the agreement
between the models based on continuum fractional Gaussian fields and their
discrete-space approximations. For ease of understanding and reproducibility,
we provide all MATLAB® codes in Appendix. The dataset can be obtained
from ftp://usgodae.org/pub/outgoing/argo/geo/indian_ocean/ and also from
the corresponding author.

2 Fractional Gaussian fields and their approximations

2.1 Fractional Gaussian fields

We follow Lodhia et al. (2016) and consider anisotropic two dimensional
fractional Gaussian fields as

$$\psi(u,v) = (-\nabla)^{-\nu/2}\xi(u,v), \quad (u,v) \in \mathbb{R}^2, \tag{1}$$

where $\xi(u, v)$, for $(u, v) \in \mathbb{R}^2$ represent Gaussian white noise with marginal
variance σ^2, ν denotes the fractional or the long range dependence parameter,
and ∇ is the anisotropic Laplacian

$$\nabla = 4\alpha\frac{\partial^2}{\partial u^2} + 4\left(\frac{1}{2} - \alpha\right)\frac{\partial^2}{\partial v^2}. \tag{2}$$

The parameter $0 < \alpha < 1/2$ controls the degree of geometric anisotropy in
both the x and y directions and the value $\alpha = 1/4$ corresponds to isotropic random fields.

It is important to note that a Gaussian white noise is not pointwise defined,
rather, it is a generalized random field (Chiles and Delfiner, 2009; Lodhia
et al., 2016; Matheron, 1973). In fact, a white noise ξ on $\mathbb{R}^2$ is defined such
that for any pair of disjoint measurable sets A and B, $\int_A \xi(u,v)dudv$ and
$\int_B \xi(u,v)dudv$ are independent Gaussian random variables with zero means
and variances $\sigma^2|A|$ and $\sigma^2|B|$, respectively, where $|A|$ and $|B|$, respectively,
denote the areas of A and B. However, either pointwise or in a distributional
sense, fractional Gaussian fields exist for all real values of ν. Fractional Gaussian fields include many important models as special cases. In particular,
$\nu = 0$ corresponds to the White noise model, $\nu = 1$ gives the de Wijs process
or the Gaussian free fields, $\nu = 2$ indicates thin plate splines (also known as
bi-Laplacian random fields) and $\nu = 3/2$ denotes the Lévy Brownian motion.
For all $\nu \leq 1$, fractional Gaussian fields correspond to generalized random fields
and are defined in a distributional sense. For $1 < \nu < 2$, fractional Gaussian fields
have stationary (zeroth-order) increments. For $2 \leq \nu < 3$, fractional Gaussian
fields have stationary first-order increments, and so on. For non-negative ν,

it can be shown that the generalized spectral density of the fractional Gaussian fields in (1) is

$$\rho(\omega, \eta) = \frac{\sigma^2}{\left(4\alpha\omega^2 + 4\left(\frac{1}{2} - \alpha\right)\eta^2\right)^\nu}, (\omega, \eta) \in \mathbb{R}^2. \tag{3}$$

Thus, for $1 < \nu < 2$, standard Fourier integral formulas give an expression for variogram of Ψ as (Dutta and Mondal, 2016a)

$$\gamma(h, k) = \frac{1}{2}\text{var}(\psi(u + h, v + k) - \psi(u, v)) = \int_{\mathbb{R}^2} \{1 - \cos(h\omega + k\eta)\}\rho(\omega, \eta)d\omega d\eta$$

$$= \frac{\sigma^2 \Gamma\left(\nu - \frac{1}{2}\right)}{16\sqrt{\pi\alpha\left(\frac{1}{2} - \alpha\right)}\Gamma(\nu)\Gamma(2\nu - 1)\sin(-\nu\pi)} \left(\frac{h^2}{4\alpha} + \frac{k^2}{4\left(\frac{1}{2} - \alpha\right)}\right)^{\nu - 1},$$

$$\tag{4}$$

for any $(h, k) \in \mathbb{R}^2$. By virtue of (4), fractional Gaussian fields, for values of $1 < \nu < 2$, correspond to widely used power variogram models in geostatistics. Fractional Gaussian fields can also be seen as a limiting case of Matérn models. The latter emerge as a solution to the stochastic partial differential equation

$$\left(\kappa^2 - \nabla\right)^{\nu/2}\psi^\dagger(u, v) = \xi(u, v), (u, v) \in \mathbb{R}^2, \tag{5}$$

where $\kappa > 0$ is the inverse range parameter. The limiting cases, as $\kappa \rightarrow 0$, provide the fractional Gaussian fields in (1).

The spectral density (3) and the variogram function (4) play an important role in all subsequent statistical computations. For example, for $1 < \nu < 2$, the variogram function (4) is key to computing the actual likelihood function, which we shall discuss in Section 3.1.1.

2.2 Lattice approximations

For $m \geq 1$, let $\mathbb{Z}_m^2$ denote the sub-lattice of the two-dimensional integer lattice $\mathbb{Z}^2$ with spacing $1/m$. Following Dutta and Mondal (2016a), let Δ_m be the Laplace difference operator on the sub-lattice $\mathbb{Z}_m^2$. Thus, for any real valued function ω defined at the lattice points of $\mathbb{Z}_m^2$, we get

$$\Delta_m w(u, v) = w(u, v) - \left[\alpha_m\left\{w\left(u + \frac{1}{m}, v\right) + w\left(u - \frac{1}{m}, v\right)\right\}\right.$$

$$\left. + \left(\frac{1}{2} - \alpha_m\right)\left\{w\left(u, v + \frac{1}{m}\right) + w\left(u, v - \frac{1}{m}\right)\right\}\right],$$

where $0 \leq \alpha_m \leq 1/2$. Next, we consider

$$\psi^{(m)}(u, v) = \Delta_m^{-\nu/2} \xi_{u,v}^{(m)}, \quad \nu \geq 0, \tag{6}$$

where $\xi_{u,\,v}^{(m)}$ is a Gaussian white noise on the sub-lattice $\mathbb{Z}_m^2$ with

$$\text{var}\xi_{u,v}^{(m)} = \sigma_m^2/m^2.$$

Then, the random field $\{\psi^{(m)}\,(u,\,v)\}$ that arises from the above fractional Laplacian differencing can be interpreted as an approximation of the fractional Gaussian random fields 1 on the sub-lattice $\mathbb{Z}_m^2$. It then follows from the standard theory on linear transformation or spectral representation that the generalized spectral density function of $\{\psi^{(m)}\,(u,\,v)\}$ has the form

$$\rho_m(\omega,\eta) = \frac{\sigma_m^2}{m^2\left[4\alpha_m\sin^2\left(\frac{1}{2m}\omega\right) + 4\left(\frac{1}{2} - \alpha_m\right)\sin^2\left(\frac{1}{2m}\eta\right)\right]^{\nu}}, \tag{7}$$

with $\omega,\ \eta \in (-\pi m,\ \pi m]$, $\sigma_m > 0$ and $\nu > 0$. Under appropriate scaling of the parameters σ_m^2 the lattice random field converges to the fractional Gaussian fields (1). That is, as $m \to \infty$,

$$4^{\nu}m^{2\nu-2}\sigma_m^2 \to \sigma^2,$$

and $\alpha_m \to \alpha$, the spectral density ρ_m converges to ρ pointwise and in L_p for all $p \le 2/[\nu - 1]$. The preceding result indicates that continuum fractional Gaussian fields are scaling limits of fractionally differenced Gaussian random fields on regular lattices and it also explicitly describes the rescaling of parameters needed.

For integer values $\nu = 1,\ 2,\ \ldots$, fractional Laplacian differencing corresponds to intrinsic autoregressions of order $\nu - 1$ on the sub-lattice $\mathbb{Z}_m^2$. Furthermore, for $1 < \nu < 2$, fractional Laplacian differencing leads to a random field with stationary (zeroth-order) increments. Similarly, for $2 < \nu < 3$, fractional Laplacian differencing gives rise to a random field with stationary first-order increments, and so on.

For $1 < \nu < 2$, the variogram function of $\psi^{(m)}$ takes the form of

$$\begin{aligned}
\gamma_m(h,k) &= \frac{1}{2}\text{var}\left(\psi^{(m)}(u+h,v+k) - \psi^{(m)}(u,v)\right) \\
&= \frac{1}{4\pi^2}\int_{-m\pi}^{m\pi}\int_{-m\pi}^{m\pi}\{1 - \cos(\omega h + \eta k)\}\rho_m(\omega,\eta)d\omega d\eta.
\end{aligned} \tag{8}$$

for $(h,\ k) \in \mathbb{Z}^2$, and one can show that $\sup_{(h,k)\in\mathbb{Z}^2}|\gamma_m(h,k) - \gamma(h,k)| \to 0$ as $m \to \infty$. However, unlike (4), there is no such exact analytic formula available for (8). Interestingly, we can apply the numerical method presented in Dutta and Mondal (2016b) to calculate (8) and assess how well γ_m in (8) approximate the limiting variogram function (4). The plots in Fig. 1 display the difference $\gamma_m\,(h,\,k) - \gamma\,(h,\,k)$ for $\sigma^2 = 1$, $\sigma_m^2 = 4^{-\nu}m^{2-2\nu}$, $\alpha_m \equiv \alpha = 0.1$ and various values of ν between 1 and 2 and, for $m = 2$, 4 and 8. We find that the difference is essentially a small constant independent of the spatial lag, but depending on ν and α. Because the variogram of a nugget effect is constant, these results thus suggest that, when augmented with a nugget effect, the fractionally differenced random field at the one-eighth lattice provides an excellent approximation of an

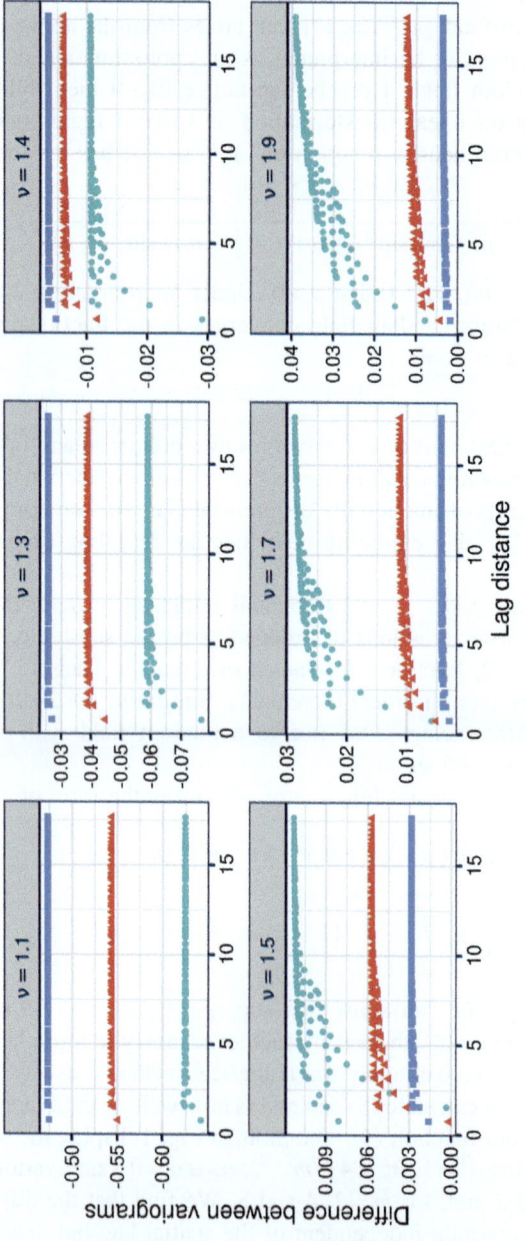

FIG. 1 Plot of $\gamma_m (h, k) - \gamma (h, k)$ against the lag distance $\sqrt{h^2/(4\alpha) + k^2/(4(\frac{1}{2} - \alpha))}$ for difference values of ν and m.

fractional Gaussian field plus a nugget effect on the original lattice. This approximation result is consistent with the isotropic case discussed in Dutta and Mondal (2016b).

3 Model based geostatistics

In practice, the spatial random fields are often observed indirectly via some noise, blurring, treatment or covariate effects. Let the available data consist of values $y_1, ..., y_n$ at respective sites or sampling stations $s_1, ..., s_n$. Here each s_i represents a small region (relative to the scale of sampling) and is often referenced by a point (u_i, v_i) in $\mathbb{R}^2$. In model based geostatistics (Diggle and Ribeiro, 2007; Diggle et al., 1998), it is assumed that the observed data values are realizations of an explicitly specified stochastic model, such as the linear mixed model

$$y_i = \mu + \psi(u_i, v_i) + \epsilon_i, \tag{9}$$

where μ is the overall mean, ψ is the underlying fractional Gaussian field (1) with the variogram function given by Eq. (4), and, independent of ψ, random errors $\epsilon_1, ..., \epsilon_n$ are iid $N(0, \tau^{-1})$ residual or nugget components. The parameter τ^{-1} is popularly known as the nugget variance. Under the intrinsic assumption, the joint distribution of the contrasts observations $y_1, y_2, ..., y_n$ are then used for estimating the mean and spatial parameters, and conditional distribution of ψ (u, v) given observed data values $y_1, ..., y_n$ is used to make predictions at an unsampled locations $(u, v) \in \mathbb{R}^2$.

For a suitable value of m, we next assume that the sampling stations s_i, $1 \le i \le$ n, can be embedded in the sub-lattice $\mathbb{Z}_m^2$. Furthermore, on the sub-lattice $\mathbb{Z}_m^2$, let the point $(u_{i,m}, v_{i,m})$ best represents the sampling station s_i. We can then consider a lattice approximation of the linear mixed model (9) by replacing ψ with $\psi^{(m)}$. This leads to an approximate model

$$y_i = \mu^{(m)} + \psi^{(m)}(u_{i,m}, v_{i,m}) + \epsilon_i^{(m)}. \tag{10}$$

In the above is now the overall mean, and random errors $\epsilon_1^{(m)}, ..., \epsilon_n^{(m)}$ are independent of $\psi^{(m)}$ and are iid $N(0, \tau_m^{-1})$. The nugget variance τ_m^{-1} is analogous to τ^{-1}.

One important aspect of model based geostatistics is that it explicitly describes the joint distribution of the observations, thus providing a likelihood for the parameters. It provides a complete approach to inference based on variograms which is primarily used by practitioners. For more detail on model based geostatistics we refer the readers to Diggle and Ribeiro (2007) and references therein.

3.1 Maximum likelihood estimation

Generally, the Gaussian linear mixed models allow maximum likelihood methods for estimating spatial parameters of interest thus facilitating model selection via information criteria, statistical inference, and more importantly, assessment of the uncertainty of the parameter estimates. However, maximum likelihood estimation for models (9) and (10) presents significant challenges. In particular, exact MLE calculations for (9) can be very challenging for any values of $\nu \geq 2$. Furthermore, $\nu \leq 1$, fractional Gaussian fields are not defined pointwise but only in a distributional sense. This also presents additional complications. For MLE calculations with $\nu = 1$, we refer to McCullagh and Clifford (2006) and Dutta and Mondal (2015). Here, for ease of exposition, we restrict our discussion to $1 < \nu < 2$. When $1 < \nu < 2$, fractional Gaussian fields have stationary increments. Thus, for this range of the fractional parameter, the marginal variances of the observations are infinite but all contrasts possess valid joint distribution. Moreover, in this case, the expected value of any contrast of the vector $y = (y_1, \ldots, y_n)^\top$ is zero. In the next subsection, we use these properties to advance MLE calculations.

3.1.1 MLE for fractional Gaussian fields

We assume $1 < \nu < 2$. In the continuum model (8), the observations themselves do not possess a regular joint distribution because the marginal variances are not finite. However, in this case, all contrasts of the observations admit a nonsingular multivariate normal distribution. To that end, suppose C is an $(n-1) \times n$ matrix of orthogonal contrasts so that $C1_n = 0$ and $CC^\top = I_{n-1}$, where 1_n is the $n \times 1$ vector of ones. Then the joint distribution of Cy is multivariate normal with zero mean vector and covariance matrix $C\Sigma C^\top + \tau^{-1} I_{n-1}$, where the (i, j)th entry of Σ arises from the variogram (4) and is given by

$$\sigma_{ij} = \frac{\sigma^2 \pi \Gamma(\nu - 1/2)}{16\sqrt{\alpha(\frac{1}{2} - \alpha)}\Gamma(\nu)\Gamma(2\nu - 1)\sin(\nu\pi)}\left(\frac{(u_i - u_j)^2}{4\alpha} + \frac{(v_i - v_j)^2}{4(1/2 - \alpha)}\right)^{\nu - 1},$$

where $\sigma^2 > 0$, $\tau > 0$, $1 < \nu < 2$ and $0 < \alpha < 1/2$. Note that, although Σ is not non-negative definite, $C\Sigma C^\top$ is positive semi-definite. Consequently, the log-likelihood of the parameter $\theta = (\tau, \sigma^2, \nu, \alpha)$ is given by

$$2\ell(\theta) = -(n-1)\log(2\pi) - \log\det\left(C\Sigma C^\top + \tau^{-1}I_{n-1}\right)$$
$$- y^\top C^\top \left(C\Sigma C^\top + \tau^{-1}I_{n-1}\right)^{-1}Cy. \tag{11}$$

This likelihood function is invariant to the choice of the orthogonal contrast matrix C because different choices change the log-likelihood by an additive constant that does not depend on the parameters. ML estimates of the parameters are obtained by maximizing ℓ within the domain. Because the parameters ν and α are constrained inside intervals, the limited memory Broyden–Fletcher–Goldfarb–Shanno algorithm with box constraints (L-BFGS-B) provides a practically useful tool for ML estimation. We also obtain the numerical hessian matrix as a byproduct of the algorithm and compute the standard errors of the parameters as the square roots of the diagonals of the inverse hessian matrix.

There are some practical drawbacks of estimating the parameters using this method. First, the method requires inversion of an $(n-1) \times (n-1)$ covariance matrix, which is typically done using the dense Cholesky factorization, that requires $O(n^2)$ storage space in memory and has $O(n^3)$ computational complexity. Thus the method is only useful for moderate sample sizes. Second, the log-likelihood is not a concave function. Thus we cannot guarantee a global maximum. At the same time, maximization can run into boundary problems, meaning that maximum value is susceptible to occur at the boundary of the parameter space. Finally, for $\nu \geq 2$, MLE calculations get exceedingly difficult, as we need to consider a different contrast matrix C that can generate all first-order increments of the observed data.

3.1.2 MLE with lattice approximations

Exact MLE calculations for the model (10) also presents challenges. This is because unlike (4), variogram calculations (8) require expensive numerical computation. However, on any finite regular lattice, (6) provides another alternative way to approximate the model (10). To that end, suppose that for a specific value of m, the spatial domain is embedded in a finite regular rectangular array with r rows and c columns (both of which depend on m). Then, under a restriction of Δ_m to the finite $r \times c$ array, a solution φ to (6) has a precision matrix $\lambda_m R^\nu$ where $\lambda_m = m^2/\sigma_m^2$ and R is the $rc \times rc$ matrix representing the restriction of Δ_m to the finite $r \times c$ array. Under a column major ordering of the entries of the $r \times c$ array, Dutta and Mondal (2015), have shown that the $rc \times rc$ matrix R admits a spectral decomposition given by

$$R = P^\top(4\alpha_m D_{01} + 4(1/2 - \alpha_m)D_{10})P,$$

with $P = P_c \otimes P_r$, $D_{10} = I_c \otimes D_r$ and $D_{10} = D_c \otimes I_r$, where for $l = r$ or c, P_l is the $l \times l$ orthogonal matrix with (i, j)th entry given by

$$p_{1j} = l^{-\frac{1}{2}}, \ p_{i,j} = (2/l)^{-\frac{1}{2}} \cos\left\{\pi(i-1)\left(j - \frac{1}{2}\right)/l\right\}, \ i = 2, ..., l, \ j = 1, ..., l,$$

and D_l is the $l \times l$ diagonal matrix with ith diagonal entry

$$d_i = \sin^2\{\pi(i-1)/(2l)\}, \ 1 \leq i \leq l.$$

Consequently, suppressing m, we revise (10) using φ as

$$y = \mu + F\varphi + \varepsilon$$

where F is the $n \times rc$ incidence matrix with ith row f_i such that $f_i^\top \varphi$ gives the φ-values at (u_{im}, v_{im}), $\varepsilon = (\varepsilon_1^{(m)}, ..., \varepsilon_n^{(m)})^\top$, and the improper density for φ is given by,

$$f(\varphi) \propto |\lambda_m R^\nu|_+^{1/2} \exp\left(-\frac{1}{2}\lambda_m \varphi^\top R^\nu \varphi\right). \tag{12}$$

In the above, we interpret the fractional power of R via its spectral density,

$$R^\nu = P^\top (4\alpha_m D_{01} + 4(1/2 - \alpha_m)D_{10})^\nu P. \tag{13}$$

In order to estimate the parameters $\theta_m = (\tau_m, \lambda_m, \nu, \alpha_m)$, Dutta and Mondal (2016a) takes an h-likelihood approach. Unlike the method described in Section 3.1.1, the above finite regular lattice approximations and the h-likelihood method are valid for all $\nu > 0$. The h-likelihood method goes as follows. Let B denote the last $rc - 1$ rows of the matrix M so that $B\varphi$ is an $rc - 1$ variate normal random vector with diagonal precision matrix G consisting of the $rc - 1$ non-zero eigen values of $\lambda_m R^\nu$. Next, define the following matrices and vectors

$$X = \begin{pmatrix} 1_n & F \\ 0 & B \end{pmatrix}, \ z = \begin{pmatrix} y \\ 0 \end{pmatrix}, \ \beta = \begin{pmatrix} \mu^{(m)} \\ \varphi \end{pmatrix}, \ Q = \begin{pmatrix} \tau_m I_n & 0 \\ 0 & G \end{pmatrix} \text{ and}$$

$$H = X(X^\top Q X)^{-1} X^\top Q.$$

Dutta and Mondal (2016a) then obtain the residual likelihood (REML) function ℓ_R given by

$$2\ell_R(\widehat{\theta}) = \log \det Q - \log |X^\top Q X|_+ - (z - X\widehat{\beta})^\top Q (z - X\widehat{\beta}) \tag{14}$$

where $\widehat{\beta}$ is the solution to

$$(X^\top Q X)\beta = X^\top Q z. \tag{15}$$

Traditional maximization of the log REML function uses score equations which are obtained by equating the gradient of ℓ_R to zero. Thus suppose $Q_1 = \partial Q/\partial \tau_m$, $Q_2 = \partial Q/\partial \lambda_m$, $Q_3 = \partial Q/\partial \nu$, and $Q_4 = \partial Q/\partial \alpha_m$. The score equations that maximize the log–REML function in (14) are then given by

$$\frac{1}{2}\mathrm{Tr}(Q^{-1}Q_i) - \frac{1}{2}\mathrm{Tr}\{(X^\top Q X)^{-1}X^\top Q_i X\} - \frac{1}{2}(z - X\widehat{\beta})^\top Q_i (z - X\widehat{\beta}) = 0$$

for $i = 1, \ldots, 4$. Note that these score equations can also be expressed succinctly as

$$\frac{1}{2}\mathrm{Tr}(I - H)Q^{-1}Q_i - \frac{1}{2}(z - X\widehat{\beta})^\top Q_i (z - X\widehat{\beta}) = 0, \ i = 1, \ldots, 4. \tag{16}$$

Typically, Fisher's scoring method is used to solve the score equations and to obtain REML estimates. However, this also requires computation of the second derivatives of the log REML function or the information matrix $\mathfrak{J}$ whose (i, j)th entry is equal to

$$\mathfrak{J}(i,j) = \frac{1}{2}\mathrm{Tr}\{(I - H)Q^{-1}Q_i(I - H)Q^{-1}Q_j\}, \tag{17}$$

which can also be used to derive standard errors of the estimates. However, computing the trace terms either in (16) or (17) are not straightforward as they require computing the diagonal entries of the hat-matrix H. For large values of n, an exact computation of these trace terms has $O(n^3)$ computation complexity and requires $O(n^2)$ memory storage space. As a practical alternative, Dutta and Mondal (2016a) then suggests instead solving the unbiased system of equations

$$g_i(\theta_m) = \frac{1}{2K} \sum_{t=1}^{K} u_t^\top Q^{-1} Q_i (I - H) u_t - \frac{1}{2}\left(z - X\widehat{\beta}\right)^\top Q_i \left(z - X\widehat{\beta}\right) = 0, \quad (18)$$

where u_t's are i.i.d Rademacher random vectors with entries ± 1 with probability 1/2 each. Here the number of Rademacher vectors, K, should be large. However, the results of Dutta and Mondal (2016a) suggest $K = 50$ retains sufficient statistical efficiency of the estimates.

Dutta and Mondal (2016a) provide a sophisticated matrix-free trust-region algorithm for solving (18) that crucially depends on the matrix-free discrete cosine transformation for computing matrix-vector multiplications of the form Pv and the matrix-free inverse discrete cosine transformation for computing $P^\top v$ for $v \in \mathbb{R}^{rc}$, (Frigo and Johnson, 2005; Rao and Yip, 1990) and a matrix-free preconditioned Lanczos algorithm for solving large system of linear Eq. (15) and those involved in (18). Furthermore, this computational framework yields the standard errors of the parameter estimates as well as the best linear unbiased predictions of the random field φ that serves as the kriged surface of the random field. Overall, in contrast to the dense-matrix computations the computational complexity of the matrix-free algorithms is essentially $O(n(\log n)^2)$ using only $O(n)$ storage in memory.

4 Simulation studies

We perform two simulation studies. The goal of the first simulation study is to derive the estimates for the fractionally differenced random field model when the data is generated from a continuum fractional Gaussian field plus a nugget effect, and to compare these estimates with the actual maximum likelihood estimates. The goal of the second simulation study is to demonstrate the scalability of statistical computation for fractional Laplacian differencing.

4.1 An experiment with power-law variogram

We generate data on 4000 randomly selected grid points in a 100×100 lattice embedding the unit square from an intrinsic Matérn random field with $\nu = 1.25$, $\tau = 1$, $\sigma^2 = 2$, and $\alpha = 0.25$. We compute the estimates of ν, τ, σ^2 and α using the method described in Section 3.1.1. Next, we fit the lattice model (10) the original 100×100 array (so that $m = 1$) and compute estimates of ν, τ_m, λ_m and α_m. We repeat this process 100 times. Overall, the h-likelihood method was between 40 and 80 times faster than the direct ML estimation of

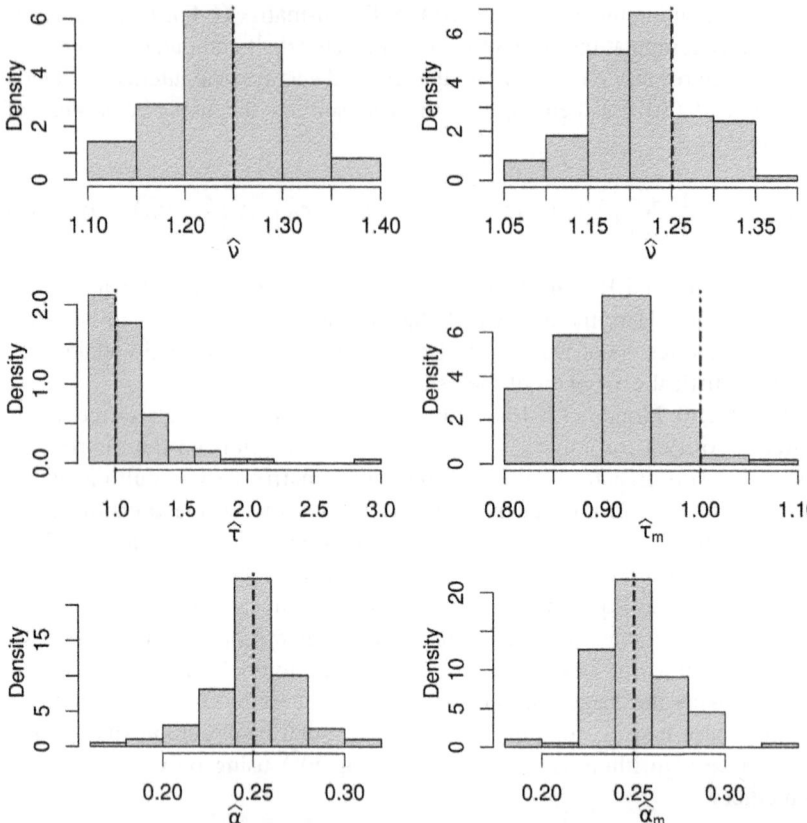

FIG. 2 Histograms of the estimates of ν (top), nugget precision (middle) and anisotropy parameters (bottom) using direct ML estimation of intrinsic Matérn model (left column) and h-likelihood method on the lattice model (right column).

intrinsic Matérn model and in one of these simulations the direct ML method failed to converge, yielding estimates on the boundary. We discard this case from our analysis. It is expected that the analysis on the original scale with the fractionally differenced model would yield biased estimate of τ_m because it compensates or absorbs the difference between the lattice variogram and the continuum variogram as seen in Fig. 1. Fig. 2 shows the histograms of these estimates from the two models along with the true values. These plots show that the lattice based fractionally differenced model provide practically useful estimates of ν and the anisotropy parameter. However, it over estimates the nugget variance (underestimates nugget precision). On the other hand, the confidence intervals and their average widths in Table 1 show that the fractionally differenced model provides shorter and more practically meaningful confidence intervals for ν and the anisotropy parameters.

TABLE 1 Coverage probabilities and mean widths of 95% confidence intervals based on normal approximations.

	ν		$\log\tau : \log\tau_m$		$\alpha : \alpha_m$	
Model	Coverage	Width	Coverage	Width	Coverage	Width
Fractional Gaussian fields	100	1.31	92.9	0.89	100	0.19
Fractional Laplacian differencing	96	0.27	60.6	0.24	94	0.09

4.2 Large scale computation with lattice approximations

In this section, we demonstrate the scalability of the likelihood computations using the fractionally differenced model. To that end, we now generate data on grid points of a 256×256 regular rectangular array from two fractionally differenced models. We keep $\tau_m = 4$, $\lambda_m = 8$ and fix $\alpha_m = 0.25$ and use two different values $\nu = 1.25$ and $\nu = 1.5$. We randomly keep 60% of the observations resulting in a sample of size around $39{,}321 \pm 125$ (mean$\pm$sd). The data is generated from the fractionally differenced model because such the method for generating from the intrinsic Matérn model runs out of the memory. Similarly, the method for fitting the intrinsic Matérn model using dense-matrix computations also fail on such large datasets. In contrast, the fractionally differenced model fits without any issue on a standard personal computer. The process is repeated 100 times for each choices of ν and the resulting boxplots of the estimates and their standard errors are shown in Fig. 3. We find that the estimates are very close to the true values of the parameters and are also unbiased. Furthermore, the standard errors are also very small suggesting the estimators are statistically efficient, a fact that is also noted in Dutta and Mondal (2016a).

5 Indian ocean surface temperature from Argo floats

The Argo Program is part of the Global Ocean Observing System from an international collaboration among more than 30 countries from all continents that provides useful data on important ocean variables. Conceived in the early 2000s, the Argo fleet now consists of more than 4000 drifting battery-powered machines called Argo *floats* that are deployed worldwide. These floats weigh around 20–30 kg each, and typically probe the drifts at a depth where they are stabilized by their buoyant (around 1 km). Every 10 days or

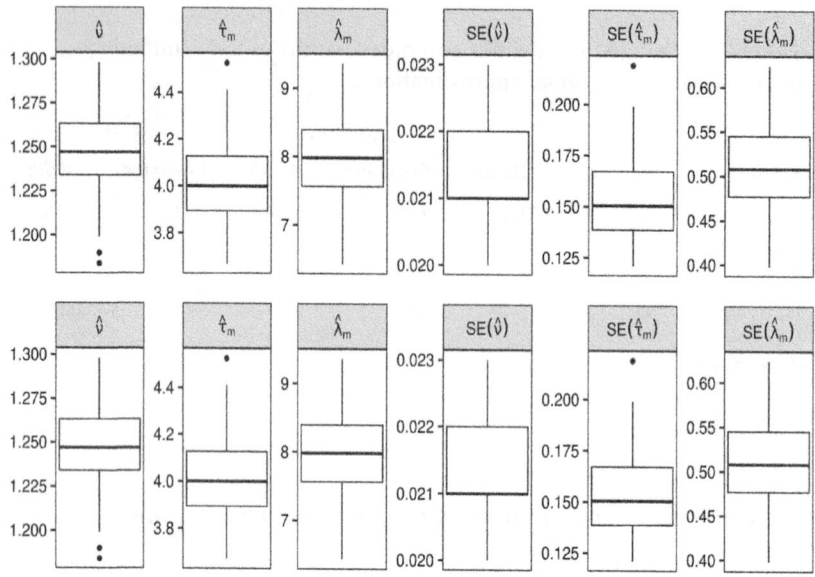

FIG. 3 Boxplots of the parameter estimates and their standard errors. True values of the parameters are: $\tau_m = 4$ and $\lambda_m = 8$ and $\nu = 1.25$ in top panel and $\nu = 1.5$ in bottom panel.

so, these floats change their buoyant and dive to a depth of 2 km and then rise to the water-surface measuring conductivity and temperature profiles as well as pressure, over about 6 h. From the surface, they transmit their location as well as the collected data to satellites and dives back to their drifting depth. In this section, we analyze the monthly data on sea-surface temperature in the Indian Ocean obtained from April 1, 2020 till April 30, 2020. These data were collected and made freely available by the International Argo Program and the national programs that contribute to it (Argo, 2020). After removing the erroneous measurements as described on the Argo website, we obtain 2525 observations of sea-surface temperature (in °C) and plot them in the bottom left panel of Fig. 4. Note that the Argo floats are quite scattered over the Indian Ocean and the temperature are clearly spatially auto-correlated. Furthermore, from the top panel of Fig. 4 we can see that the temperature variation seems to be more along the latitude compared to the longitude, as one would naturally expect. In fact, this suggests that an (intrinsically) stationary model may not be accurate. To account for this trend along the latitude, we fit a quadratic mean model

$$\mu(l) = a_0 + a_1 l + a_2 l^2 \tag{19}$$

where l denotes the latitude using ordinary least squares. Next, we obtain the residuals from this quadratic mean model and use them as response for the following spatial analysis.

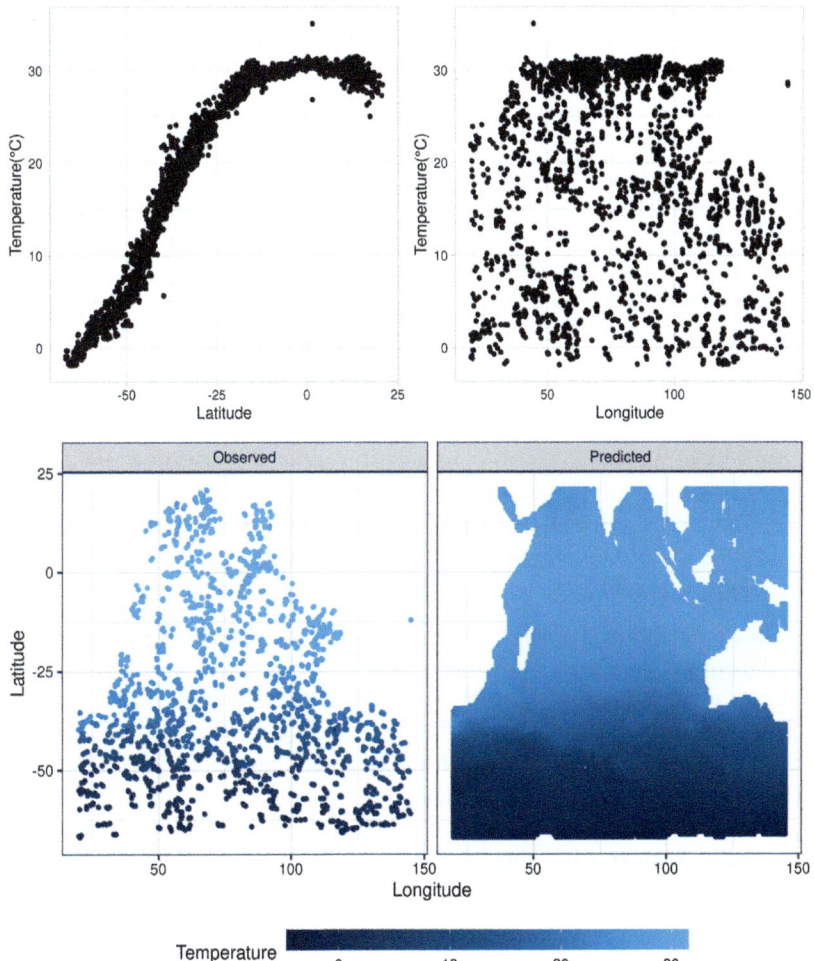

FIG. 4 Sea-surface temperature (in °C) measured by the Argo floats in the Indian Ocean during April, 2020. Top: Observed temperature against the geographic coordinates. Bottom left: Image of observed temperature values and right: Krigged sea-surface temperature.

First we run some exploratory analyses. We compute and plot the empirical variogram using the R package geoR along the four directions and plot it using the R package ggplot2.

```
library (geoR)
library (dplyr)
library (ggplot2)
library (RColorBrewer)
```

```
temp = read.table("april-data.txt")
names(temp)=c("Latitude","Longitude","Temperature","Resid.quad")
temp.geo = as.geodata (temp, data.col = 4)
# Using the residuals from quadratic model as response
vg4 = variog4 (temp.geo)
vg.gg=data.frame (h=vg4$'0'$u, v=vg4$`0`$v, Direction='0⁰') %>%
   rbind (., data.frame (h = vg4$'45'$u, v = vg4$`45`$v, Direction =
'45⁰')) %>%
   rbind (., data.frame (h = vg4$'90'$u, v = vg4$`90`$v,
Direction= '90⁰')) %>%
   rbind (., data.frame (h = vg4$'135'$u, v = vg4$`135`$v,
Direction= '135⁰'))
```

```
ggplot (vg.gg, aes (x=h, y=v, group=Direction, color=Direction,
   lty=Direction)) + geom_line (size=1.5) + xlab ("Spatial lag") +
   ylab ("Variogram") + theme_light (base_size = 16) +
   scale_color_brewer (palette="Dark2") + theme
   (legend.position = "bottom", legend.key.width = unit(2.5,"cm"))
```

The dataset is also available by an email request to the corresponding author. The directional variograms are shown in Fig. 5. We see that the variogram increases more along the 90° and the 45° directions supporting that there is more spatial variability across the latitude. Furthermore, the variograms along these directions do not seem to reach a sill, suggesting that an intrinsic model could be more appropriate for the data.

We first fit the intrinsic Matérn model to the data using the method described in Section 3.1.1. The estimates of ν, α and τ are shown in the first

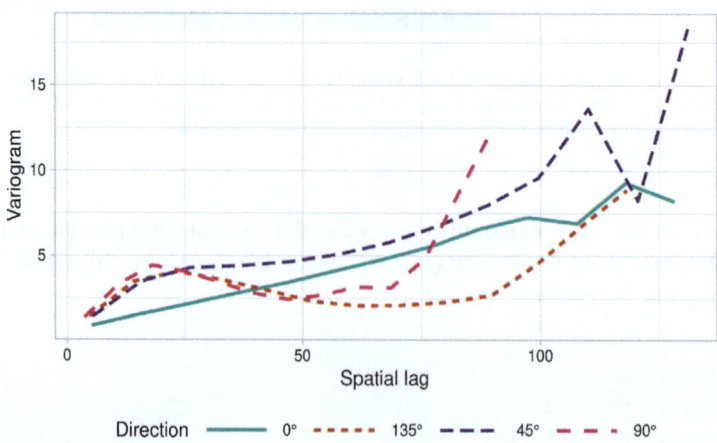

FIG. 5 Directional variograms of the residual temperature values.

TABLE 2 Estimates of the spatial parameters from the spatial linear mixed model based on fractional Gaussian field (FGF) and its lattice approximation (FLD).

Model	$\widehat{\nu}$	$\widehat{\alpha}:\widehat{\alpha}_m$	$\widehat{\tau}:\widehat{\tau}_m (°C^{-2})$
FGF	1.400 (0.114)	0.058 (0.0314)	3.59 (0.285)
FLD	1.426 (0.051)	0.074 (0.012)	3.11 (0.267)

Standard errors are shown in parentheses.

row of Table 2 and the estimate of σ^{-2} is $\widehat{\sigma}^{-2} = 8.987°C^{-2}(\text{s.e.}1.19°C^{-2})$. The estimate of α corroborates the observation that the temperature varies more across the latitude than the longitude.

Next, we fit the fractionally differenced process to the data using the method described in Section 3.1.2. To that end, we embed the region bounded between by 21° to the North, 67° to the South, 20° to the West and 145° to the East in a 128×180 regular rectangular array so that each pixel is approximately 0.6875° latitude by 0.694° longitude. Next we average the residuals from the quadratic model falling inside the same lattice pixel, resulting in around 7.43% observed pixels. We use 50 Rademacher variables for stochastically approximate the score equations. Abusing the notation, we drop the subscript m from τ_m, α_m and λ_m, as m is implicitly chosen via the array dimensions. The estimates of ν, α and τ are shown in the second row of Table 2 and the estimate of λ is $\widehat{\lambda} = 15.381°C^{-2}$ (s.e.$3.35°C^{-2}$). The results largely agree with the findings in Section 4.1. In particular, although the estimates of ν and the anisotropy parameters are very close from the two methods, the fractionally differenced model yields smaller standard errors of these estimates. The slight discrepancy in the estimates of α occurs because the pixels are not exact squares. On the other hand, the estimate of the nugget precision is lower from the fractionally differenced model than the intrinsic Matérn model.

Note that as a byproduct of fitting the fractionally differenced model, we also obtain the best linear prediction $\widehat{\varphi}$ of the underlying spatial random field φ. The image $\widehat{\varphi}$ plus the quadratic mean from (19) is shown on the right panel of Fig. 4. Note that the fine scale features of the temperature gradient are more prominent in the krigged map. Such interpolated maps are often useful in studying other oceanic and atmospheric activities.

6   Concluding remarks

This book chapter presents a brief review on fractional Gaussian fields, their lattice based approximations, and connections to the intrinsic and stationary

Matérn, power-law and other generalized random fields. Likelihood based inference methods have been developed for spatial linear mixed models based on the fractional Gaussian random fields and fractional Laplacian differencing on regular lattice. Computational methods for maximum likelihood estimation of parameters have been described and compared. Using both simulation and data examples, it is demonstrated that the lattice based model facilitate faster and more stable statistical computations of the maximum likelihood estimators than the model based on fractional Gaussian fields, while providing practically close estimates of dependence and anisotropy parameters. Moreover, the h-likelihood method for the model on regular lattice provide more useful estimates of uncertainty and confidence intervals for the aforementioned parameters than the maximum likelihood method for the geostatistical model based on the fractional Gaussian fields.

It must be stressed that there are definite advantages in discretizing the space using a regular lattice instead of other well-known ideas such as triangulation (Lindgren et al., 2011) and neighborhood selection (Datta et al., 2016) of irregularly distributed sampling stations. One advantage is the explicit spectral decomposition that allows for the use of fractional values ν and provides fast matrix-free computation in terms of discrete cosine transformation. Another advantage is accommodation of geometric anisotropies. It must be noted that irregular discretizations do not permit us to accommodate geometric anisotropies in any obvious way.

Our presentation has focused on fractional models with long range dependence. To study short range dependence, we can consider stationary Matérn covariance models (Guinness and Fuentes, 2017; Haskard et al., 2007; Stein, 2012). As fractional Gaussian fields are limiting cases of Matérn models, we can also obtain lattice approximations of the latter. Under the same setup as Section 3.1.2, the inverse variance covariance matrix of this approximate Matérn model takes the form

$$\lambda_m R^\nu = \lambda_m P^\top \left(\kappa_m + 4\alpha_m D_{01} + 4\alpha_m' D_{10} \right)^\nu P \tag{20}$$

where κ_m, α_m, α_m' are non-negative and $\kappa_m + 4\alpha_m + 4\alpha_m' = 2$. Thus, both Matérn models and their lattice approximations contain an additional range parameter, and at first it may appear that Matérn models and their lattice approximations have added flexibility due to the extra range parameter. However, inclusion of an unknown finite range parameter often leads to long flat ridges in the likelihood function, which in turn incur substantial numerical instability in the MLE computations. This, for example, has been observed in Lim et al. (2017) and also in our own experiments with the lattice approximation (20). Interestingly, the work of Zhang (2004) suggest that the scale and the range cannot both be estimated consistently. In fractional fields, we set the range parameter at infinity. In the short range dependence case, we can also fix the range parameter to a finite number to lessen numerical instabilities and to enhance interpretability. We can then proceed computation as presented in Section 3 of this chapter.

Acknowledgments

The authors thank an anonymous referee for helpful comments. Dutta's research was supported in part by the United States Department of Agriculture (USDA) National Institute of Food and Agriculture (NIFA) Hatch project IOW03617. Mondal's research was supported by the National Science Foundation (NSF) award DMS-1916448. The content presented in this chapter are those of the authors and do not necessarily reflect the views of NIFA, USDA and NSF.

Appendix

MATLAB codes for Section 3.1.1

```
function [x, se] = mlFGF (row, col, y, initial)
% INPUT:
% row: n x 1 vector of x-coordinates
% col: n x 1 vector of y-coordinates
% y: n x 1 vector of observations
% initial: Starting values [tau, lambda, nu, alpha] where lambda is
1/sigma^2
%
% OUTPUT:
% x: estimate of the parameters in otder [tau, lambda, nu, alpha]
%    where lambda = 1/sigma^2
% se: standard error of the parameters
n = length (row);
if n ~= length (y) || n ~= length (col)
    error('lengths of the three vectors must be equal');
end
if (initial(4) > 0.5) || (initial(4) < 0)
    error('Initial  for  anisotropy  parameter  must  be  between
0 and 0.5');
end
diffrow2 = (row - row'). ^2;
diffcol2 = (col - col'). ^2;
% orthogonal contrast matrix
[Cmat, ~] = qr(eye(n) - ones(n)/n);
Cmatt = Cmat(:,1:n-1);
Cmat = Cmatt';
Cy = Cmat * (y-mean(y));
f = @(logx) -loglikFGF(logx,diffrow2,diffcol2,Cmat,Cmatt,Cy,n);
logx0 = log(initial);
logx0(3) = log(initial(3)-1);
logx0(4) = log(2*initial(4)/(1-2*initial(4)));
[x, ~, ~, ~, ~, hess] = fminunc(f,logx0);
x(1:2) = exp(x(1:2));
```

```
x(3) = exp(x(3)) + 1;
se = sqrt(diag(inv(hess)));
se(1:3) = x(1:3).*se(1:3);
beta = exp(x(4));
x(4) = 0.5*beta/(1+beta);
se(4) = sqrt(beta)/(1+beta) * se(4)/2;
end
% function for computing the fractional Gaussian field log-
likelihood
function v = loglikFGF(logx,diffrow2,diffcol2,Cmat,Cmatt,Cy,n)
ly = exp(logx(1));
lp = exp(logx(2));
nu = exp(logx(3)) + 1;
beta = exp(logx(4));
beta = 0.5*beta/(1+beta);
const = pi^1.5*gamma(nu-0.5)/(16*sqrt(pi) * gamma(nu)* gamma
        (2*nu-1))/... (sin(nu*pi) * sqrt(beta*(0.5-beta)));
h = diffrow2/(4*beta) + diffcol2/(4*(0.5-beta));
ucont = const * (h.^(nu-1));
Sigma = (Cmat*ucont*Cmatt)/lp + eye(n-1)/ly;
R = chol(Sigma);
z = R'\Cy;
v = -sum(log(diag(R))) - 0.5*sum(z.^2);
end
```

MATLAB codes for Section 3.1.2

This code requires two functions dct2mod and idct2mod which takes input a matrix or vector with mn entries and m and n and computes the discrete cosine transformation and the inverse discrete cosine transformation of the $m \times n$ matrix using the column-major format. The returned value must be a $mn \times 1$ vector.

```
function [x, se, psi] = fracdiffML(y,w,initial,nseed)
% [x, se, psi] = iMaternRemIisotropic(y,w,initial,nseed)
% y = r x c matrix of observations
% w = r x c incidence matrix (0 = missing pixel, 1 = observed
pixel)
% initial guess for [lambda_y;lambda_psi;nu; alpha]
% nseed : number of Rademacher variables (optional).
% OUTPUT:
% x: REML estimate of precision parameters [lambda_y; lambda_psi ;nu]
% se: standard errors of x
% psi: BLUP of the random effects (r x c matrix)
if nargin == 3
```

```
 nseed = 50;
end
[r, c] = size(y);
q = r*c;
yield = y(w>0);
n = length(yield);
idx = find(w>0);
F = sparse(1:n,idx,1,n,q);
xr = sin(0.5*pi*(0:(r-1))'/r).^2;
xc = sin(0.5*pi*(0:(c-1))'/c).^2;
xxr = kron(ones(c,1),xr);
xxc = kron(xc,ones(r,1));
FF = full(diag(F'*F));
Z = F'*yield;
eve = rng; % backup the random number generator
rng(2441139);
RadVar = 2*(rand(n+q-1,nseed) < 0.5) - 1;
rng(eve);
options = optimset('Display','iter','TolFun',0.01,...
        'TolX',0.001,'MaxFunEvals',500,'MaxIter',40);
x0 = [log(initial(1)); log(initial(2)); log(initial(3));...
      log(2*initial(4)/(1-2*initial(4)))];
gr = @(pars) gradfunAniso (pars,F,FF,Z,yield,xxr,xxc,r,c,n,q,
RadVar,nseed);
[x, fval, exitflag, output, hess] = fsolve(gr,x0,options);
se = sqrt(diag(inv(-hess)));
ly = exp(x(1));
lp = exp(x(2));
nu = exp(x(3));
beta = 0.5/(1 + exp(-x(4)));
se(1) = se(1)*ly;
se(2) = se(2)*lp;
se(3) = se(3)*nu;
se(4) = 0.5*se(4) * exp(x(4))/(1 + exp(x(4)))^2;
x = [ly;lp;nu;beta];
L = lp*(4*beta*xxr + 4*(0.5-beta)*xxc).^nu;
PtZ = dct2mod(Z,r,c);
precon_mat = 1./(ly + L);
mxf = @(vv) precon_mat.*(ly*dct2mod(FF.*idct2mod(precon_mat.*
      vv,r,c),r,c) + ... L.*precon_mat.*vv);
[psi0, flag, rel, iter] = symmlq(mxf,ly*precon_mat.*PtZ,1e-12,q);
psi1 = idct2mod(precon_mat.*psi0,r,c);
psi = reshape(psi1,r,c);
end
```

```
% Subfunction for computing the score equations.
function grad = gradfunAniso(pars,F,FF,Z,yield,xxr,xxc,r,c,n,q,
RadVar,nseed)
ly = exp(pars(1));
lp = exp(pars(2));
nu = exp(pars(3));
beta = 0.5/(1 + exp(-pars(4)));
lastelt = @(v) v(2:q);
Xt = @(vv) F'*vv(1:n) + idct2mod([0;vv(n+1:n+q-1)],r,c);
X = @(vv) [F*vv; lastelt(dct2mod(vv,r,c))];
Q = [ly*ones(n,1); lp*(4*beta*xxr(2:q) +
    4*(0.5-beta)*xxc(2:q)).^nu];
dQ1 = [ones(n,1);zeros(q-1,1)];
dQ2 = [zeros(n,1);(4*beta*xxr(2:q) + 4*(0.5-beta)*xxc(2:q)).^nu];
dQ3 = [zeros(n,1);Q(n+1:n+q-1).*log(4*beta*xxr(2:q) + 4*(0.5-beta)
*xxc(2:q))];
dQ4 = [zeros(n,1); 4*lp*nu*(xxr(2:q) - xxc(2:q)).*(4*beta*xxr(2:q) +
    ... 4*(0.5-beta)*xxc(2:q)).^(nu-1)];
L = lp*(4*beta*xxr + 4*(0.5-beta)*xxc).^nu;
PtZ = dct2mod(Z,r,c);
precon_mat = 1./(ly + L);
mxf = @(vv) precon_mat. * (ly*dct2mod(FF.*idct2mod(precon_mat.*
    vv,r,c),r,c) + ... L.*precon_mat.*vv);
% Compute the BLUP
[psi0, flag, rel, iter] = symmlq(mxf,ly*precon_mat.*PtZ,1e-12,q);
psi = idct2mod(precon_mat.*psi0,r,c);
res2 = ([yield; zeros(q-1,1)] - X(psi)).^2;
g1 = 0; g2 = 0; g3 = 0; g4 = 0;
% Computing the score function
parfor t=1:nseed
    v0 = Xt(Q.*RadVar(:,t));
    v0 = dct2mod(v0,r,c);
    [v1, flag] = symmlq(mxf,precon_mat.*v0,1e-12,q);
    v = idct2mod(precon_mat.*v1,r,c);
    v = RadVar(:,t) - X(v);
    g1 = g1 + sum(RadVar(:,t).*dQ1.*v./Q)/nseed;
    g2 = g2 + sum(RadVar(:,t).*dQ2.*v./Q)/nseed;
    g3 = g3 + sum(RadVar(:,t).*dQ3.*v./Q)/nseed;
    g4 = g4 + sum(RadVar(:,t).*dQ4.*v./Q)/nseed;
end
g1 = g1 - sum(res2.*dQ1);
g2 = g2 - sum(res2.*dQ2);
g3 = g3 - sum(res2.*dQ3);
```

```
g4 = g4 - sum(res2.*dQ4);
grad = [g1;g2;g3;g4];
grad=0.5*grad.*[ly;lp;nu;0.5*exp(pars(4))/(1+exp(pars(4)))^2];
end
```

References

Argo, 2020. Argo Float Data and Metadata From Global Data Assembly Centre (Argo GDAC)—Snapshot of Argo GDAC of April 9st 2020. Available from https://doi.org/10.17882/42182#72592.

Beran, J., 1994. Statistics for Long-Memory Processes. vol. 61 CRC press.

Besag, J., Kooperberg, C., 1995. On conditional and intrinsic autoregressions. Biometrika 82 (4), 733–746.

Besag, J., Mondal, D., 2005. First-order intrinsic autoregressions and the de wijs process. Biometrika 92 (4), 909–920.

Chiles, J.P., Delfiner, P., 2009. Geostatistics: Modeling Spatial Uncertainty. John Wiley & Sons.

Cohen, S., Istas, J., 2013. Fractional Fields and Applications. Springer.

Cressie, N., 2015. Statistics for Spatial Data. John Wiley & Sons.

Datta, A., Banerjee, S., Finley, A.O., Gelfand, A.E., 2016. Hierarchical nearest-neighbor gaussian process models for large geostatistical datasets. J. Am. Stat. Assoc. 111 (514), 800–812.

Diggle, P.J., Ribeiro Jr., P., 2007. Model Based Geostatistics. Springer, New York.

Diggle, P.J., Tawn, J.A., Moyeed, R.A., 1998. Model-based geostatistics. J. R. Stat. Soc. Ser. C. Appl. Stat. 47 (3), 299–350.

Dobrushin, R.L., 1979. Gaussian and their subordinated self-similar random generalized fields. Ann. Probab. 7 (1), 1–28.

Dutta, S., Mondal, D., 2015. An h-likelihood method for spatial mixed linear models based on intrinsic auto-regressions. J. R. Stat. Soc. Ser. B (Stat Methodol.) 77 (3), 699–726. [Online]. Available from https://doi.org/10.1111/rssb.12084.

Dutta, S., Mondal, D., 2016a. Reml estimation with intrinsic matérn dependence in the spatial linear mixed model. Electron. J. Statist. 10 (2), 2856–2893. [Online]. Available from https://doi.org/10.1214/16-EJS1125.

Dutta, S., Mondal, D., 2016b. Variogram calculations for random fields on regular lattices using quadrature methods. Environmetrics 27 (7), 380–395. [Online]. Available from https://doi.org/10.1002/env.2390.

Frigo, M., Johnson, S.G., 2005. The design and implementation of FFTW3. Proc. IEEE 93 (2), 216–231.

Gangolli, R., 1968. Asymptotic behaviour of spectra of compact quotients of certain symmetric spaces. Acta Math. 121 (1), 151–192.

Gay, R., Heyde, C., 1990. On a class of random field models which allows long range dependence. Biometrika 77, 401–403.

Granger, C.W., Joyeux, R., 1980. An introduction to long-memory time series models and fractional differencing. J. Time Ser. Anal. 1 (1), 15–29.

Gu, C., Wahba, G., 1993. Semiparametric analysis of variance with tensor product thin plate splines. J. R. Stat. Soc. B. Methodol. 55 (2), 353–368.

Guinness, J., Fuentes, M., 2017. Circulant embedding of approximate covariances for inference from gaussian data on large lattices. J. Comput. Graph. Stat. 26 (1), 88–97.

Haskard, K.A., Cullis, B.R., Verbyla, A.P., 2007. Anisotropic matérn correlation and spatial prediction using reml. J. Agric. Biol. Environ. Stat. 12 (2), 147–160.

Hosking, J.R.M., 1981. Fractional differencing. Biometrika 68 (1), 165–176. [Online]. Available from https://doi.org/10.1093/biomet/68.1.165.

Kelbert, M.Y., Leonenko, N.N., Ruiz-Medina, M., 2005. Fractional random fields associated with stochastic fractional heat equations. Adv. Appl. Probab. 37 (1), 108–133.

Künsch, H.R., 1987. Intrinsic autoregressions and related models on the two-dimensional lattice. Biometrika 74 (3), 517–524.

Lim, C.Y., Chen, C.H., Wu, W.Y., 2017. Numerical instability of calculating inverse of spatial covariance matrices. Statist. Probab. Lett. 129, 182–188.

Lindgren, F., Rue, H., Lindstrom, J., 2011. An explicit link between gaussian fields and gaussian markov random fields: the spde approach (with discussion). J. R. Stat. Soc. Series B Stat. Methodology 73 (4), 423–498.

Lodhia, A., Sheffield, S., Sun, X., Watson, S.S., 2016. Fractional gaussian fields: a survey. Probab. Surv. 13, 1–56.

Ma, C., 2003. Power-law correlations and other models with long-range dependence on a lattice. J. Appl. Probab. 40 (3), 690–703.

Mandelbrot, B.B., Van Ness, J.W., 1968. Fractional brownian motions, fractional noises and applications. SIAM Rev. 10 (4), 422–437.

Matheron, G., 1970. Random functions and their application in geology. In: Geostatistics. Springer, pp. 79–87.

Matheron, G., 1973. The intrinsic random functions and their applications. Adv. Appl. Probab. 5 (3), 439–468.

McCullagh, P., Clifford, D., 2006. Evidence for conformal invariance of crop yields. Proc. Roy. Soc. Lond. A: Math. Phys. Eng. Sci. 462 The Royal Society, pp. 2119–2143.

McKean Jr., H., 1963. Brownian motion with a several-dimensional time. Theory Probab. Appl. 8 (4), 335–354.

Mondal, D., 2015. Applying dynkin's isomorphism: an alternative approach to understand the markov property of the de wijs process. Bernoulli 21 (3), 1289–1303.

Mondal, D., 2018. On edge correction of conditional and intrinsic autoregressions. Biometrika 105 (2), 447–454.

Rao, K.R., Yip, P., 1990. Discrete Cosine Transform: Algorithms, Advantages, Applications. Academic, Boston, MA.

Rue, H., Held, L., 2005. Gaussian Markov Random Fields: Theory and Applications. CRC Press.

Sheffield, S., 2007. Gaussian free fields for mathematicians. Probab. Theory Relat. Fields 139 (3–4), 521–541.

Stein, M.L., 2012. Interpolation of Spatial Data: Some Theory for Kriging. Springer Science & Business Media.

Whittle, P., 1962. Topographic correlation, power-law covariance functions, and diffusion. Biometrika 49 (3–4), 305–314.

Yaglom, A.M., 1957. Some classes of random fields in n-dimensional space, related to stationary random processes. Theory Probab. Appl. 2 (3), 273–320.

Yaglom, A., 1987. Correlation Theory of Stationary and Related Random Functions: Supplementary Notes and References. Springer Science & Business Media.

Zhang, H., 2004. Inconsistent estimation and asymptotically equal interpolations in model-based geostatistics. J. Am. Stat. Assoc. 99 (465), 250–261.

Chapter 6

Estimating individual-level average treatment effects: Challenges, modeling approaches, and practical applications

Victor B. Talisa[a],* and Chung-Chou H. Chang[a,b]
[a]*Department of Biostatistics, University of Pittsburgh, Pittsburgh, PA, United States*
[b]*Department of Medicine, University of Pittsburgh, Pittsburgh, PA, United States*
Corresponding author: e-mail: vit13@pitt.edu

Abstract

Good prediction models are practical tools that take advantage of the heterogeneity of individual traits in a population, exploiting the natural laws that connect them. But if generating a useful approximation for the natural processes linking the present to the future is helpful for understanding nature, often the ultimate goal of the applied scientist is to manipulate it to control an outcome in desirable ways. Just as we may expect some members of a population to have better outcomes than others following an infection, or be less likely to fall into poverty following an economic recession compared with others, we may also expect heterogeneity in the causal effect of taking some action, or treatment, to manipulate these outcomes. The causal effect is the difference between the outcome that would have occurred had the treatment been taken, and the outcome had a treatment not been taken. It may be very useful in the applied sciences to be able to distinguish between those members who benefit from a treatment and those who may be harmed, or between members benefitting to a great degree from those who benefit little. In this chapter we review the highly active literature merging ideas from supervised prediction modeling and causal inference to estimate the causal treatment effect at the individual level, called the conditional average treatment effect (CATE). After brief reviews of supervised statistical learning and the assumptions necessary for estimation and inference on causal treatment effects, we review a set of challenges inherent to the CATE estimation problem, and we demonstrate empirically how two naive approaches can fail. We then review the current state of the literature regarding model training and selection strategies tailored to CATE estimation. Finally, we cover some

Handbook of Statistics, Vol. 44. https://doi.org/10.1016/bs.host.2020.09.001

155

higher-level applications of CATE models that target specific features of the CATE function, such as the rank order of the true CATEs among the individuals in a dataset and identification of a subgroup of treatment responders.

Keywords: Conditional average treatment effects, Causal inference, Prediction modeling, Supervised learning

---

1 Introduction

Although it may seem trivial in hindsight, the idea that individual variability is real and not simply apparent was not taken seriously until as recently as Charles Darwin's lifetime (Xie, 2013). This perspective provides the foundation for the modern applied sciences, summarized by Simon (2002) as the prediction of unknown values of some variables from the known values of other variables, made possible by natural laws connecting them. Questions of prediction are a cornerstone of the scientific method: theories are evaluated on the basis of their ability to make falsifiable predictions about future observations (Hofman et al., 2017).

Prediction models are a necessity in many settings: a doctor may need to predict the course of a particular patients' medical condition; an economist may want to predict whether a particular child is likely to become impoverished; a developer of consumer products needs to understand whether a particular type of consumer demographic is most likely to purchase a product. But if generating a useful approximation for the natural processes connecting the present to the future is helpful for *understanding* nature, often the ultimate goal in applied science is to *manipulate* it to control the outcome in desirable ways. For the doctor, the relevant question usually is "how can I treat my patient to improve the course of their condition?" The economist may wonder whether there is evidence that a particular economic policy could improve the probability of children falling into poverty, and the product marketer would like to understand the sales impact of a particular consumer retention campaign. The doctor, economist, product marketer, and many others pose questions about *causality*, or whether an outcome occurred *because* a particular action was taken, and would have occurred differently otherwise.

The real variability present within members of a population may manifest in heterogeneity in the *causal effect* of a specific treatment (henceforth used generically to refer to any action, policy decision, etc.) among its members: the difference in the outcome that would have occurred had the treatment been taken, and the outcome had the treatment not been taken. Fundamentally, the doctor would seem to benefit from having an accurate prediction of the causal effect of a treatment on a specific individual in order to decide whether to treat.

In this chapter we review the highly active literature that merges supervised prediction modeling and causal inference to tackle the problem of estimating causal treatment effects at the individual level. To simplify discussion we mostly limit the scope of discussion to binary treatment decisions, and also because at this point in time most research in this area is focused on this setting. In Section 2, we offer an introduction to some important concepts in supervised statistical learning and decision theory for the traditional prediction setting, and we fix some notation and definitions for use throughout the chapter. Estimable quantities for causal inference and the conditions under which they are identifiable from data are presented in Section 3. In Section 4 we review a set of challenges that are unique to the causal inference setting, leading to biases incurred by the naive approach. In Section 5 we summarize modeling strategies for training models designed to estimate individual-level treatment effects, and in Section 6 we describe tailored model evaluation strategies for the causal inference problem. Finally, we review two applications of individual-level treatment effect estimates in Section 7: subgroup identification and treatment effect stratification. We close with a summary in Section 7.

2 Introduction to statistical learning

It will be necessary to review some fundamental concepts in supervised statistical learning, which serve as the basis for many of the approaches for developing and evaluating treatment effect models. In addition to a high-level review of important concepts (see Hastie et al. (2009) or James et al. (2014) for more comprehensive treatments), we also establish some necessary notation that we will use throughout later sections of this chapter.

2.1 Preliminaries

Let X and Y be a pair of jointly distributed random variables, where X is a p-vector of *input variables* or *covariates*, and Y is the *target, output* or *outcome* variable. We use uppercase letters to denote random variables, such as Y and X, and lower-case letters to denote realizations of these variables, such as y and x. The expectation of some arbitrary function $h(\cdot)$ of random variables A and B is taken with respect to their joint distribution with density function $p(a, b)$. For example, $E_A[h(A, B)|B = b]$ is the expectation of $h(\cdot)$ taken with respect to the conditional distribution of random variable A conditional on B held fixed at value b: $\int_A h(a, b)\, p(a|b)\, da$, where $p(a|b) = p(a, b)/p(b)$.

Define $\mathcal{X}$ as the space spanned by X. The objective of the standard prediction problem is to learn an approximating function, $f(X)$, that can be used to predict Y before it is measured, on the basis of the information contained in X. In general, there are many methods available for learning $f(X)$, each one leading to a different model specification. Given a set of K models $f_k(\cdot)$, $k = 1, \ldots, K$ taking realizations from X as inputs, we would like to rank models based on some performance criterion matching the objective at hand: usually predicting the value of Y itself, but sometimes on other features of Y such as its sign or the rank ordering of a set of realizations from Y. Evaluating the prediction errors of a given $f_k(\cdot)$ with respect to the goals of the analysis is called *model evaluation* and requires the specification of a *loss function* taking realizations of Y and X as inputs: $L(y, f_k(x))$.

For reasons that will be explored below, the most common way to learn $f(X)$ and evaluate its performance is to consider two independent sets of draws from the joint distribution of (Y, X), a training set and a test set. We index observations in a dataset using subscript i. Let S^{tr} be the set of indices i belonging to the training set, and S^{te} be the set of indices from the test set. We indicate the number of observations in arbitrary dataset S as $|S|$. The components of p-vector X are indexed using subscript j, for $j = 1, ..., p$. Lastly, let x_{ij} indicate the realization of the jth component of X for the ith observation, and let x_i indicate the ith p-vector realization of X.

In general, $f(X)$ is learned from the observations in S^{tr}, so to avoid confusion moving forward we explicitly denote this dependency by writing, e.g., $f(x; S^{tr})$. The test set is used for evaluation of the loss function, given $f(x; S^{tr})$. We treat any dataset S of observations itself as a random variable, and often take expectations over the *sampling distribution*: the distribution of all possible sets of draws from the joint distribution of (Y,X) that could have resulted in the sample S. For example, $E_{S^{tr}}[L(Y, f(X; S^{tr}))|Y = y, X = x]$ denotes the expectation of the loss function $L(Y, f(X; S^{tr}))$ over all possible training sets S^{tr} of $|S^{tr}|$ observations, holding values of Y and X fixed.

One of the most commonly used loss functions is the squared error loss: $L(y, f(x; S^{tr})) = (y - f(x; S^{tr}))^2$, which we will focus on for the sake of analysis and decomposition. Note that for a given data point (y, x), the squared error loss is the squared L2 norm, or squared Euclidean distance, between y and $f(x; S^{tr})$. The squared error loss penalizes functions $f(x; S^{tr})$ that produce predictions that are far from the true value y. Since the magnitude of errors made by $f(x; S^{tr})$ in predicting y is likely to vary when evaluated at different points (y, x) in S^{te}, a common criterion for model selection is to identify the function minimizing the expected loss over the joint distribution of (Y, X). For squared error loss, this is the conditional mean squared error, also sometimes called the conditional prediction error:

$$\text{MSE}_Y(S^{tr}) = E_{X,Y}\left[(Y - f(X; S^{tr}))^2 \big| S^{tr}\right]. \tag{1}$$

In this notation it is clear that the expectation is taken over the joint distribution of Y and X, but that the prediction function $f(\cdot)$ was derived from a fixed training sample. In words, smaller values of (1) indicate that the function f is more generalizable to new realizations from (Y, X) that may not have been available in the training set. After all, in most prediction settings we want the model to predict Y well at points $X = x$ that are potentially not in the training set.

2.2 Bias-variance decomposition

We generally cannot estimate (1) directly; instead we usually estimate the unconditional MSE_Y:

$$MSE_Y = E_{\mathcal{S}^{tr},X,Y}\left[(Y - f(X;\mathcal{S}^{tr}))^2\right]. \tag{2}$$

A brief analysis of (2) will reveal several concepts essential to prediction modeling. Consider MSE_Y conditional on X:

$$MSE_Y(x) = E_{\mathcal{S}^{tr},Y}\left[(Y - f(x;\mathcal{S}^{tr}))^2 \Big| X = x\right].$$

In words, this is the expected prediction error at a fixed point $X = x$, where the expectation is taken over the distribution of all training sets and the conditional distribution of $Y|X = x$. First of all, the function $f(x;\mathcal{S}^{tr})$ minimizing $MSE_Y(x)$ is the conditional expectation $\mu(x) = E[Y|X = x]$. Thus, if $\mu(x)$ were known for all x, MSE_Y would be minimized and we would have our optimal prediction function for Y. However, we only ever observe noisy variates from $\mu(x)$ in our data, so at best we can hope to generate an imperfect (but hopefully useful) estimate from our training data, $\hat{\mu}(x;\mathcal{S}^{tr})$. Therefore, moving forward we refer to the prediction function $f(\cdot)$ as $\hat{\mu}(\cdot)$. The bias-variance decomposition is an important tool for analyzing the sources of error when doing prediction with $\hat{\mu}(x)$. To start, let us assume an additive errors model with constant variance: $Y = \mu(X) + \epsilon$, where ϵ has mean 0 and variance σ_ϵ^2:

$$MSE_Y(x) = E_{\mathcal{S}^{tr},Y}\left[(Y - \hat{\mu}(x;\mathcal{S}^{tr}))^2 \Big| X = x\right]$$

$$= \sigma_\epsilon^2 + (E_{\mathcal{S}^{tr}}[\hat{\mu}(x;\mathcal{S}^{tr})] - \mu(x))^2 + E_{\mathcal{S}^{tr}}\left[(\hat{\mu}(x;\mathcal{S}^{tr}) - E_{\mathcal{S}^{tr}}[\hat{\mu}(x;\mathcal{S}^{tr})])^2\right] \tag{3}$$

This decomposition explicitly shows how bias and variance contribute to $MSE_Y(x)$, and therefore MSE_Y (since $MSE_Y = E_X[MSE_Y(X)]$). The second term in (3) is the square of the bias of $\hat{\mu}(x;\mathcal{S}^{tr})$; if we fit a new model to all possible training samples, the bias would be the difference between the averaged predictions from each fit evaluated at $X = x$ and the true conditional mean at $X = x$. This bias is the result of misspecifying our model $\hat{\mu}(\cdot)$. The third term is the estimation variance of $\hat{\mu}(x;\mathcal{S}^{tr})$, the result of using a random sample to estimate $\mu(x)$. The first term is the *irreducible error* associated with the stochastic target Y. Even if we had the true $\mu(x)$ as our prediction function, and thus no bias or estimation variance, the minimum MSE_Y could not be reduced below the value of σ_ϵ^2.

The second and third terms in (3) combine to form the MSE of $\hat{\mu}(x;\mathcal{S}^{tr})$ for estimating $\mu(x)$:

$$MSE_\mu(x) = E_{\mathcal{S}^{tr}}\left[(\hat{\mu}(x;\mathcal{S}^{tr}) - \mu(x))^2 \Big| X = x\right], \tag{4}$$

where $\text{MSE}_\mu = E_X[\text{MSE}_\mu(X)]$. Thus, the standard prediction problem (at least when using squared error loss) can be expressed as an estimation problem, where the parameter of interest is the function $\mu(x)$. The advantage of using MSE_Y for model selection is that we do not need access to $\mu(x)$, unlike MSE_μ, making estimation more straightforward. As we will see in concrete examples below, in many cases the bias-variance decomposition reveals that the prediction modeler seeking to minimize MSE_Y may not find it among the class of unbiased, "correct" estimators of $\mu(x)$, unless it also achieves lower variance than other unbiased models. Often the "wrong" model (a biased one) is preferable for having a lower MSE_Y, and therefore better prediction performance as a whole. Although it may be very unlikely that a modeling procedure captures every nuance of nature necessary to produce truly unbiased estimates of $\mu(x)$, the hope is that some will provide a good enough approximation as to be useful in predicting Y.

2.3 Examples of the bias-variance tradeoff

A classical example of the bias-variance tradeoff makes use of the class of models that take local averages. The k-nearest neighbors (k-nn) models can be expressed as follows:

$$\hat{\mu}_k(x; \mathcal{S}^{tr}) = \frac{1}{k} \sum_{i \in \mathcal{N}_k(x, \mathcal{S}^{tr})} y_i,$$

where $\mathcal{N}_k(x, \mathcal{S}^{tr})$ is the set of k closest observations in terms of Euclidean distance from x among the observations in the training set. In this notation, k is a *tuning parameter* that indexes the set of models in the space of k-nn fits given $\mathcal{S}^{tr}$, $\{\hat{\mu}_k(x; \mathcal{S}^{tr}); \text{ for } k = 1, ..., |\mathcal{S}^{tr}|\}$.

When $k = |\mathcal{S}^{tr}|$ is the size of the entire training sample, the model estimates $\mu(x)$ as the overall mean $\bar{y} = (1/|\mathcal{S}^{tr}|)\sum_{i \in \mathcal{S}^{tr}} y_i$ for all points x. Over all possible training samples $\mathcal{S}^{tr}$, $\bar{y}$ is an unbiased estimator for $\mu = E_X(\mu(x))$. Because $\bar{y}$ uses all of the data in the training sample, its variance is small. However, for an arbitrary x, the distance between μ and the conditional mean $\mu(x)$ is likely to be large (e.g., high bias), unless Y and X are independent (in which case $\mu(x) = \mu$ for all x). If we could take k to be 1, $\hat{\mu}(x; \mathcal{S}^{tr})) = \{y_i : i = \mathcal{N}_1(x, \mathcal{S}^{tr})\}$. In other words, the prediction is equal to the outcome of the single nearest observation in $\mathcal{S}^{tr}$. This model will generate predictions that are on average (over all $\mathcal{S}^{tr}$) equal to $\mu(x)$ (low bias), but since they are trying to estimate $\mu(x)$ based on a sample size of 1 they have very high variance. The tradeoff between bias and variance for the k-nearest neighbors model is illustrated via simulated data in Fig. 1. The two models in Fig. 1 have differing *complexity* or *smoothness*. The 1-nn model has higher complexity, as seen by the roughness or lack of smoothness of the prediction function in panel A. The 100-nn model has the lowest complexity, as it predicts Y

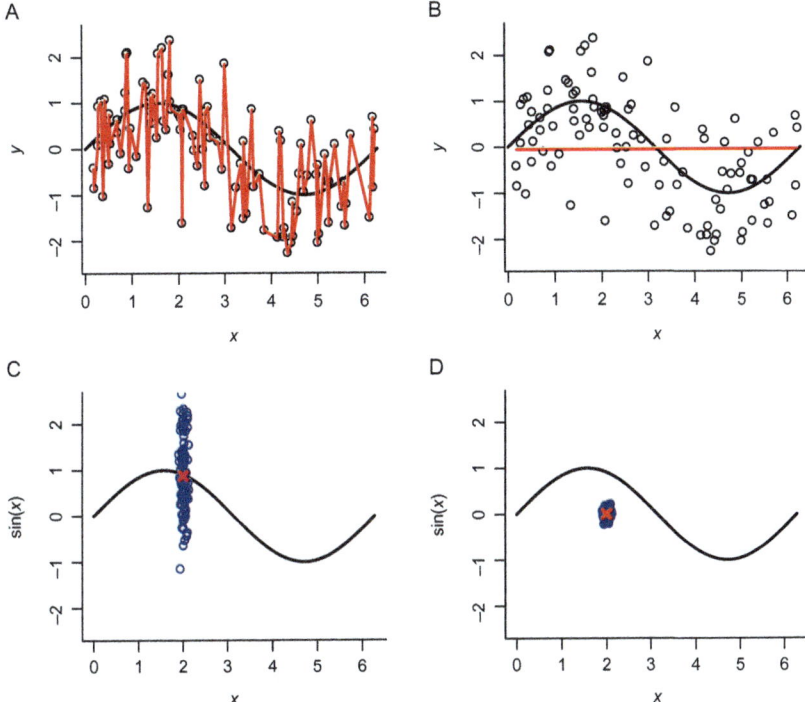

FIG. 1 Illustration of the bias and variance of predictions from two k-nn models fit to 100 data points (Y, X), where $Y \sim N(\sin(X), 0.75^2)$, $X \sim N(0, 2\pi)$. The *solid black curve* shows the true $E[Y|X]$. In panels A and B, 100 draws of (Y, X) are shown as hollow points, the model fit of a 1-nn model (*red curve*, panel A), and the model fit of a 100-nn (100 nn) model (*red curve*, panel B). In panels C and D, separate models were fit to 100 independent datasets, each of 100 draws from (Y, X). *Blue hollow circles* show predictions from these models (1 nn in panel C; 100-nn in panel D) at $X = 2$, and *red crosses* show the mean of these predictions over the 100 datasets. Predictions from 1-nn are essentially unbiased (*red cross* is on the *black curve*), but are highly variable, while predictions from 100-nn are very biased but have low variance.

using a smooth, perfectly straight line in panel B. When a model is too complex we say that it is *overfit* to the training data; lack of sufficient complexity is called underfitting. The decomposition of $\mathrm{MSE}_Y(x, \mathcal{S}^{tr})$ for the additive errors model, fixing x and $\mathcal{S}^{tr}$, shows this relationship between choice of k, bias and variance:

$$\mathrm{MSE}_Y(x, \mathcal{S}^{tr}) = E_Y\left[(Y - \hat{\mu}_k(x; \mathcal{S}^{tr}))^2 | X = x, \mathcal{S}^{tr}\right]$$

$$= \sigma_\epsilon^2 + \left(\frac{1}{k}\sum_{i \in \mathcal{N}_k(x_i, \mathcal{S}^{tr})} y_i - \mu(x)\right)^2 + \frac{\sigma_\epsilon^2}{k} \tag{5}$$

Thus, as k increases, the variance component decreases but the bias term increases for values of x corresponding to $\mu(x)$ that are very distant from the sample average of observations from $\mathcal{S}^{tr}$.

For predicting the output values in an arbitrary test set, the best choice of k is the choice that leads to the combination of bias and variance minimizing $\text{MSE}_Y(\mathcal{S}^{tr})$. However, empirically the performance of such a model tends to deteriorate as p (the dimension of X) increases relative to the number of observations in the training sample. This is due to the *curse of dimensionality*, which decreases the likelihood of finding 2 points in a small neighborhood of each other in $\mathcal{X}$ as p increases. As an example, consider taking 1000 random draws from a 10-dimensional X following a multivariate standard normal distribution with diagonal covariance matrix. Separately for D from 2 to 10, compute the average distance from the D-dimensional origin to the coordinates taken as the first D dimensions of each draw from X. When we did this, the average distance was about 1.2 for $D = 2$; 1.9 for $D = 5$, and 2.5 when $D = 10$. In this case, as the dimensionality increases, the average distance of any point from the D-dimensional origin increases. For fixed k, the average distance between any point and its k-nearest neighbors would also increase, increasing the variance of the k-nn estimator for all x.

If it were possible to identify the dimensions of X that best describe the variation in $\mu(x)$, intuitively the k nearest neighbors identified in the reduced-dimensionality input space would produce better estimates of $\mu(x)$. For example, say that $k = 2$, and we had 3 observations of a 3-dimensional $X = (U, V, W)$ and scalar Y in the training set, each component of X distributed standard uniform. Say that for observation $i \in \{1, 2, 3\}$, $x_i = (u_i, v_i, w_i)$; then consider the 3 points $x_1 = (0, 0, 0)$, $x_2 = (0.6, 0.6, 0.2)$, and $x_3 = (0.3, 0.3, 0.9)$. Assuming that the U, V, and W components of X are important predictors of Y, x_2 would be the nearest neighbor of x_1. However, if $Y \perp W$, then W contains no information about Y and can be ignored. In this case, we can reduce the dimensionality of X to 2, so that $X = (U, V)$, $x_1 = (0, 0)$, $x_2 = (0.6, 0.6)$, and $x_3 = (0.3, 0.3)$; hence, x_3 would be the nearest neighbor of x_1. A 1-nn model fit to a larger dataset ignoring random variable W would have lower variance for predictions taken at points close to x_1, because the nearest neighbor computed this way would on average be closer to x_1. Thus, by ignoring W we would be able to reduce the variance by reducing the dimensionality of input vector X.

The relationship between the dimension of X and the tradeoff between bias and variance is explicitly evident for another class of models, the linear regression model. For linear models, $\hat{\mu}(x; \mathcal{S}^{tr}) = \hat{\alpha} + \sum_{j=1}^{p} \hat{\beta}_j x_j$. Parameter estimation proceeds via ordinary least squares (OLS) by minimizing an estimate of $\text{MSE}_Y(\mathcal{S}^{tr})$ evaluated on the training data themselves:

$$(\hat{\alpha}, \hat{\beta}_1, \ldots, \hat{\beta}_p) = \underset{\alpha, \beta_1, \ldots, \beta_p}{\arg\min} \frac{1}{|\mathcal{S}^{tr}|} \sum_{i \in \mathcal{S}^{tr}} \left(y_i - \alpha - \sum_{j=1}^{p} \beta_j x_{ij} \right)^2. \qquad (6)$$

Note that when $p = 0$, the regression estimate of $\hat{\alpha}$ is equal to the sample mean of Y. This mirrors the k-nearest neighbors case in which the lowest complexity model is equivalent to applying the overall mean as a prediction for all future observations of Y. The Gauss–Markov theorem shows that unbiased solutions to this problem have minimum achievable variance. However, unbiasedness depends on the assumption that the true $\mu(x)$ is linear in the inputs x. We can see the dependence of MSE_Y on p through the average of $\mathrm{MSE}_Y(x)$ evaluated over points x_i from a fixed $\mathcal{S}^{te}$ (see Hastie et al. (2009) for details):

$$\frac{1}{|\mathcal{S}^{te}|} \sum_{i \in \mathcal{S}^{te}} \mathrm{MSE}_Y(x_i) = \frac{1}{|\mathcal{S}^{te}|} \sum_{i \in \mathcal{S}^{te}} E_{Y, \mathcal{S}^{tr}} \left[(Y - \hat{\mu}(x_i; \mathcal{S}^{tr}))^2 \Big| X = x_i \right]$$

$$= \sigma_\epsilon^2 + \frac{1}{|\mathcal{S}^{te}|} \sum_{i \in \mathcal{S}^{te}} \left[\mu(x_i) - E_{\mathcal{S}^{tr}} \left(\hat{\alpha} + \sum_{j=1}^{p} \hat{\beta}_j x_{ij} \right) \right]^2 + \frac{p}{|\mathcal{S}^{te}|} \sigma_\epsilon^2.$$

(7)

The first term in (7) is the usual error variance, while the second is the squared bias of the estimated regression function, averaged over points in the test sample. In the last term of (7) we can see explicitly how the estimation variance of the OLS estimator increases with the dimension of X. In high-dimensional problems with relatively low sample size, prediction performance may be improved by "enriching" the input covariate space by ignoring variables that would be relatively ineffective for prediction of Y.

2.4 Regularization

As we have argued through the classical examples of k-nearest neighbors and linear regression models, one way to improve prediction performance is to limit the effective dimension of the input space. Thus, at the potential cost of increasing the bias of the predictor, the variance component can be reduced along with MSE_Y. There are several approaches to this task, which we will refer to generally as *regularization*. Below are three ways to achieve regularization:

1. Variable selection: The simplest form of regularization is to exclude variables based on prior information, or based on analyses conducted in the training sample. Several algorithmic methods exist for the latter, including best subset selection, and forward or backward stepwise selection. Note that for a linear regression model, excluding a covariate from analysis is equivalent to setting its coefficient β_j to 0 in the full model. Over many independent $\mathcal{S}^{tr}$, a covariate X_j may sometimes have a nonzero coefficient (if the algorithm used for variable selection did not exclude it), or a coefficient of 0 (if the algorithm excluded it). Thus, the average of β_j coefficient estimates across many independent training analyses is likely to be biased toward 0, compared to the coefficients from the "unregularized" estimator.

2. Loss penalization: In the case of the linear regression model, we may achieve lower model complexity (desirable in high dimensions) by optimizing an objective function that penalizes the complexity of the model. The class of generalized additive models provides an example of a penalized generalized additive model (Hastie and Tibshirani, 1990):

$$\hat{\mu}(x; \mathcal{S}^{tr}) = \hat{\alpha} - \sum_{j=1}^{p} \hat{g}_j(x_j)$$

$$(\hat{\alpha}, \hat{g}_1, ..., \hat{g}_p) = \arg\min_{\alpha, g} \frac{1}{|\mathcal{S}^{tr}|} \sum_{i \in \mathcal{S}^{tr}} \left(\left(y_i - \alpha - \sum_{j=1}^{p} g_j(x_{ij}) \right)^2 + \lambda J(g_1, ..., g_p) \right).$$

$$(8)$$

Notice that the first term in the optimization function in (8) is the squared error loss; if we take $\lambda = 0$, the function we are optimizing is an estimate of $MSE_Y(\mathcal{S}^{tr})$ taken in the training sample. Regularization occurs when we limit the complexity of the $g(\cdot)$ functions by taking $\lambda > 0$, through some penalty function $J(\cdot)$. Many model fitting algorithms can be specified this way, including penalized spline models, and regularized linear regression models. For example, if we restrict $\hat{\mu}(x; \mathcal{S}^{tr})$ to be linear in X, then $g_j(X_j) = \beta_j X_j$, and popular choices for $J(\cdot)$ include the L1 norm where $J(\beta) = \sum_{j=1}^{p} |\beta_j|$, and the L2 norm where $J(\beta) = \sum_{j=1}^{p} \beta_j^2$. Using the latter penalty is called ridge regression, and for solutions to the corresponding optimization function the bias can be shown to increase with increasing λ. Ridge regression assumes Y varies most in the dimensions of high variance of X, and shrinks the dimensions with low variance toward 0. By shrinking coefficients toward 0, the effective dimension of X is reduced. Other forms of penalization achieve dimension reduction in broadly similar fashion. Common to all methods is the need to select a value for the tuning parameter λ, controlling the extent of penalization. For regularized linear regression models, when $\lambda = 0$ there is no penalization and objective function becomes the OLS objective; when $\lambda = \infty$, all parameters are shrunken to 0 leading to the overall mean estimator. Thus, the complexity of the model is a function of tuning parameter λ, and so the value of λ minimizing our training objective must be determined separately (see Section 2.5).

3. Representation learning: This is a more general class of approach that includes variable selection. A representation is a vector-valued function of inputs, $\phi(x)$. For variable selection, $\phi(x)$ is the function that takes the original p-dimensional vector x and outputs a new vector of dimension h, where h is the number of variables selected, and $\phi(x)$ outputs the untransformed values for the selected variables. More generally, the dimensions of $\phi(x)$ could represent weighted combinations of the original inputs, along with transformation or selection. A classical method for representation learning is principal components analysis (PCA) (Bengio et al., 2013). PCA learns

an h-dimensional representation as a linear transformation of inputs x, where $\phi(x) = W^T x + b$. The $h \times p$-dimensional weight matrix W has columns that form an orthonormal basis for the h directions of greatest variance in $\mathcal{X}$ apparent from the training data. For highly correlated X's, we could find $h \ll p$ dimensional representation using PCA, and then fit a model of Y on the transformed inputs $\phi(x)$. When the model is a linear regression, the prediction error from this strategy is comparable to ridge regression (Hastie et al., 2009). PCA fits into a more general class of representation learning that is not constrained to be linear in the inputs, such as the representations learned as part of deep learning.

2.5 Evaluating a model's prediction performance

Thus far we have only briefly addressed how to assess a model's prediction performance from data. We saw in (8) that some approaches are fit by optimizing an objective that, given tuning parameters, resembles an estimate of $\mathrm{MSE}_Y(\mathcal{S}^{tr})$ evaluated at observations in the training sample. However, estimates of MSE taken from the training data alone may be overly optimistic of the performance of a model for making predictions using new draws from X not present in the training data. In general, there are 2 related but separate motivations for assessing the performance of a given modeling strategy:

1. *Model selection:* for any given problem, we may have a large number of competing modeling strategies. This could include making a choice of tuning parameter for a given model specification (e.g., ridge regression models with different λ tuning parameters; see Section 2.4), or a choice between two completely different modeling approaches (e.g., a penalized linear regression vs k-nn model type).
2. *Model evaluation:* given a model $\hat{\mu}(x, \mathcal{S}^{tr})$, we would like to estimate its expected error in predicting a new sample, $\mathrm{MSE}_Y(\mathcal{S}^{tr})$.

A necessary tool for accomplishing both of these tasks is the following estimator of $\mathrm{MSE}_Y(\mathcal{S}^{tr})$:

$$\widehat{\mathrm{MSE}}_Y(\mathcal{S}^{tr}) = \frac{1}{|\mathcal{S}^{te}|} \sum_{i \in \mathcal{S}^{te}} (y_i - \hat{\mu}(x_i, \mathcal{S}^{tr}))^2. \tag{9}$$

This estimator takes the average of the squared prediction errors over the observations in the test set. Because we are able to directly observe our target Y, the estimator in (9) is model-free (does not depend on nuisance parameters) and has a simple justification: we compare our estimator $\hat{\mu}(x, \mathcal{S}^{tr})$ with noisy empirical realizations y from some unknown distribution with mean $\mu(x)$.

Selecting among several modeling choices is essentially a problem of choosing a strategy that minimizes $\mathrm{MSE}_Y(\mathcal{S}^{tr})$. However, we usually cannot get a good estimate of $\mathrm{MSE}_Y(\mathcal{S}^{tr})$ from $\mathcal{S}^{tr}$ itself. Instead, the strategy involves finding the modeling strategy minimizing an estimate of MSE_Y from $\mathcal{S}^{tr}$ through a procedure called *cross-validation* (CV). To perform CV, $\mathcal{S}^{tr}$ is split into non-overlapping subsets $\mathcal{S}_v^{CVtr}$ and $\mathcal{S}_v^{CVte}$, so that $\mathcal{S}^{tr} = \mathcal{S}_v^{CVtr} \cup \mathcal{S}_v^{CVte}$ for all $v = 1, ..., V$. The subscript v indexes across a set of splits of $\mathcal{S}^{tr}$. For example, a common variant is called Monte Carlo CV, where the observations in $\mathcal{S}^{tr}$ are randomly split into two subsets of fixed (but possibly unequal) size. Randomly splitting $\mathcal{S}^{tr}$ V times would lead to V pairs of datasets: $(\{\mathcal{S}_1^{CVtr}, \mathcal{S}_1^{CVte}\}, \{\mathcal{S}_2^{CVtr}, \mathcal{S}_2^{CVte}\}, ..., \{\mathcal{S}_D^{CVtr}, \mathcal{S}_V^{CVte}\})$. Other splitting schemes are described in Yang (2007). Given a set of pairs $\{\mathcal{S}_v^{CVtr}, \mathcal{S}_v^{CVte}\}_{v=1}^V$, and a procedure for estimating $\hat{\mu}(\,\cdot\,)$ from each $\mathcal{S}_v^{CVtr}$, we can define the CV estimator of MSE_Y as follows:

$$\widehat{MSE}_Y = \frac{1}{V} \sum_{v=1}^V \frac{1}{|\mathcal{S}_v^{CVte}|} \sum_{i \in \mathcal{S}_v^{CVte}} \left(y_i - \hat{\mu}(x_i, \mathcal{S}_v^{CVtr})\right)^2. \tag{10}$$

The CV estimator of $\widehat{MSE}_Y$ is thus an average of the estimates of $\widehat{MSE}_Y(\mathcal{S}_v^{CVtr})$ over all V splits of $\mathcal{S}^{tr}$. Model selection can be performed by minimizing $\widehat{MSE}_Y$.

Performing a final evaluation of the model selected in $\mathcal{S}^{tr}$ usually involves estimating $\widehat{MSE}_Y(\mathcal{S}^{tr})$ via (9) using the independent test set $\mathcal{S}^{te}$. However, it is also common for the analyst not to have access to enough data to justify separation into independent $\mathcal{S}^{tr}$ and $\mathcal{S}^{te}$; in these cases the estimate of $\widehat{MSE}_Y$ via (10) is often reported as an evaluation of the predictive performance. For a survey of the statistical properties of cross-validation estimators see Arlot and Celisse (2010).

3 Causal inference

The main subject of this article is on prediction of treatment effects, not factual outcomes Y. However, the predominant approach to estimation of treatment effects stems directly from that of the standard problem of predicting Y. In this section we expand the notation from the previous section, define treatment effects, and state the usual assumptions necessary for estimating them.

3.1 Notation and counterfactual theory

The problem of causal inference is usually stated in terms of the Neyman–Rubin potential outcomes framework, which we follow throughout (Rubin, 1974, 2005). The quantity we would like to estimate is usually some function of a vector-valued random outcome variable $[Y^k]_{k \in 0...K-1}$, where K is the number of possible treatments (aka interventions, actions, etc.). Define the vector of realizations from $[Y^k]_{k \in 0...K-1}$ for individual i to be $[y_i^k]_{k \in 0...K-1}$. The *fundamental problem of causal inference* states that only one element of $[y_i^k]_{k \in 0...K-1}$ can be observed for a given observation i, corresponding to the treatment that was actually taken. This is in contrast to the standard prediction problem, where we wanted to predict scalar (or even vector-valued) Y, where all the components are potentially observable simultaneously. To facilitate discussion throughout most of this chapter, we limit the complexity of the problem by taking $K = 2$, representing a binary treatment space. This scenario is common in practice, e.g., in a standard clinical trial testing the efficacy of experimental therapy over the standard of care.

In the binary treatment scenario, the Neyman–Rubin potential outcomes model specifies the following complete vector of random variables: (Y^0, Y^1, T, X), where X is a p-dimensional vector input variables, T is a (usually scalar) treatment assignment variable taking values 0 or 1, and (Y^0, Y^1) are the potential outcomes. Let the treatment effect be defined as a random variable $D = Y^1 - Y^0$, and let $d_i = y_i^1 - y_i^0$ be the unobservable realization of D for patient i. Common estimands in causal inference include the average treatment effect (ATE) $E_{Y^1,Y^0}[Y^1 - Y^0]$, and the average treatment effect among the treated (ATT) $E_{Y^1,Y^0}[Y^1 - Y^0|T = 1]$. For individual-level treatment effect predictions, our primary estimand is the conditional average treatment effect (CATE):

$$\tau(x) = E_{Y^1,Y^0}[Y^1 - Y^0|X = x].$$

Analogous to the standard prediction problem where we wanted to predict Y using an estimate of $\mu(x)$, in the individual treatment effect prediction problem we would like to learn an estimator for $\tau(x)$ from training data, $\hat{\tau}(x; S^{tr})$, with which to predict $Y^1 - Y^0$ at new draws of X such as those in a test set.

We are guaranteed by the fundamental problem of causal inference to be missing the value of at least one potential outcome per individual in our training set. The missing potential outcome is often referred to as the *counterfactual*, while the observed one is called the *factual* outcome. In general, at least three untestable assumptions are necessary in order for $\tau(x)$ to be estimable:

1. *Unconfoundedness*: Also referred to as *exogeneity*, or *ignorability*. This assumption is satisfied if potential outcomes are independent of treatment assignment given X: $(Y^0, Y^1) \perp T|X$.
2. *Overlap*: The conditional probability of being assigned to either treatment given X is bounded away from 0 and 1: $0 < Pr(T = 1|X = x) < 1$ for all x. Under this assumption, treatment assignment is not allowed to be deterministic given covariates X.
3. *Consistency*: The factual outcome is equal to the potential outcome associated with the actual treatment assignment: $Y = TY^1 + (1 - T)Y^0$.

Under these assumptions, $\tau(x)$ is identifiable from observed data. Under unconfoundedness and consistency,

$$E[Y^t|X = x] = E[Y^t|T = 1, X = x] = E[Y|T = t, X = x], \text{ for } t = 0, 1. \quad (11)$$

The motivation behind the unconfoundedness assumption is that with it, we can treat nonrandomized studies similarly to randomized experiments. In other words, for individuals with very similar values of X, differences in Y between treatment groups could not be due to differences in X (because X is essentially held constant) or any other variables (because we assume no other variables simultaneously cause Y and T). Let $\mu_t(x) = E[Y \,|T = t, X = x]$ for $t = 0, 1$. Following from the three assumptions and (11) we can estimate $\tau(x) = \mu_1(x) - \mu_0(x)$ from observed data.

The vast majority of the literature on prediction of individualized causal effects is limited to scenarios where causal assumptions 1–3 are met and the treatment is binary. For a brief mention of the literature covering other scenarios see Section 8.

3.2   Theoretical MSE criteria for causal inference

Recall from Section 2.1 that in the standard prediction problem with squared error loss, we sought to fit a model in $\mathcal{S}^{tr}$ that minimizes $\mathrm{MSE}_Y(\mathcal{S}^{tr})$, but that we typically could only estimate the unconditional MSE_Y from our data. Additionally, MSE_Y is equal to MSE_μ plus a constant that does not depend on any estimate, so that the model minimizing one quantity would minimize the other.

It is useful to define similar quantities for estimation of $\tau(x)$. Define the following two quantities:

$$
\begin{aligned}
\mathrm{MSE}_D(x) &= E_{D,\mathcal{S}^{tr}}\left[(D - \hat{\tau}(x; \mathcal{S}^{tr}))^2 \Big| X = x\right] \\
&= E_{Y^1,Y^0,\mathcal{S}^{tr}}\left[((Y^1 - Y^0) - \hat{\tau}(x; \mathcal{S}^{tr}))^2 \Big| X = x\right]
\end{aligned}
\tag{12}
$$

$$
\mathrm{MSE}_\tau(x) = E_{\mathcal{S}^{tr}}\left[(\tau(x) - \hat{\tau}(x; \mathcal{S}^{tr}))^2 \Big| X = x\right].
\tag{13}
$$

These two MSE functions are thus analogous to quantities (3) and (4), and could be subject to equivalent decompositions. We occasionally refer to their unconditional counterparts MSE_D and MSE_τ, or counterparts conditional on the training sample $\mathrm{MSE}_D(\mathcal{S}^{tr})$ and $\mathrm{MSE}_\tau(\mathcal{S}^{tr})$, which are defined analogously to the quantities in Section 2.1.

4   Challenges in learning the CATE function

In Section 3.1, we established the conditions necessary to decompose $\tau(x)$ into the difference in conditional means $\mu_1(x) - \mu_0(x)$. As a starting point for our discussion of modeling strategies for $\tau(x)$, in this section we explore what can go wrong if we translate the approach from standard prediction modeling over to causal inference without modification. We hope to make clear the various sources of error of the naive approach. These errors stem from a seemingly trivial observation: the target from the standard prediction problem, Y, is not the same as the unobservable target for causal inference, $D = Y^1 - Y^0$.

4.1 The naive optimization strategy fails

In this section we look at what can happen if our strategy for predicting D (or equivalently, estimating $\tau(x)$) is to fit models to the training set that are designed for the another purpose: predicting Y. The logic behind this naive strategy is at first glance conceptually sound: if $\tau(x) = \mu_1(x) - \mu_0(x)$, then good estimators of $\mu_1(x)$ and $\mu_1(x)$ should produce a good estimator for $\tau(x)$. As it turns out, in finite samples where regularization is most important, this approach can fail.

4.1.1 Generalized random forests

Before we can offer a concrete example, we first introduce the generalized random forest (GRF) algorithm, a general method for using recursively partitioned decision trees to learn heterogeneity (in X) of a parameter of interest as long as it can be specified in terms of estimating equations (Athey et al., 2019). Recursive binary partitioning is a general approach for dividing $\mathcal{X}$ into a set of subspaces called nodes. At each step of the algorithm, each node (called the parent, P) is divided into two by a single covariate axis-aligned split. The algorithm selects a single covariate, X_j, and a subset A of the support values of X_j, and splits P into two child subgroups, $C_1 = \{x_j \in A\}$, $C_2 = \{x_j \notin A\}$, such that $C_1 \cup C_2 = P$. Different splitting criteria can be used to identify the optimal pair (X_k, A) for splitting using data points in the training set. For ordinal-scale biomarkers, the rule is usually restricted so that A is of the type $\{x_j < a\}$ or $\{x_j \geq a\}$ for fixed threshold value a. Recursive splitting of child nodes proceeds until exhaustion of the data, or when a predefined stopping rule is satisfied (e.g., when a minimum sample size is reached in each child node). Nodes that are not split are referred to as "terminal" nodes. The completed algorithm generates a model structure called a tree. Given a vector of covariates x_i, observation i can be classified into a unique terminal node and assigned a predicted value calculated from the data in that node. For the standard prediction problem, decision tree models have relatively low variance but high bias due to the fact that the model for $\mu(x)$ is a piecewise constant. Breiman et al. (1984) proposed an improvement to the predictive performance of single decision trees via a *random forest* algorithm. In a random forest, separate trees are grown on random samples of the original dataset, and the algorithm for growing each individual tree is slightly modified in that only a random subset of variables is considered when splitting each parent node.

The GRF method uses the basic elements of Breiman's random forest algorithm to adaptively obtain nearest neighbor weights, and estimating parameters of interest by solving weighted estimating equations. Parameters of interest $\theta(x)$ are identified via local estimating equations of the form:

$$E_O\left[\psi_{\theta(x),\nu(x)}(O)\Big|X=x\right] = 0 \text{ for all } x \in \mathcal{X}, \tag{14}$$

where $O = Y$ for standard prediction and $O = (Y, T)$ for treatment effect modeling, and $\psi_{\theta(x), \nu(x)}(O)$ is a possibly vector-valued estimating equation depending on O, X-dependent parameters of interest $\theta(x)$, and possibly X-dependent nuisance parameters $\eta(x)$. Using Breiman's random forest algorithm along with problem-specific splitting criterion $\Delta(C_1, C_2)$, weights $\alpha_i(x)$ are generated for each observation in S^{tr}. Specifically, a set of B trees are grown, indexed by $b = 1, ..., B$, by splitting nodes using criterion $\Delta(C_1, C_2)$. Let $L_b(x)$ be the set of training observations falling in the same terminal node as x; weights $\alpha_i(x)$ are then defined as:

$$\alpha_i(x) = \frac{1}{B} \sum_{b=1}^{B} \frac{1(i \in L_b(x))}{|L_b(x)|}. \tag{15}$$

In other words, when $B = 1$, the weights $\alpha_i(x)$ are equal to the inverse of the proportion of observations in the training sample falling into the same terminal node as point x. When B is large, $\alpha_i(x)$ captures the probability that training observation i falls into the same terminal node as x across all trees in the forest. Given these weights, $(\hat{\theta}(x), \hat{\nu}(x))$ solves an empirical version of (14):

$$\sum_{i \in S^{tr}} \alpha_i(x) \psi_{\hat{\theta}(x), \hat{\eta}(x)}(O_i) = 0.$$

In the GRF methodology, regularization occurs adaptively as part of the calculation of the weights; dimensions of $\mathcal{X}$ that are most important in describing heterogeneity in $\theta(x)$ are identified by the problem-specific splitting criterion $\Delta(C_1, C_2)$. Splitting is *greedy*, in the sense that splits are taken that immediately increase the difference in estimates of $\theta(x)$ between child nodes as much as possible. Athey et al. (2019) provide a generalizable scheme for constructing $\Delta(C_1, C_2)$ given $\psi_{\hat{\theta}(x), \hat{\eta}(x)}(O)$. In the example that follows, we will consider two ψ functions:

$$\psi_{\mu(x)}(O) = Y - \mu(x), \tag{16}$$

$$\psi_{\tau(x), \mu_0(x)}(O) = \begin{pmatrix} Y - \mu_0(x) - \tau(x)T \\ T(Y - \mu_0(x) - \tau(x)T) \end{pmatrix}. \tag{17}$$

Using (16) and (17), we can derive respective criteria for splitting arbitrary parent node P into child nodes C_1 and C_2:

$$\Delta_\mu(C_1, C_2) = \sum_{k=1}^{2} |C_k| (\overline{Y}_{C_k} - \overline{Y}_P)^2 \tag{18}$$

$$\Delta_\tau(C_1, C_2) = \sum_{k=1}^{2} \frac{1}{|C_k|} \left[\sum_{i \in C_k} \frac{(T_i - \overline{T}_P)^2}{\frac{1}{|C_k|} \sum_{i \in P} (T_i - \overline{T}_P)^2} (Y_i - \overline{Y}_P - (T_i - \overline{T}_P)\hat{\tau}_P) \right]^2 \tag{19}$$

In the expressions above, notation of the type $\overline{S}_L$ is short-hand for the sample average of arbitrary random variable S among the observations in node L: $\frac{1}{|L|}\sum_{i \in L} S_i$. In (19), $\hat{\tau}_P$ is the unadjusted OLS estimator for the treatment effect in node P. The Δ_μ criterion in (18) is maximized when the difference between $\overline{Y}_{C_1}$ and $\overline{Y}_{C_2}$ is maximized, satisfying the heuristic objective of finding splits leading to maximum heterogeneity in $\theta(x)$, which in this case is $\mu(x)$. Note that when treatment is randomly assigned with equal probability (as in the example to follow in Section 4.1.2), $\overline{T}_P \approx 1/2$ for all P, and 19 can be approximated as follows:

$$\Delta_\tau(C_1, C_2) \approx \sum_{k=1}^{2} |C_k| \left(\overline{Z}_{C_k} - \hat{\tau}_P \right)^2, \tag{20}$$

where $\overline{Z}_{C_k} = (1/|C_k|) \sum_{i \in C_k} (Y_i - \overline{Y}_P)/(T_i - \overline{T}_P)$. In this expression, we point out the similarity with (18), and note that $\overline{Z}_{C_k}$ is an estimate of treatment effect in C_k. Thus, Δ_τ is designed to find splits maximizing heterogeneity in $\tau(x)$. As a final note, Athey et al. (2019) suggest using independent subsamples for fitting trees and estimating parameters in the terminal nodes. They refer to this as *honest* estimation, as it is free of the *optimism bias* associated with using a single sample to adaptively construct the tree and estimate the target parameters $\theta(x)$ given the tree structure.

4.1.2 Example: Difference of independently fit $\hat{\mu}_1(x)$ and $\hat{\mu}_0(x)$ does not achieve performance of single fit $\hat{\tau}(x)$

We are now ready to look at an example comparing two different ways we could use GRFs to estimate $\tau(x)$, inspired by the example in Radcliffe and Surry (2011). The naive approach builds two independent models to the treated and untreated samples, each fitting an objective that makes no reference to the other sample. Using (16) and (18), we estimate $\mu_1(x)$ using the sample of treated observations, and estimating $\mu_0(x)$ in the sample of nontreated observations. We call this the "Double ψ_μ" approach, because we estimate $\tau(x)$ by subtracting our estimates of $\mu_0(x)$ from the estimates of $\mu_1(x)$. This approach is based on the sound observation that, assuming the conditions in Section 3.1, the difference in two correct models $\mu_1(x) - \mu_0(x)$ equals the correct model for $\tau(x)$. However, it does not follow that two incorrect models can be expected to be combined to form a decent estimator of $\tau(x)$. As discussed above, we usually consider the effects of regularization to render any model incorrect, which we are typically satisfied with as long as predictions are good. We compare the Double ψ_μ approach with a "Single ψ_τ" approach, which uses (17) and (19) to build a model for $\tau(x)$ directly.

We simulated 500 training datasets, each with 500 independent draws from the following model for Y:

$$Y = 2X_1 + 2X_2 + TX_3 + TX_4 + \epsilon,$$

where covariates X_1, X_2, X_3, X_4 are independently distributed standard uniform, ϵ is normally distributed with mean 0 and variance 0.5, and treatment T is randomly assigned values 0 or 1 with equal probability. We complete each dataset by simulating 10 more standard uniform covariates independent of all other variables. In this model, the third and fourth terms are interaction effects, in that they create heterogeneity in $\tau(x)$. However, these effects are comparatively much weaker compared to the other terms in the model. For each dataset, we fit 3 single-tree ($B = 1$) GRFs:

1. A model using (16) and (18) fit to observations with $T = 0$, leading to estimator $\hat{\mu}_0(x; \mathcal{S}^{tr})$
2. A model using (16) and (18) fit to observations with $T = 1$, leading to estimator $\hat{\mu}_1(x; \mathcal{S}^{tr})$
3. A model using (17) and (19) fit to all observations, leading to estimator $\hat{\tau}(x; \mathcal{S}^{tr})$

Models for $\mu_0(x)$ and $\mu_1(x)$ were fit using the `regression_forest` command in R package `grf`, with all default settings except for setting the minimum node size to 10. Models for $\hat{\tau}(x)$ were fit using `causal_forest` in the same package, with minimum node size of 10, and without local centering (not needed for randomized data; see Athey et al. (2019) for details). We conducted *honest* estimation by randomly splitting each simulated dataset into two equally sized halves, constructing trees in one half and estimating parameters (given the fixed tree structure) in the other. From each of the two modeling strategies, we generated two sets of predictions for the observations in each simulated training dataset: (1) $\hat{\mu}(t, x; \mathcal{S}^{tr}) = t\hat{\mu}_1(x; \mathcal{S}^{tr}) + (1 - t)\hat{\mu}_0(x; \mathcal{S}^{tr})$, the estimated conditional mean $\mu(t, x) = E[Y \mid T = t, X = x]$; (2) $\hat{\tau}(x; \mathcal{S}^{tr})$, the estimated $\tau(x)$, which was estimated directly in Single ψ_τ and as $\hat{\mu}_1(x; \mathcal{S}^{tr}) - \hat{\mu}_0(x; \mathcal{S}^{tr})$ from Double ψ_μ. We then calculated the following metrics:

$$\widehat{\text{MSE}}_\mu = \frac{1}{|\mathcal{S}^{tr}|} \sum_{i \in \mathcal{S}^{tr}} (\mu(t_i, x_i) - \hat{\mu}(t_i, x_i; \mathcal{S}^{tr}))^2$$

$$\widehat{\text{MSE}}_\tau = \frac{1}{|\mathcal{S}^{tr}|} \sum_{i \in \mathcal{S}^{tr}} (\tau(x_i) - \hat{\tau}(x_i; \mathcal{S}^{tr}))^2$$

In other words, $\widehat{\text{MSE}}_\mu$ captures the performance of each approach for estimating $\mu(x, t)$, while $\widehat{\text{MSE}}_\tau$ measures performance in estimating $\tau(x)$. Note that although these metrics are not calculable in practice because we do not observe the true values of $\mu(x, t)$ or $\tau(x)$, we do have them available in our simulations. For each analysis, we also calculated the proportion of splits in each tree that were made on X_1 or X_2, compared with X_3 or X_4. We did this separately for the two trees built as part of the Double ψ_μ approach as well as the tree from the Single ψ_τ approach. Results are shown in Fig. 2.

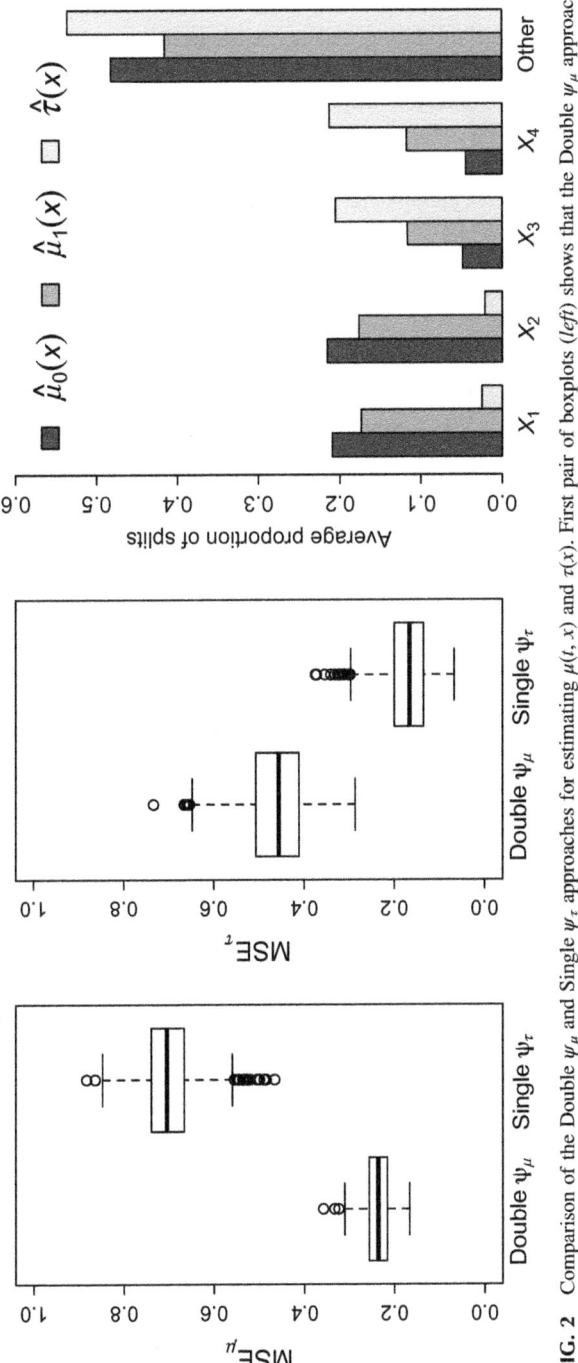

FIG. 2 Comparison of the Double ψ_μ and Single ψ_τ approaches for estimating $\mu(t, x)$ and $\tau(x)$. First pair of boxplots (*left*) shows that the Double ψ_μ approach largely achieves better estimation of $\mu(t, x)$ compared to Single ψ_τ. The next set of boxplots (*middle*) shows that Single ψ_τ outperforms Double ψ_μ in estimating $\tau(x)$. The bar plot (*right*) shows that $\hat{\mu}_0(x)$ and $\hat{\mu}_1(x)$, fit as part of Double ψ_μ, are far more likely to split on variables X_1 and X_2 which do not contribute to heterogeneity in $\tau(x)$, compared to X_3 and X_4. The opposite is true for Single ψ_τ which split more often on variables that interact with treatment.

From the values of $\widehat{\text{MSE}}_\mu$ and $\widehat{\text{MSE}}_\tau$ from the 500 simulated datasets, we can see that the choice of modeling strategy leads to differential strengths and weaknesses. The Single ψ_τ approach, by making splitting choices that target treatment effect heterogeneity, produces much better estimates of $\tau(x)$ than the naive approach. However, better estimation of $\tau(x)$ comes at the cost of worse estimation of $\mu(t, x)$. An explanation for the strengths and weaknesses of each approach can be gleaned from the bar plot. A majority of the time the two tree models fit as part of the Double ψ_μ approach chose splits on variables X_1 and X_2, the strongest predictors of Y. The strength of these variables' effects was likely to have overwhelmed the relatively weak contributions of X_3 and X_4 for predicting Y in the treated group, even though these are the most important variables for estimating $\tau(x)$. In comparison, the Single ψ_τ approach was able to identify this relatively weak signal much of the time, resulting in a much lower $\widehat{\text{MSE}}_\tau$. As a final comment, we note that the differential strengths and weaknesses persisted when setting $B = 2000$, although in general $\widehat{\text{MSE}}_\mu$ and $\widehat{\text{MSE}}_\tau$ were lower compared to when $B = 1$.

In summary, this example shows how choice of objective can lead to differences in how regularization is carried out. Decision tree models perform regularization by making discrete choices about which variables to perform splits on. The Single ψ_τ succeeds in modeling $\tau(x)$ by trying to avoid splits on variables that appear to be independent of $\tau(x)$. However, doing so involves an opportunity cost, in that the same set of variables that are independent of $\tau(x)$ may not be independent of $\mu(x)$.

4.2 Selection bias

The treatment assignment mechanism, also called the propensity score $e(x) = Pr(T = 1|X = x)$, is known in experimental studies, and is usually independent of X. In observational settings it is typically unknown, and assumed to be a function of X. Because of the dependence between T and X, the distribution of X differs between the treatment groups: $Pr(X|T = 1) \neq Pr(X|T = 0)$. This leads to a sampling bias called "selection bias" because the observations selected to receive treatment are likely to be systematically different compared to those who do not receive the treatment. For classical problems of causal inference where the estimand is the ATE or ATT, estimation via the difference in sample averages leads to biased estimates. To see why, consider the following linear model for the conditional expectation of Y_i:

$$E_Y[Y|T = t, X = x] = \tau(x)t + \sum_{j=1}^{p} \alpha_j x_j,$$

where $\tau(x)$ the CATE. Taking expectations with respect to the distribution of X given T, we can write the ATE as:

$$E_Y[Y|T=1] - E_Y[Y|T=0] = E_X[\tau(X)|T=1]$$

$$+ \sum_{j=1}^{p} \alpha_j \left(E_{X_j}[X_j|T=1] - E_{X_j}[X_j|T=0] \right). \qquad (21)$$

When $e(x) = Pr(T=1|X=x) = 0.5$, as in many RCTs, the distribution of all X_j's is the same for each treatment group, and the expectations in the sum in (21) cancel out. Also, notice that the term $E_X[\tau(X)|T=1]$ is the ATT, which is equal to ATE when $T \perp X$ (implying $E_X[\tau(X)|T=1] = E_X[\tau(X)]$). Under selection bias, the second term in (21) is not equal to 0, leading to biased estimates.

If we make the assumptions in Section 3.1, we might expect not to have problems with selection bias in estimating the CATE function, because our estimand $\tau(x) = E[Y^1 - Y^0|X = x]$ does not appear to involve expectations over the distribution of X. This reasoning would be correct if the true functional relationship between Y and X were known and did not need to be estimated from a finite sample. Of course, this is not the case, and selection bias continues to present an issue due to the curse of dimensionality. Selection bias can be described in terms of *covariate shift*, a term from the machine learning literature describing scenarios where the distribution of X in the training set is different from the distribution in the test set (Bickel et al., 2009). All CATE models are trained on S^{tr}, which consists of draws from the factual distribution of X and T: $\{x_i, t_i\}_{i \in S^{tr}}$. However, to estimate $\tau(x)$, we must also be able to predict values of the outcome Y for points in the counterfactual set $\{x_i, 1 - t_i\}_{i \in S^{tr}}$. Define random variable $R = 1 - T$. Under selection bias, the joint distributions of (X, T) and (X, R) are unequal, and so the problem is one of training a prediction model on draws of (X, T) that also perform well at draws from (X, R). Asymptotically, this is not an issue: as long as the supports of (X, T) and (X, R) are equal we would have some data at every point (x, t), and so we could rely on predictions at $(x, 1 - t)$ to be just as accurate. However, for finite samples the disparity between (X, T) and (X, R) means that there will be some regions in $\mathcal{X}$ that are not well-represented by observations from both treatment groups. Therefore, to make predictions at the counterfactual treatment assignment value, we need to rely on extrapolations from our model into these underrepresented regions. The *curse of dimensionality* says that higher the dimension of $\mathcal{X}$, the more distant points will be; extrapolations will thus tend to be less and less accurate with increasing dimensionality of $\mathcal{X}$. As we will see in Section 5, the general strategy for approaching this issue is to encourage similarity (e.g., balance) between treated and control populations, either by sample weighting or by learning an appropriate representation of X.

4.3 Regularization biases

We saw in Section 4.1 that learning a model for $\tau(x)$ by fitting models to the data from each treatment group independently can fail, even under randomized treatment. In this section we explore the sources of bias of another naive approach, where a single model is fit using T as just another predictor variable like any other in X. Like the approach in Section 4.1, part of the reason why this approach also fails is because it is regularized by appealing to the model's predictiveness of Y. However, by allowing regularization of the effect of T on Y, we open up unique opportunities for bias. We refer to these biases collectively as *regularization biases*.

To demonstrate regularization bias, let us assume that $\tau(x)$ is constant in x, $\tau(x) = \tau$, all causal inference assumptions from Section 3.1 are satisfied, and that Y is correctly assumed to be linearly related to T and X. We consider $\tau(x)$ to be a constant in order to simplify the discussion and examples, but the sources of bias considered in this section are also applicable to nonconstant $\tau(x)$ (Imai and Ratkovic, 2013; Hahn et al., 2019). Since in our example unconfoundedness is satisfied $(Y^1, Y^0) \perp T|X$, the OLS estimator of τ is unbiased no matter the dimension of X_i, as long as all possible causes of treatment assignment T are included on the right-hand side. Let us consider the ramifications of taking a naive approach to estimating τ by fitting a linear regression model regularized by penalizing the squared error loss function as in (8), under two scenarios: (1) treatment with equal probability $e(x) = 0.5$, and (2) selection bias resulting from *targeted selection*. Targeted selection is a form of selection bias common especially in medicine when doctors prescribe treatments, and it describes a monotonic relationship between the propensity score and $\mu_0(x)$ (i.e., patients understood by their doctors to be very sick will be more likely to receive treatment).

We generated outcomes Y from the following linear model:

$$Y = -2 - T + \sum_{j=1}^{6} X_j + \epsilon,$$

where all covariates X_j and residuals ϵ are independent standard normal variables, and T is distributed Bernoulli with mean $1/(1 + \exp(-\eta\sum_{j=1}^{6} X_j))$. Thus, parameter η controls the strength of targeted selection, and $e(x) = 0.5$ when $\eta = 0$. Also note that the true $\tau = -1$. We generated samples of varying sizes from the joint distribution of (Y, X, T), for a set of η values, and amended the datasets with 44 additional standard normally distributed covariates independent of T and Y. For each dataset, we fit a ridge regression model, setting the penalization parameter λ as the value minimizing test set error via CV (by minimizing $\widehat{\text{MSE}}_Y$ as in Section 2.5). For each sample size and η parameter setting, we generated and fit models to 50 independent datasets, and reported the average of the estimate $\hat{\tau}$ minus true τ over all each simulation runs. A plot of the results is shown in Fig. 3. The simulations show that when $e(x) = 0.5$, the treatment effect estimate has positive bias that diminishes with sample size. As expected, the ridge regression penalty shrinks $\hat{\tau}$ toward 0, which creates the positive bias with respect to the true value of τ at -1. A model that is

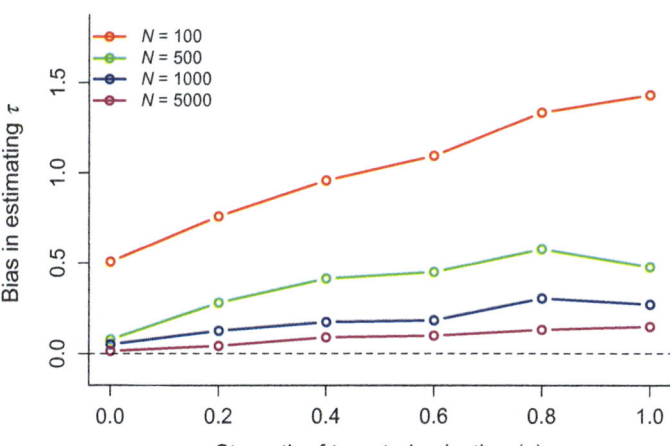

FIG. 3 Example simulation showing bias in estimating treatment effects under treatment randomized with equal probability ($\eta = 0$) and targeted selection ($\eta > 0$).

not subject to this bias would have been one that does not penalize the coefficient on T. As the strength of targeted selection increases, the estimator exhibits additional bias on top of shrinkage: *regularization-induced confounding* (RIC). It can be shown that the bias of ridge regression estimator of τ is a function of (1) the strength of targeted selection and (2) every element of the unknown vector of β, the linear coefficient of X, in the conditional mean model of Y (Hahn et al., 2018).

Some insight on RIC can be gained by considering the way in which covariate adjustment is intended to be used for controlling selection bias. Consider the scenario where both Y^1 and Y^0 follow additive errors models with constant variances. Under the unconfoundedness and consistency assumptions, we can write the model for each potential outcome as:

$$Y^1 = \mu_1(X) + \epsilon^1$$

$$Y^0 = \mu_0(X) + \epsilon^0,$$

where for simplicity both error terms are considered standard normally distributed. In this case, we can rewrite the unconfoundedness assumption in terms of the error terms (see Section 5 of Morgan and Winship (2007) for a more detailed discussion): $(\epsilon^0, \epsilon^1) \perp T|X$. Given estimators for $\mu_0(x)$ and $\mu_1(x)$, write a new pair of error terms as:

$$\hat{\epsilon}^1 = Y^1 - \hat{\mu}_1(X; \mathcal{S}^{tr})$$

$$\hat{\epsilon}^0 = Y^0 - \hat{\mu}_0(X; \mathcal{S}^{tr})$$

These are the residuals expressed in terms of our approximations of $\mu_0(x)$ and $\mu_1(x)$. As we have mentioned earlier, we may expect these estimators to be biased unless they are correctly specified, which is unlikely in the case of regularized fits. Now imagine that the effect of the bias can lead these estimators to be "bent" away from the true values of $\mu_0(x)$ and $\mu_1(x)$ in different directions, perhaps more drastically in specific regions of $\mathcal{X}$. If, through selection bias, values of X in these regions also predict T, then $(\hat{\epsilon}^0, \hat{\epsilon}^1) \perp T | X$ will not hold. Thus, estimates of treatment effect quantities will not be unconfounded, leading to selection bias.

In summary, the naive approach leading to RIC ignores the three roles played by X in estimation of τ or $\tau(x)$: removing confoundedness and estimating $\mu_0(x)$ and $\mu_1(x)$. As we will explore below, modern modeling approaches can reduce the impact of RIC by decoupling the regularization of the relationship between T and X and regularization of the models for $\mu_0(x)$ and $\mu_1(x)$ (or alternatively, $\tau(x)$ directly).

5 Survey of CATE modeling approaches

In this section we review strategies for modeling the CATE function $\tau(x)$. We focus specifically on the various approaches to model fitting given a training sample, and how various modeling strategies approach the task of estimating $\tau(x)$ from a high level. We specifically omit any discussion of hyperparameter selection for discussion in Section 6, because model selection in the context of estimating $\tau(x)$ poses some unique problems that deserve separate discussion.

5.1 Meta-learners

Following Künzel et al. (2019), we refer to meta-learners as the class of CATE estimators that can be fit using off-the-shelf machine learning or statistical regression modeling procedures. Common "base" models for CATE meta-learners include generalized additive models (Hastie and Tibshirani, 1990), random forests (Breiman et al., 1984), boosted trees (Chen and Guestrin, 2016), Bayesian additive regression trees (Chipman et al., 2010), and neural networks (Goodfellow et al., 2016).

Two classes of meta-learners were first described in the causal inference literature, where the ultimate aim was "back-door" adjustment for confounders toward the aim of estimating ATE or ATT via linear regression methods (Rubin, 1977) or other nonparametric estimators (Imbens, 2004). Following the appellations given by Künzel et al. (2019), we will refer to these as S- or T-type meta-learners. In the S-learner (S for single model), T is treated like any other covariate in X, and we estimate $\mu(t, x) = E[Y | T = t, X = x]$ using a single model; refer to this model as $\hat{\mu}(t, x; \mathcal{S}^{tr})$. The S-learner estimate of $\tau(x)$ is taken to be the difference $\hat{\mu}(1, x; \mathcal{S}^{tr}) - \hat{\mu}(0, x; \mathcal{S}^{tr})$. On the other hand, T-learners are constructed from two models fit to independent samples. Define $\mathcal{S}^{tr}_{t=0}$ and $\mathcal{S}^{tr}_{t=1}$ as the subsets of treated and nontreated observations in $\mathcal{S}^{tr}$, respectively. In the T-learner, $\mu_0(x)$ is estimated from $\mathcal{S}^{tr}_{t=0}$, and $\mu_1(x)$

is estimated independently from $\mathcal{S}^{tr}_{t=1}$; refer to the models generated by the T-learner as $\hat{\mu}_0(x;\mathcal{S}^{tr}_{t=0})$ and $\hat{\mu}_1(x;\mathcal{S}^{tr}_{t=1})$. CATEs are then estimated as $\hat{\mu}_1(x;\mathcal{S}^{tr}_{t=1}) - \hat{\mu}_0(x;\mathcal{S}^{tr}_{t=0})$ by the T-learner.

Although originally described in the context of estimating ATE or ATT, these algorithms have also been used for the ultimate objective of estimating $\tau(x)$ (Foster et al., 2011; Radcliffe, 2007; Jaskowski and Jaroszewicz, 2012; Hill, 2011; Lu et al., 2018). However, both algorithms can be problematic in finite samples for different reasons. Firstly, models are trained to minimize MSE_Y in the training sample, without attempt at appealing to the real objective for CATE prediction, MSE_D. This becomes an issue in high dimensions where the fit is regularized, even when treatment is completely randomized, because the bias-variance tradeoff that applies to the prediction of Y does not necessarily lead to the same solutions that we would reach if we were able to fit the model to realizations from $Y^1 - Y^0$ directly. Because the S-learner treats T like any other variable in X, there is a chance of the treatment effects getting "washed out" via regularization, or shrunken toward 0 as we show in the example of Section 4.3 (an S-learner where homogeneity of CATE is assumed). As seen in Section 4.1.2, T-learners have somewhat the opposite problem in that by regularizing two prediction functions independently, they may pick up differences between the conditional means attributable not to a heterogeneous $\tau(x)$ but rather to differences in how the two estimators are regularized (Powers et al., 2018; Radcliffe and Surry, 2011). In the presence of treatment selection, both S-learner and the T-learner algorithms are prone to RIC (Hahn et al., 2018, 2019; Powers et al., 2018).

The X-learner is a two-stage algorithm that takes a much more tailored approach to estimating CATEs (Künzel et al., 2019). The first stage proceeds exactly as the T-learner, producing separate model estimators for $\hat{\mu}_0(x; \mathcal{S}^{tr}_{t=0})$ and $\hat{\mu}_1(x; \mathcal{S}^{tr}_{t=1})$. These are then used to impute the missing counterfactual outcomes for each individual. By imputing the counterfactuals, an estimate of d_i is created for every observation in the dataset:

$$\hat{d}_i = \begin{cases} y_i - \hat{\mu}_0(x; \mathcal{S}^{tr}_{t=0}), & \text{for } i \in \mathcal{S}^{tr}_{t=1} \\ \hat{\mu}_1(x; \mathcal{S}^{tr}_{t=1}) - y_i, & \text{for } i \in \mathcal{S}^{tr}_{t=0} \end{cases}$$

In the second stage, the data are again split into the treated sample and the nontreated sample. Separate standard supervised learning models are fit to the $\hat{d}_i$'s for observations in $\mathcal{S}^{tr}_{t=0}$ and $\mathcal{S}^{tr}_{t=1}$ separately, yielding two estimators of $\tau(x)$: $\hat{\tau}_0(x; \mathcal{S}^{tr}_{t=0})$ and $\hat{\tau}_1(x; \mathcal{S}^{tr}_{t=1})$. In the final step, the final estimator of $\tau(x)$ is taken as a weighted sum of the two:

$$\hat{\tau}(x; \mathcal{S}^{tr}) = e(x)\hat{\tau}_1(x; \mathcal{S}^{tr}_{t=1}) + (1 - e(x))\hat{\tau}_0(x; \mathcal{S}^{tr}_{t=0}),$$

where the propensity score weights are estimated in cases of selection bias. Note that in the second stage, the models are essentially fit to minimize a plug-in version of MSE_D. The authors show that the X-learner converges faster than S- or T-learners do in scenarios where the true $\tau(x)$ is smoother than $\mu_1(x)$ or $\mu_0(x)$, which they suspect is often the case in real datasets. They also show that the X-learner shows considerable improvement over the T-learner when the treatment groups are severely imbalanced.

5.2 Transformed outcome methods

Another class of CATE estimation methods can also be performed using off-the-shelf prediction methods, by transforming the outcome so that its weighted mean difference from the true $\tau(x)$ is equal to 0 (Knaus et al., 2018). These estimators result from a weighted optimization function:

$$\hat{\tau}(x, \mathcal{S}^{tr}) = \arg\min_{\tau(x)} \sum_{i \in \mathcal{S}^{tr}} w_i \left(y_i^* - \tau(x_i)\right)^2, \tag{22}$$

with weights w_i and transformed outcome variables y_i^*. Table 2 shows several versions of (22) reported in the literature, differing in the weight and transformation functions.

Note that for the transformations involving nuisance parameters, Chernozhukov et al. (2018) suggest they be estimated in a different sample than the CATEs to avoid overfitting. They describe a procedure called *cross-fitting* to do this, using repeated sampling without replacement to (1) estimate nuisance parameters via machine learning in the subsample, and (2) estimate CATEs in the out-of-sample observations, given nuisance parameter predictions. The Horvitz–Thompson transformation (Hirano et al., 2003) has been used as the basis for many approaches tailored for CATE estimation (Powers et al., 2018; Athey and Imbens, 2016; Sugasawa and Noma, 2019; Jaskowski and Jaroszewicz, 2012). While y_i^* via Horvitz-Thompsom transformation is unbiased for CATE, its variance can be large due to the propensity score in the denominator (Powers et al., 2018). The Doubly Robust transformation was adapted from (Robins and Rotnitzky, 1995) by (Knaus et al., 2018), and attempts to stabilize the variance of the Horvitz–Thompson transformation. The efficiency of the Modified Covariates method (Tian et al., 2014) can be improved via local centering of y_i at $\mu(x_i)$. This local centering is referred to as orthogonalization (Chernozhukov et al., 2018). Orthogonalized variables are also a part of the R-learner method, which deliberately separates the elimination of correlations between $e(x)$ and $\mu(x)$ from the task of regularizing a fit for estimating $\tau(x)$ (Nie and Wager, 2017). In the case where $\tau(x)$ is a linear function of X, R-learner can achieve performance on par with an oracle estimator in which nuisance parameters $e(x)$ and $\mu(x)$ are considered to be known.

Knaus et al. (2018) compare these approaches in simulated datasets and conclude that the Horvitz–Thompson and Modified Covariates methods perform much more poorly than the others in terms of $MSE_\tau(\mathcal{S}^{tr})$. However, they note that partialling out the effect of covariates on y_i as part of the Modified

Covariates method (so that $y_i^* = 2(2t_i = 1)(y_i - \mu(x_i)))$ improves its performance considerably. They did not compare a similarly modified version of the Horvitz–Thompson transformation.

5.3   Shared basis methods

Shared basis models of $\tau(x)$ can be written in terms of *linear basis expansions*:

$$\sum_{m=1}^{M} \beta_m g_m(x),$$

where $g_m(X)$ is the mth scalar-valued transformation of X, $m = 1, ..., M$. As we reviewed in Section 4.1.1, the GRF methodology combines an adaptive nearest neighbors with an estimating equations approach to estimating $\theta(x)$ (Athey et al., 2019). When $\theta(x) = \tau(x)$, the resulting GRF is called a causal forest. For the bth tree in a causal forest, $b = 1, ..., B$, we can view the resulting estimate of $\tau(x)$ as an additive model:

$$\hat{\tau}_b(x) = \sum_{m=1}^{M} \hat{\tau}_m g_m(x),$$

where the set of functions $g_m(X)$ are transformations of X induced by the tree structure, M is equal to the number of terminal nodes, and $\hat{ta\textit{u}}_m$ is the treatment effect estimate for the mth node. For example, consider a tree with 3 terminal nodes: the subspaces of $\mathcal{X}$ where, (1) $X_1 \le a_1$; (2) $X_1 > a_1$ and $X_2 \le a_2$; and (3) $X_1 > a_1$ and $X_2 > a_2$. This tree structure could be written as a basis expansion with the following transformations:

$$g_1(x) = \mathbf{1}(x_1 \le a_1)$$
$$g_2(x) = \mathbf{1}(x_1 > a_1)\mathbf{1}(x_2 \le a_2)$$
$$g_3(x) = \mathbf{1}(x_1 > a_1)\mathbf{1}(x_2 > a_2)$$

The causal forest estimator can be thought of as being equivalent to taking the average of $\hat{\tau}_b(x)$ over all B trees in the forest. Because the two treatment arms are modeled using the same basis expansions, Powers et al. (2018) refer to causal forest as a *shared basis model*. As in the example of Section 4.1.2, the advantage of shared basis models is that they perform regularization by choosing basis functions that are sensitive to heterogeneity of the target parameter of interest for CATE modeling, $\tau(x)$. The causal forest's approach to handling selection bias follows closely the orthogonalization approach of the R-learner from Section 5.2. The relative simplicity of fitting piecewise constants make causal forests very competitive in terms of computational speed, at the expense of some bias resulting from the piecewise constant basis functions. Friedberg et al. (2019) propose a modification of the GRF algorithm using local linear regression adjustments to reduce this bias.

A few other shared basis models are developed in Powers et al. (2018). The Causal Boosting algorithm applies the concept of *boosting* (see for example Chen and Guestrin (2016)) using Causal Forests with $B = 1$ as the base learner. Causal Multivariate Adaptive Regression Splines (MARS) is an adaptation of the MARS algorithm for the standard problem of predicting Y (see Section 9.4 in Hastie et al. (2009)). Causal MARS considers input variable transformations of linear regression splines of the form $\max(0, x_j - c)$ or $\max(0, c - x_j)$. Higher order products of these spline terms can also be considered. The algorithm starts by fitting an intercept term $g_1(x) = 1$, where corresponding β_1 corresponds to the overall ATE; further terms are then added to the sum which best explain the treatment effect. The algorithm proceeds by adding basis functions to the sum until there are a prespecified number of terms M, at which point the algorithm stops. The authors recommend using bootstrapping to reduce the variance of causal MARS, and offer a propensity-stratified version of their algorithm for use under treatment selection.

5.4 Neural networks

Several methods leverage neural networks for modeling $\tau(x)$, not only because of the flexibility of the functional forms with which they can model $\mu_1(x)$ and $\mu_0(x)$, but also because it is possible to colearn useful representations of X to overcome selection bias. The balanced neural network (BNN) of Johansson et al. (2018) estimates $\tau(x)$ using a two-stage approach. In the first stage, they impute a counterfactual outcome for each observation in S^{tr} using the 1-nn match from the opposite treatment group. They then cooptimize a representation $\phi(x)$ and an estimator for the conditional mean of factual and imputed counterfactual Y given this representation and T, $\hat{\mu}(\phi(x), t; S^{tr})$. The representation is designed to minimize the discrepancy distance, which captures the distance between the distributions of X among the treated and nontreated observations. In stage 2, a penalized linear model is fit to the factual outcomes Y given T and $\phi(X)$ using a neural network independent from that used in stage 1. Thus, the model fit in stage 2 is equivalent to an S-learner.

Shalit et al. (2017) note that the BNN method suffers from the problem typical of the S-learner: that the treatment effects can be shrunken toward 0. Their proposed counterfactual regression (CFR) eliminates this limitation by jointly training the representation model, and models estimating $\mu_1(x)$ and $\mu_0(x)$, in a single neural network. Their representation $\phi(x)$ is trained to minimize the Wasserstein distance between treated and nontreated samples, while $\mu_1(x)$ and $\mu_0(x)$ are modeled using $\phi(x)$ as the inputs, and using separate parameterizations. The deep counterfactual networks of Alaa et al. (2017) take a similar approach and parameterize models for $\mu_1(x)$ and $\mu_0(x)$ separately in a single multitask network, while simultaneously generating a model for the propensity score $e(x)$. Selection bias is handled using a propensity-dropout regularization scheme, the analog of propensity score weighting. Yao et al. (2018) point out that the representations learned as part of the BNN and CFR methods improve the similarity between treated and nontreated populations globally, but do not try to preserve potentially small distances of nearest neighbors from assigned to opposite treatments. They train a representation $\phi(x)$ preserving these local similarities, and simultaneously learn a model for the factual outcomes Y given T and $\phi(x)$. Finally, the GANITE method of (Yoon et al., 2018) draws counterfactuals from a distribution trained via generative adversarial networks, and learns a model for $\tau(x)$ by combining the factual and generated counterfactuals.

5.5 Bayesian nonparametric methods

Nonparametric Bayesian methods are a natural choice for modeling $\tau(x)$ since the Bayesian posterior incorporates uncertainty in regions of $\mathcal{X}$ with poor overlap between treated and nontreated samples. Hill (2011) use Bayesian additive regression trees (Chipman et al., 2010) in an S-learner style approach to generate draws from the posterior predictive distribution of Y given $t = 0, 1$ for each point x in a test set. Under selection bias, they show that the posterior predictive distribution for x in regions of poor overlap have appropriately high variance. However, under strong confounding their method can be badly biased; moreover it inherits the regularization bias common to the S-learner (Hahn et al., 2019). The Bayesian Causal Forest method separately regularizes propensity score $e(x)$, nontreated conditional mean $\mu_0(x)$ and $\tau(x)$ functions. By separating the regularization of $\mu_0(x)$ and $\tau(x)$ functions, their method shrinks estimates of $\tau(x)$ toward constant τ instead of toward 0; further separating regularization of $e(x)$ is designed to alleviate bias due to of RIC.

Alaa and van der Schaar (2018) study the information rates (Bayesian analogue of convergence rates) of Bayesian nonparametric estimators of $\tau(x)$. They show that the rate is mainly limited by the complexity or smoothness of the more complex of $\mu_1(x)$ and $\mu_0(x)$. They use Gaussian processes to model these two quantities jointly in a way that allows their complexities to differ freely. Their algorithm minimizes an objective function comprised of two terms: (1) the sum of squared differences between the observed factual outcomes and the expectation of the posterior predictive distribution for the factual outcomes, and (2) the sum of the variances of the posterior predictive distribution for the counterfactuals. Thus, the model posterior is regularized by the sum of the uncertainty of the counterfactual outcomes, which will be higher for more complex fits to the factual outcomes.

6 CATE model selection and assessment

In Section 2.5, we described the CV procedure for selecting an optimal model for predicting Y on the basis of estimates of MSE_Y, which itself was an average of estimates of $MSE_Y(S^{tr})$ taken over many different CV-training and CV-test samples. These estimators had the advantage of being free of nuisance parameters; all one needs is prediction function $\hat{\mu}(x; S^{tr})$ and an independent test set S^{te}. An equivalent estimator for evaluating models of $\tau(x)$ would be based on:

$$\widehat{MSE}_D(S^{tr}) = \frac{1}{|S^{te}|} \sum_{i \in S^{te}} (d_i - \hat{\tau}(x_i; S^{tr}))^2, \tag{23}$$

where d_i is a hypothetical realization of random variable $D = Y^1 - Y^0$. However, it should be immediately clear that we cannot evaluate (23) because we do not directly observe realizations of D; we only observe realizations from Y^1 for observations i for whom treatment was administered, and realizations from Y^0 otherwise (assuming consistency).

Thus, in practice, we must find a substitute for (23) that we can use for model selection in the training set (e.g., fixing tuning parameters, selecting between model specifications), and model evaluation in the test set. The naive approach would be to simply select the model minimizing the CV estimator of $\widehat{MSE}_Y$; in other words, find the best model for predicting Y, given X and T. However, this approach can fail to identify the best estimator of $\tau(x)$ for the same reasons that led the Double ψ_μ approach to fail in the example of Section 4.1.2. See Rolling and Yang (2014) for an example where the use of $\widehat{MSE}_Y$ will always lead to selection of the wrong model for estimating $\tau(x)$ as long as sample size is large enough.

6.1 Survey of evaluation methods for estimators of $\tau(x)$

Rolling and Yang (2014) propose a plug-in estimate of d_i. For each observation i, let $\hat{d}_i = (2t_i - 1)(y_i - y_{\bar{i}})$, where $\bar{i}$ is the index of the nearest observation in the treatment group opposite that of observation i:

$$\bar{i} = \arg\min_{i'} \sqrt{\sum_{j=1}^{p} (x_{ij} - x_{i'j})^2}, \text{ among all } i' \text{ in } S^{te} \text{ for which } t_{i'} = 1 - t_i.$$

Model selection then proceeds by plugging $\hat{d}_i$ into (23) to estimate MSE_D via cross-validation.

Several of the transformed outcome methods in Table 1 have also been proposed as the basis for model selection metrics, based on (22). For a given estimator $\hat{\tau}(x; S^{tr})$ learned from the training data, calculate $\sum_{i=1}^{|S^{te}|} w_i (y_i^* - \hat{\tau}(x_i; S^{tr}))^2$ in the test dataset. Gutierrez and Gerardy (2016) use the Horvitz–Thompson approach, while Nie and Wager (2017) suggest using the weights and variable transformations as in the R-learner. Schuler et al. (2018) compare Horvitz–Thompson and R-learner-based model evaluation metrics, concluding in their simulations that the latter more consistently chooses the truly optimal models.

TABLE 1 Formulas for weights and outcome transformations for estimation of CATEs via weighted MSE in (22).

Name	w_i	y_i^*
Horvitz–Thompson	1	$y_i\left(\frac{t_i - e(x_i)}{e(x_i)(1-e(x_i))}\right)$
Doubly Robust	1	$\mu_1(x_i) - \mu_0(x_i)$
		$+\ (y_i - \mu(t_i, x_i))\left(\frac{t_i - e(x_i)}{e(x_i)(1-e(x_i))}\right)$
Modified Covariates	$\frac{(2t_i-1)(t_i-e(x_i))}{4e(x_i)(1-e(x_i))}$	$2(2t_i - 1)y_i$
R-learner	$(t_i - e(x_i))^2$	$\frac{y_i - \mu(x_i)}{t_i - e(x_i)}$

Notation: $e(x_i) = E[T|X = x_i]$; $\mu(t_i, x_i) = E[Y\,|T = t_i, X = x_i]$; $\mu(x_i) = E[Y\,|X = x_i]$; $\mu_0(x_i) = E[Y\,|T = 0, X = x_i]$; $\mu_1(x_i) = E[Y\,|T = 1, X = x_i]$

TABLE 2 Evaluation of the models fit on the training data in Fig. 4.

Scoring system	ABUC	$COR_\tau(\mathcal{S}^{tr}, F)$
Causal Forest	0.157	0.599
Random Forest S-learner	0.107	0.302
Random Forest T-learner	0.113	0.261

Note: ABUC is the Area Between the Uplift Curves generated from $\mathcal{S}^{tr}$ alone. $COR_\tau(\mathcal{S}^{tr}, F)$ is the Pearson correlation between $\tau(x)$ and the CDF of each scoring system evaluated at points from a test set of 10,000 observations.

To our knowledge the performance of adaptations of the other methods in Table 1 have not been studied.

Alaa and van der Schaar (2019) take the approach of estimating $MSE_\tau(\mathcal{S}^{tr})$ using influence functions. They begin by expanding the true $MSE_\tau(\mathcal{S}^{tr})$ in a first-order Taylor-like expansion. Since the expansion depends on unknown $\tau(x)$ they use a plug-in estimator $\widetilde{\tau}(x)$:

$$\widehat{MSE}_{\widetilde{\tau}}(\mathcal{S}^{tr}) = MSE_{\widetilde{\tau}}(\mathcal{S}^{tr}) + \frac{1}{|\mathcal{S}^{tr}|}\sum_{i \in \mathcal{S}^{tr}}\widehat{MSE}_{\widetilde{\tau},i}^{(1)}(\mathcal{S}^{tr}). \tag{24}$$

where the last term is the unique first-order influence function of $MSE_\tau(\mathcal{S}^{tr})$, and

$$MSE_{\widetilde{\tau}}(\mathcal{S}^{tr}) = E_X[\widetilde{\tau}(X) - \hat{\tau}(X; \mathcal{S}^{tr})]^2. \tag{25}$$

They show that the first-order influence function in (24) can be written as follows:

$$
\widetilde{\text{MSE}}_{\tilde{\tau},i}^{(1)}(\mathcal{S}^{tr}) = \left(1 - \frac{2t_i(t_i - e(x_i))}{e(x_i)(1 - e(x_i))}\right)\tilde{\tau}^2(x_i)
$$

$$
+ \frac{2t_i(t_i - e(x_i))}{e(x_i)(1 - e(x_i))}y_i(\tilde{\tau}(x_i) - \hat{\tau}(x_i; \mathcal{S}^{tr}))
$$

$$
- (t_i - e(x_i))(\tilde{\tau}(x_i) - \hat{\tau}(x_i; \mathcal{S}^{tr}))^2 + \hat{\tau}^2(x_i; \mathcal{S}^{tr}) - \text{MSE}_{\tilde{\tau}}(\mathcal{S}^{tr})
$$

Since the influence function depends on unknown $\tau(x)$, they establish the conditions under which a plug-in estimator $\tilde{\tau}(x)$ can be used to yield efficient estimates of $\text{MSE}_\tau(\mathcal{S}^{tr})$. They suggest calculating $\tilde{\tau}(x)$ using tree-based methods such as XGBoost (Chen and Guestrin, 2016). Although others have suggested plug-in estimators of $\text{MSE}_\tau(\mathcal{S}^{tr})$ derived from (25) (Powers et al., 2018), such estimates are model-dependent and can be biased. Alaa and van der Schaar (2019) developed the *unplugged* estimator in (24) to reduce bias, and show that it increases the chances of selecting the same model as an oracle $\text{MSE}_\tau(\mathcal{S}^{tr})$ compared to the plug-in across many benchmark datasets.

7 Practical application of CATE estimates

So far, the statistical or machine learning approaches we have considered have been designed for the explicit objective of CATE estimation or treatment effect prediction. This objective is primarily motivated by the desire to inform treatment decisions for individuals in the population under study. In practice, however, this objective can often be a step removed from actionable information that might result if we further refined our objective. For example, once one has a CATE estimate for every x, what does one do with them? In this section we consider 2 *features* of the CATE function that are often of higher-level interest. In some cases these features can be effectively learned in a two-stage analysis, where the first stage involves learning CATEs using the modeling strategies in Section 5, followed by a separate analysis motivated by the specific higher-level objective. In other cases, it is possible to achieve optimal performance with respect to the higher-level objective without explicitly estimating the CATE function.

7.1 Stratification on the CATEs

As we've seen, it is difficult enough to design methods to evaluate the accuracy of predictions from a CATE model, let alone provide an interpretable and transparent evaluation. Existing methods for doing so rely on relatively complex approaches that present the evaluation through a level of abstraction quite removed from the observable data. As a result, some analysts may feel hesitation when considering models for the specific purpose of predicting D, or estimating $\tau(x)$, in the same way that models might be used to predict Y.

Instead of choosing a model based on its inferred ability to estimate $\tau(x)$ itself, it may be attractive to choose a model, or *scoring system* $s(x; \mathcal{S}^{tr})$, from the observations in the training set, producing *scores* that can be used to sort a new dataset in order of increasing $\tau(x)$. We call this objective *stratification on the CATEs*, and we emphasize that scoring systems optimizing this objective do not necessarily minimize MSE_τ, or even estimate $\tau(x)$ at all. Scoring systems could include estimates of $\tau(x)$, or even estimates of $\mu_0(x)$ as we will explore below in the case of binary Y. Stratification on the CATEs can lead to easily interpretable and transparent evaluation metrics, and therefore may be more immediately appealing in practice compared with prediction accuracy.

A typical analysis of this sort could be described in general terms as follows. Given test set observations $\mathcal{S}^{te}$, and a scoring function $s(x; \mathcal{S}^{tr})$ assumed to be positively associated with $\tau(x)$:

1. Sort the observations in the test set based on their scores $s(x_i; \mathcal{S}^{tr})$, for all $i \in \mathcal{S}^{te}$.
2. Define intervals I_q, $q = 1, \ldots, Q$ in the support of $s(x; \mathcal{S}^{tr})$. Construct an ordered set of groups U_q, $q = 1, \ldots, Q$, where U_q is the set of indices i falling into interval I_q.
3. Using the data from each group U_q, estimate the group ATE (GATE) using the data in $\mathcal{S}^{te}$:

$$\gamma_q = E[Y^1 - Y^0 | s(X; \mathcal{S}^{tr}) \in I_q].$$

The intuition is that we would like to find a scoring system such that the estimated CATEs are monotonically increasing and maximally heterogeneous. The advantage of this method is that if the groups are large enough, it is straightforward to estimate CATEs (especially under randomized treatment assignment), by the difference in average outcomes between treatment groups within each U_q. Furthermore, presentation of the results can often be paired with an informative visualization. Because the training set (used to fit the score model $s(x; \mathcal{S}^{tr})$), and the test set (used to estimate CATEs) are independent, the CATEs should have low bias in finite samples (Abadie et al., 2018).

There are at least two procedures in the literature that follow the structure of the 3-step analysis outline presented above. These methods differ in the way in which intervals I_q and groups U_q are constructed in step 2, but they both have been shown to estimate a quantity that is proportional to the correlation between $\tau(x)$ and a monotonic function of $s(x; \mathcal{S}^{tr})$. Let us define a model evaluation criterion from this correlation:

$$\text{COR}_\tau(\mathcal{S}^{tr}, F) = \frac{E_X[AB|\mathcal{S}^{tr}]}{\sqrt{E_X[A^2|\mathcal{S}^{tr}]E_X[B^2|\mathcal{S}^{tr}]}},$$

$$A = \tau(X) - E_X[\tau(X)],$$

$$B = F(s(X; \mathcal{S}^{tr})) - E_X[F(s(X; \mathcal{S}^{tr}))|\mathcal{S}^{tr}],$$

(26)

where $F(\cdot)$ is some monotonic function of its argument. Thus, (26) captures the ability of a scoring system $s(x; \mathcal{S}^{tr})$ to stratify a sample on $\tau(x)$ by measuring the association between the two functions. The quantity in (26) can be viewed as an alternative to $\mathrm{MSE}_\tau(\mathcal{S}^{tr})$ for evaluating models of $\tau(x)$. Whereas $\mathrm{MSE}_\tau(\mathcal{S}^{tr})$ penalizes the error of an estimator $\hat{\tau}(x; \mathcal{S}^{tr})$ for estimating $\tau(x)$, (26) does not. Therefore, maximizers of (26) does not necessarily need to be estimators of $\tau(x)$, hence our preference for the generic term *scoring systems* to describe the functions $s(x; \mathcal{S}^{tr})$ to be evaluated.

The simplest way to construct groups in step 2 is to divide the support of $s(x_i; \mathcal{S}^{tr})$, $i \in \mathcal{S}^{te}$, into nonoverlapping intervals $I_q = [l_{q-1}, l_q)$ for $q = 1, ..., Q$ with equal masses, e.g., quintiles with $Q = 5$. Step 3 would then involve estimation of CATEs γ_q using the samples in each quantile group U_q, using the usual difference in treatment group averages:

$$\hat{\gamma}_q = \frac{\sum\limits_{i \in U_q} t_i y_i}{\sum\limits_{i \in U_q} t_i} - \frac{\sum\limits_{i \in U_q} (1 - t_i) y_i}{\sum\limits_{i \in U_q^{(d)}} (1 - t_i)}. \tag{27}$$

Chernozhukov et al. (2018) show that choosing a scoring system maximizing $\frac{1}{Q}\sum_{q=1}^{Q} \hat{\gamma}_q^2$ is equivalent to maximizing (26) with $F(\cdot)$ as the identity function. They also provide an estimation procedure for the best linear predictor of $\tau(x)$, given $s(x; \mathcal{S}^{tr})$, from which they further develop inference methods for testing the null hypothesis of homogeneity of treatment effect associated with the scores. This method is capable of directly estimating (26), though it has the disadvantage of losing the accessibility inherent in evaluating CATEs.

Another popular method involves constructing a sequence of $Q = |\mathcal{S}^{te}|$ nested groups in step 2 (Radcliffe, 2007; Jaskowski and Jaroszewicz, 2012). Let $U_q \subseteq \mathcal{S}^{te}$ for all $q = 1, ..., Q$. Then, a sequence of groups is obtained,

$$U_1 \subset U_2 \subset \cdots \subset U_q \subset \cdots \subset U_{Q-1} \subset U_Q, \tag{28}$$

such that U_{q+1} is the set of observations in U_q excluding the observation in U_q with the minimum score. Intuitively, in order for a scoring system to provide a good basis for stratification on $\tau(x)$, we would expect that estimates of γ_q increase monotonically in q. Given a scoring system $s(x; \mathcal{S}^{tr})$, we can plot $\hat{\gamma}_q$ as a function of the proportion $|U_q|/|\mathcal{S}^{te}| = q/Q$, for all $q = 1, ..., Q$. Such plots are referred to as *uplift curves* or *qini curves* (Radcliffe, 2007; Jaskowski and Jaroszewicz, 2012). Note that the uplift curve calculated for a sequence of groups constructed via random exclusion (instead of excluding the observation with the smallest score) should be approximately constant at $\hat{\gamma}_Q$, the estimate of overall ATE in the test set. Under mild conditions, as $|\mathcal{S}^{tr}| \to \infty$, the weighted area between the uplift curve for scoring system $s(x; \mathcal{S}^{tr})$ and the curve constructed from random exclusion converges in probability to the product of (26) and a constant, with $F(\cdot)$ as the empirical cumulative distribution function (CDF) (Zhao et al., 2013). Similar convergence results have been shown for the analogous procedure for evaluating the association of scoring systems with $\mu(x)$ (Denuit et al., 2019). Using the set of estimates $[\hat{\gamma}_q]_{q=1}^{Q}$, the ABUC is calculated as:

$$\text{ABUC} = \frac{1}{Q}\sum_{q=1}^{Q}\frac{q}{Q}(\hat{\gamma}_q - \hat{\gamma}_Q). \tag{29}$$

Note that the weights reduce the contribution of $\hat{\gamma}_q$ to the ABUC as q approaches 1; these estimates are likely to be volatile due to sample size.

In cases where the available sample size is modest, scoring systems can be evaluated and compared in the training set by plotting their uplift curves from a Monte Carlo CV procedure (Zhao et al., 2013). As in Section 2.5, we start by splitting $\mathcal{S}^{tr}$ many times into equal-sized $\mathcal{S}_v^{CVtr}$ and $\mathcal{S}_v^{CVte}$, for $v = 1,...,V$. Using the data from each of the $\mathcal{S}_v^{CVtr}$, we learn scoring system $s(x, \mathcal{S}_v^{CVtr})$. Then, in each $\mathcal{S}_v^{CVte}$, the sequence of groups is generated as in (28). Let $U_q^{(v)}$ be the qth group generated from the vth $\mathcal{S}_v^{CVte}$. Then, let the estimate of γ_q from $\mathcal{S}_v^{CVte}$ be:

$$\hat{\gamma}_q = \frac{1}{V}\sum_{v=1}^{V}\hat{\gamma}_q^{(v)},$$

$$\hat{\gamma}_q^{(v)} = \frac{\sum_{i \in U_q^{(v)}} t_i y_i}{\sum_{i \in U_q^{(v)}} t_i} - \frac{\sum_{i \in U_q^{(v)}}(1-t_i)y_i}{\sum_{i \in U_q^{(v)}}(1-t_i)}.$$

Because $|\mathcal{S}_v^{CVte}|$ is equal for all v, for each value of q the $|U_q^{(v)}|$ will also be equal for all v. The CV uplift curves plot the CV estimates $\hat{\gamma}_q$ as a function of proportion of observations in each $\mathcal{S}_v^{CVte}$ falling into subgroup $U_q^{(v)}$: $|U_q^{(v)}|/|\mathcal{S}_v^{CVte}|$. Similar to the CV procedure used to estimate MSE_Y, the CV procedure for uplift curves is likely to be a better estimate of $E_{\mathcal{S}^{tr}}[COR_\tau(\mathcal{S}^{tr}, F)]$ times a constant, where F is the CDF, compared to (26) times a constant. To our knowledge the properties of the CV estimator have not been studied extensively.

We now provide an example of the uplift curves and ABUC. We simulated a single training dataset with 500 observations from the same generative distribution as the example in Section 4.1.2, and plotted uplift curves from the Monte Carlo CV procedure with $V = 100$ in Fig. 4. Three scoring systems are compared: (1) causal forest, (2) a random forest S-learner, and (3) a random forest T-learner. All three models were fit to each $\mathcal{S}_v^{CVtr}$ for $v = 1, ..., 100$ using the `grf` package in R using functions `causal_forest`

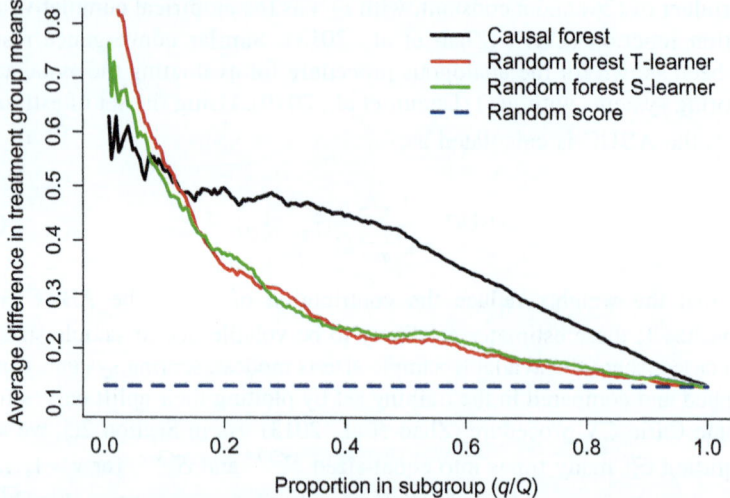

FIG. 4 Uplift curves from 3 models fit to a single training dataset of 500 observations drawn from the example setup from Section 4.1.2. Curves were generated from the Monte Carlo CV procedure with $V = 100$. The weighted area between each solid curve and the dotted line (the ABUC metric) estimates a quantity proportional to the correlation between the true $\tau(x)$ and the rank order of the scores from each model evaluated at points x in a test set. In this dataset, the uplift curves suggest that the rank ordering of scores generated from the causal forest are more associated with the true $\tau(x)$. ABUC values reported in Table 2 corroborate this.

and `regression_forest` at their default settings. Note that the 3 curves should only diverge as $|U_q^{(d)}|$ decreases (from right to left on the x-axis). Recall that U_q is generated by removing the observation with the lowest predicted score from U_{q+1}. The dotted blue line in Fig. 4 was generated by randomly excluding observations to generate the sequence of U_q's; we refer to this as the uplift curve for a random scoring system. For the random scoring system, the averaged estimates of γ_q's remain constant for all q.

The ABUCs for the scoring systems in Fig. 4 were calculated as in (29) using cross-validated estimates of each $\hat{\gamma}_q$, and are shown in Table 2. We also evaluated (26) via simulation from the true $\tau(x)$, for each scoring system fit to the entire training dataset. To do this, we generated a test set $\mathcal{S}^{te}$ comprised of 10,000 observations from the same data generating distribution as the training sample, and calculated the Pearson correlation between the true $\tau(x_i)$ and the CDF of scores $s(x_i; \mathcal{S}^{tr})$ for each observation x_i in $\mathcal{S}^{te}$. These correlations are shown for the 3 scoring systems in Table 2. The ABUC metric correctly identifies the causal forest scoring system for stratifying on the true $\tau(x)$ from this data generating distribution.

As mentioned previously, it is possible that scoring systems that are not estimators of $\tau(x)$ could perform very well in terms of the optimization criterion in (26). This phenomenon has been exploited in medical research, where it is common to *risk stratify* a sample based on scores from a model estimating $\mu_0(x)$, where Y_i is a binary outcome often representing mortality (Ioannidis and Lau, 1997; Kent and Hayward, 2007; Kent et al., 2018; Abadie et al., 2018). We will subsequently show an example scenario where $\mu_0(x)$ may be expected to outperform a good estimator of $\tau(x)$, or vice versa, in terms of (26).

Consider the following simulated example with randomized treatment assignment T, two independent covariates X_1 and X_2 distributed uniformly, and outcomes Y distributed Bernoulli with mean:

$$E[Y|T,X] = \text{expit}(-2 + \log(0.7)T + X_1 + X_2 + T + T\beta(X_1 - X_2)), \quad (30)$$

where $\text{expit}(\cdot)$ is the inverse logit function. We generated training datasets ranging in size from 500 to 3000, and magnitudes of $\exp(\beta)$ (the odds ratios for the interaction term) ranging from 0.3 to 1.7. Note that for values of $\exp(\beta)$ close to 1, the true $\tau(x)$ is approximately monotonic in $\mu_0(x)$. From each training dataset we estimated $\mu_0(x)$ and $\mu_1(x)$ functions using logistic regression models fit separately to each treatment sample, estimating separate coefficients for X_1 and X_2. Along with each training dataset, we generated a test set with 10,000 observations from the same data generating distribution and used it to evaluate $\text{COR}_\tau(\mathcal{S}^{tr}, F)$ for each scoring system. In Fig. 5 we show, for each combination of sample size and $\exp(\beta)$, the

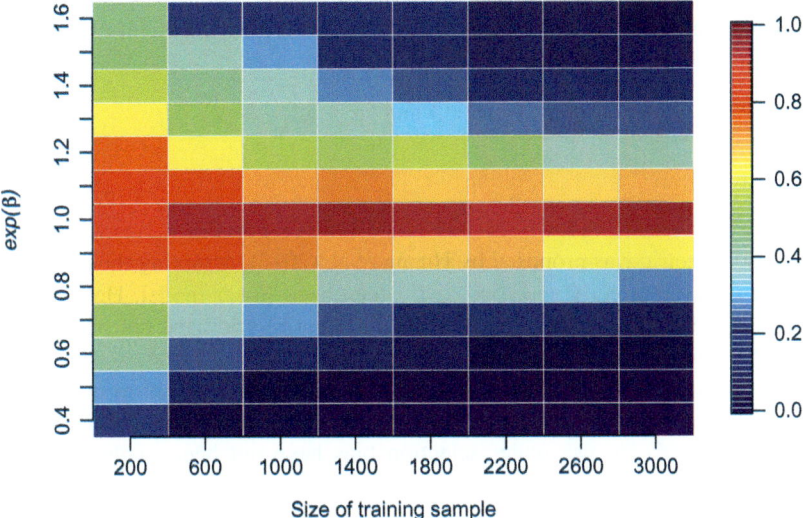

FIG. 5 Probability that the CDF of a logistic regression estimator of $\mu_0(x)$ achieves higher correlation with the true $\tau(x)$ (specifically, $\text{COR}_\tau(\mathcal{S}^{tr}, F)$) compared with logistic regression T-learner estimator of $\tau(x)$. Outcomes Y were distributed Bernoulli with mean as in (30). Randomness in the relative performance of one estimator over the other is due to finite training sample sizes. For $\exp(\beta)$ near 1, the superiority of the logistic regression estimator of $\mu_0(x)$ does not diminish with training sample sizes up to 3000.

proportion of simulation runs where $\hat{\mu}_0(x;\mathcal{S}^{tr})$ was superior to $\hat{\tau}(x;\mathcal{S}^{tr})$ in terms of $\widehat{COR}_\tau(\mathcal{S}^{tr})$. When sample sizes were small, $\hat{\mu}_0(x;\mathcal{S}^{tr})$ was likely to be optimal even for values of $\exp(\beta)$ far from 1. As sample size increased, $\hat{\tau}(x;\mathcal{S}^{tr})$ was more likely to be optimal for values of $0.8 < \exp(\beta) < 1.2$, but inferiority to $\hat{\mu}_0(x;\mathcal{S}^{tr})$ persists for values closer to 1 even when $|\mathcal{S}^{tr}| = 3000$. This example suggests that when the goal is stratification on the CATEs, a variety of estimators for $\mu_0(x)$ and $\tau(x)$ should be considered and compared using an appropriate evaluation approach such as uplift curves and ABUC, at least when Y is binary.

Another metric was proposed by van Klaveren et al. (2018) to evaluate $s(x;\mathcal{S}^{tr})$ for the objective of stratifying a sample on $\tau(x)$, when Y is binary. In terms of the 3-step analysis structure outlined above, their procedure generates the groups in step 2 as pairs of observations from opposite treatment groups matched on $s(x;\mathcal{S}^{tr})$. Letting w index the treated observations in $\mathcal{S}^{te}$, $\{w : t_w = 1\}$, let $m(w)$ indicate the corresponding match from the nontreated group:

$$m(w) = \underset{m:t_m=0}{\arg\min} \ (s(x_i;\mathcal{S}^{tr}) - s(x_m;\mathcal{S}^{tr}))^2 (.)$$

They then calculate an imputed treatment effect for each observation w as $\hat{d}_w = y_w - y_m$, and the predicted score for subject w as $s_w = (s(x_w;\mathcal{S}^{tr}) + s(x_m;\mathcal{S}^{tr}))/2$. The *c-for-benefit* metric is calculated as the proportion of all possible pairs, $\{(w, w') : w \neq w'\}$, for which imputed treatment effects and predicted scores are concordant:

$$\hat{d}_w > \hat{d}_{w'} \text{ and } s_w > s_{w'},$$

or

$$\hat{d}_w < \hat{d}_{w'} \text{ and } s_w < s_{w'}.$$

A similar metric was proposed by Huang et al. (2012), assuming the true $\mu_1(x)$ and $\mu_0(x)$ are correctly specified as a generalized linear model. However, to our knowledge the statistical properties of either metric have not been rigorously studied.

As a closing remark to this section, we note that the procedures in this section all assume randomized treatment assignment. To our knowledge equivalent procedures for use under selection bias have not been explored in the literature.

7.2 Subgroup identification

Randomized experiments, such as those carried out to test a new medical treatment, are often carried out in targeted populations satisfying the characteristics which in theory could confer benefit. However, at the start of many trial programs this population may not be known, and the investigators may seek to learn about the population of patients who may benefit in order to refine the enrollment criteria for future study. This objective, where it is necessary to learn a population of treatment responders adaptively from observed data, is often referred to as *subgroup identification*. This objective is a subset of the broader *subgroup analysis*, which includes the scenario not considered here where the subgroups of interest are predefined. In subgroup identification analyses, uncertainty quantification plays a crucial role. Often, the objective is to find a population in whom the ATE is greater than a predetermined threshold representing a minimum clinical benefit established by the investigators a priori, a claim that must be accompanied by the evaluation of a suitable null hypothesis. Sometimes, but not always, it is of high importance to restrict the representation of the learned population to be as parsimonious as possible. For example, it is much easier for a clinician to identify a patient as belonging to the identified population, and therefore potentially eligible for treatment, if it can be represented as a simple rule (hypertensive patients aged <50) rather than a complex one that requires the use of a computer to interpret. The clinical picture of an individual satisfying a complex rule such as $\hat{\tau}(x; \mathcal{S}^{tr}) > 0.07$ is difficult to communicate, and likely requires a computer for evaluation of $\hat{\tau}(x; \mathcal{S}^{tr})$.

Foster et al. (2011) identify parsimonious subgroups by first predicting CATEs, then further regressing the predicted CATEs onto X using a single decision tree. The likelihood is then considered that any apparent treatment benefit within a given terminal node is not due to chance alone. Morita and Müller (2017) describe a procedure that takes estimates of $\tau(x)$, evaluated for each patient i, as the inputs. They then evaluate a utility function for the set of subgroups in a space of parsimonious subgroup representations. Their proposed utility function penalizes lower sample sizes and higher complexity, and subgroups associated with higher values of the calculated utility function are suggested as likely candidates for follow-up study. A Bayesian approach is taken by Schnell et al. (2018), who seek a *credible subgroup pair $Pr[A \subseteq B \subseteq C|\mathcal{S}^{tr}] > 1 - \alpha$*, where with $1 - \alpha$ posterior probability the subgroup A only contains points in $\mathcal{X}$ where there is evidence of benefit, and the subgroup C contains all points where there is insufficient evidence of no benefit. The target is the unknown subgroup B which is composed of the subspace of $\mathcal{X}$ indicating benefit (e.g., x satisfying $\tau(x) > 0$ when large Y is better for the patient; $\tau(x) < 0$ otherwise). Their procedure takes as inputs draws from the posterior distribution for $\tau(x)$ generated from a Bayesian CATE model.

Li et al. (2016) describe a procedure for subgroup identification methodology similar to that used for uplift curves (see Section 7.1). Instead of calculating a simple estimate of the CATE for each of the subgroups in the sequence in (28), other statistics are calculated that are more relevant to the subgroup identification problem; specifically, they use the data in each group U_q to calculate statistics that are related to the null hypothesis $H_0 : \gamma_q = 0$, or alternatively the one-sided version of this test. These could include a relevant test statistic or P-value for testing this null hypothesis. This idea could be coupled with the CV procedure from Section 7.1 to generate plots from the training set that could be used for selecting a well-performing scoring system for the purposes of subgroup identification; given the best-performing model, the plots could be used to identify a threshold score c such that the optimal subgroup could be identified as observations x_i from the test set satisfying $s(x_i; \mathcal{S}^{tr}) > c$.

As an example, we took a single sample of 500 observations from the same data generating distribution used for the examples in Sections 4.1.2 and 7.1. As in the analysis presented in Fig. 4, we used a Monte Carlo CV procedure with $V = 100$ random splits. Using the data from $U_q^{(v)}$, the qth group in the vth test set, we could calculate a statistic related to our hypothesis test of interest, such as:

- $T_q^{(v)}$: the T-statistic testing the difference in group means in (27), from Satterswaithe's t-test.
- $p_0(T_q^{(v)})$: the P-value resulting from using $T_q^{(v)}$ to test one-sided null hypothesis $H_0 : \gamma_q \leq 0$
- $\mathbb{1}(p_0(T_q^{(v)}) < 0.05)$: An indicator of whether or not P-value $p_0(T_q^{(v)})$ was less than 0.05, the usual threshold for rejecting a null hypothesis.

As before, we can average statistics calculated within the qth group over CV replications indexed by v. Note that doing so will allow us to compare scoring systems on the same plot, since the x-axis values only depend on q. In panel A of Fig. 6, we compare the same 3 models from Fig. 4 by plotting the average $\frac{1}{V}\sum_v T_q^{(v)}$ as a function of the proportion $|U_q^{(v)}|/|\mathcal{S}_v^{CVte}|$, which is invariant across CV replications. As one might expect from Fig. 4, the plot in Fig. 6 suggests that the causal forest model produces scores (black curve) that are more effective in stratifying on $\tau(x)$ compared with the other two models (red and green curves). Panel A also suggests that for the average sample drawn from this population, the subsample with the top 50% of causal forest scores will have the largest T-statistic compared with other subsamples. Selecting a more exclusive group (say, with 20% of the observations in a given sample), might have a larger CATE, but the data in Panel A demonstrates that a lower sample size makes it more difficult to generate evidence in support of the alternative hypothesis that this CATE is greater than 0.

Plotting our statistics in terms of the proportion of the population chosen for membership into the treatment benefit subgroup may be difficult to interpret if what we want to do is identify a score threshold value c to complete a subgroup identification rule of the type $s(x; \mathcal{S}^{tr}) > c$. To facilitate identification of this threshold, we can transform the x-axis as follows. Let $W_c^{(v)}$ be equal to the group $U_q^{(v)}$ from the vth CV split composed of observations x_i

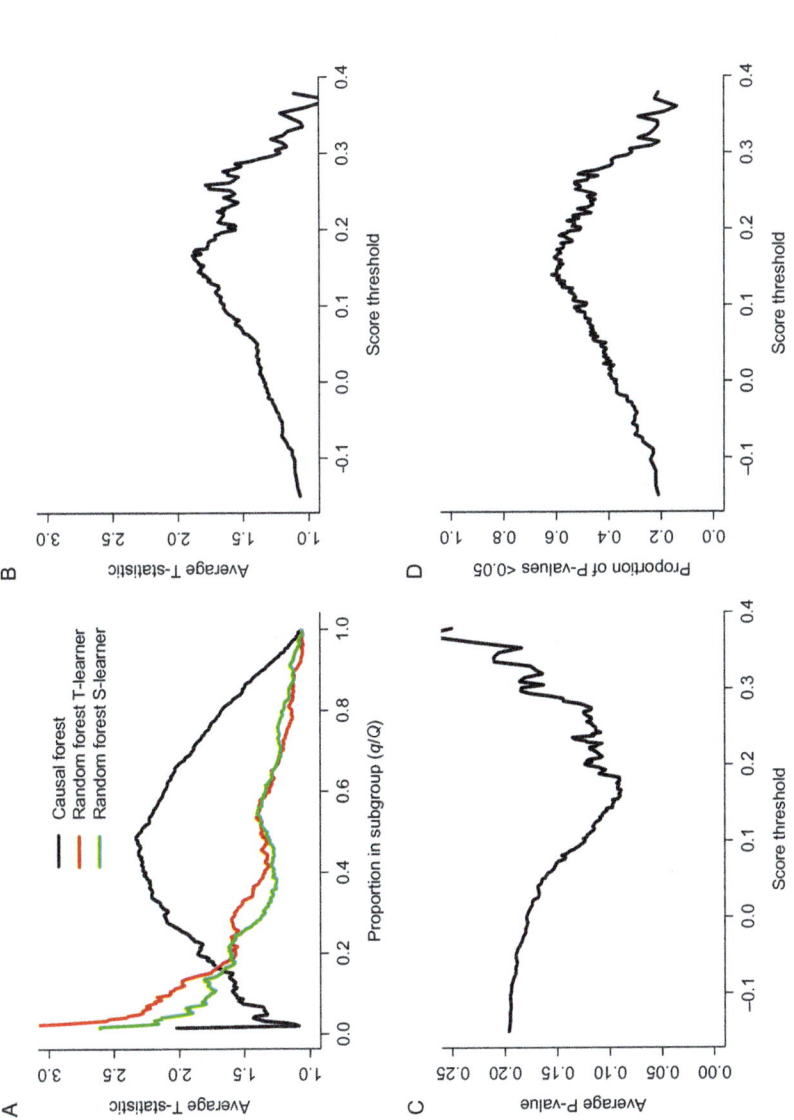

FIG. 6 Example showing the use of Monte Carlo CV for identifying a subgroup of treatment responders, using a training dataset drawn from the same distribution as in Section 7.1. In panel A, 3 scoring systems are compared using the same x-axis as in Fig. 4, but plotting average of $T_q^{(v)}$ over the $V = 100$ CV replicates on the y-axis. In Panel B, the causal forest curve from Panel A is redrawn after transforming the x-axis to be in terms of the minimum causal forest score threshold defining each subgroup, suggesting that the average T-statistic in a new sample will be maximized among subgroups with causal forest scores at around 0.15. Equivalent suggestions are made Panels C and D, which show average P-value and P-value significance indicator over CV replicates, respectively for the subgroups indexed by threshold values on the x-axis.

whose scores $s(x_i; S^{tr})$ are greater than c. Then, for each value of c, we can generate and plot the average of statistics calculated in $W_c^{(v)}$ for $v = 1, ..., V$. Correspondingly, let $T_c^{(v)}$, $p_0(T_q^{(v)})$, and $\mathbf{1}(p_0(T_q^{(v)}) < 0.05)$ be the T-statistic, P-value, and statistically significant P-value indicator statistics calculated from the data in group $W_c^{(v)}$. Panel B shows that in our example the T-statistic is approximately maximized by taking a subgroup with scores no lower than $c = 0.15$. In Panel C we show the average $p_0(T_q^{(v)})$ as a function of the score threshold c. Again, it shows that the P-value is approximately minimized in subgroups with threshold scores around $c = 0.15$. Finally, in Panel D we show the proportion of hypothesis tests over all CV splits that led to rejection, as a function of threshold value c; it, too, suggests taking $c = 0.15$ leads to a subgroup for which the chances of null hypothesis rejection are highest.

Once a scoring system $s(x; S^{tr})$ and accompanying score threshold value c are identified in the training set, a confirmatory analysis can be performed in the test set. The identified scoring system is first derived from the entire training set. Then, in the test set, the observations i satisfying $s(x; S^{tr}) > c$ would be identified, and a confirmatory test of the hypothesis of no treatment effect would be carried out in this group, for example via one-sided t-test. Such a confirmatory analysis is necessary for control of Type I error.

We also remark here, as in Section 7.1, that to our knowledge the subgroup identification problem has not been extensively discussed in the literature for settings involving selection bias.

8 Summary

In this chapter we have presented a relatively high-level discussion of CATE estimation procedures. We showed empirically how several naive strategies can fail to produce the best estimators, and review the current state of the literature regarding model training and selection strategies tailored to the CATE estimation problem. Finally, we covered some higher-level applications of CATE models that target specific features of the CATE function, such as the ranks of the true CATEs for the observations in a test set and a confirmable subgroup of treatment responders.

We have focused our discussion on modeling strategies for use in settings where the treatment option is binary, the unfoundedness assumption is met, and the outcome is a scalar. For settings with nonbinary treatment options, see Grimmer et al. (2017) and Li et al. (2019). Athey et al. (2019) describe a version of the GRF methodology that can be used with an instrumental variable to estimate heterogeneous local average treatment effects in settings where there are suspected to be unobserved confounders. Other methods for use when there are unobserved or latent confounders are described in Louizos et al. (2017) and Lee et al. (2018).

References

Abadie, A., Chingos, M.M., West, M.R., 2018. Endogenous stratification in randomized experiments. Rev. Econ. Stat. 100 (4), 567–580.

Alaa, A.M., van der Schaar, M., 2018. Bayesian nonparametric causal inference: information rates and learning algorithms. IEEE J. Sel. Top. Sign. Proces. 12 (5), 1031–1046.

Alaa, A.M., van der Schaar, M., 2019. Validating causal inference models via influence functions. In: Proceedings of the 34th International Conference on Machine Learning.

Alaa, A.M., Weisz, M., Van Der Schaar, M., 2017. Deep counterfactual networks with propensity-dropout. arXiv preprint arXiv:1706.05966.

Arlot, S., Celisse, A., 2010. A survey of cross-validation procedures for model selection. Stat. Surv. 4, 40–79. https://doi.org/10.1214/09-SS054.

Athey, S., Imbens, G., 2016. Recursive partitioning for heterogeneous causal effects. Proc. Natl. Acad. Sci. U.S.A. 113 (27), 7353–7360.

Athey, S., Tibshirani, J., Wager, S., 2019. Generalized random forests. Ann. Stat. 47 (2), 1148–1178.

Bengio, Y., Courville, A., Vincent, P., Washington, DC, USA, 2013. Representation learning: a review and new perspectives. IEEE Trans. Pattern Anal. Mach. Intell. 35 (8), 1798–1828. https://doi.org/10.1109/TPAMI.2013.50.

Bickel, S., Brückner, M., Scheffer, T., Zadrozny, B., 2009. Discriminative learning under covariate shift. J. Mach. Learn. Res. 10, 2137–2155.

Breiman, L., Friedman, J., Stone, C.J., Olshen, R.A., 1984. Classification and Regression Trees. Taylor & Francis Group.

Chen, T., Guestrin, C., 2016. XGBoost: a scalable tree boosting system. In: KDD '16. Proceedings of the 22Nd ACM SIGKDD International Conference on Knowledge Discovery and Data Mining, New York, NY, USAACM, pp. 785–794.

Chernozhukov, V., Demirer, M., Duflo, E., Fernandez-Val, I., 2018. Generic machine learning inference on heterogenous treatment effects in randomized experiments. National Bureau of Economic Research.

Chipman, H.A., George, E.I., McCulloch, R.E., 2010. BART: Bayesian additive regression trees. Ann. Appl. Stat. 4 (1), 266–298.

Denuit, M., Sznajder, D., Trufin, J., 2019. Model selection based on Lorenz and concentration curves, Gini indices and convex order. Insurance Math. Econom. 89, 128–139.

Foster, J.C., Taylor, J.M.G., Ruberg, S.J., 2011. Subgroup identification from randomized clinical trial data. Stat. Med. 30 (24), 2867–2880.

Friedberg, R., Tibshirani, J., Athey, S., Wager, S., 2019. Local linear forests. arXiv:1807.11408v4 [stat.ML].

Goodfellow, I., Bengio, Y., Courville, A., 2016. Deep Learning. MIT Press. http://www.deeplearningbook.org.

Grimmer, J., Messing, S., Westwood, S.J., 2017. Estimating heterogeneous treatment effects and the effects of heterogeneous treatments with ensemble methods. Polit. Anal. 25 (4), 413–434.

Gutierrez, P., Gerardy, J.Y., 2016. Causal inference and uplift modeling: a review of the literature. JMLR: Workshop Conf. Proc. 67, 1–13.

Hahn, P.R., Carvalho, C.M., Puelz, D., He, J., 2018. Regularization and confounding in linear regression for treatment effect estimation. Bayesian Anal. 13 (1), 163–182. https://doi.org/10.1214/16-BA1044.

Hahn, P.R., Murray, J.S., Carvalho, C., 2019. Bayesian regression tree models for causal inference: regularization, confounding, and heterogeneous effects. arXiv preprint arXiv:1706.09523v4 [stat.ME].

Hastie, T.J., Tibshirani, R.J., 1990. Generalized Additive Models. Chapman & Hall, London, . ISBN: 0412343908335.

Hastie, T., Tibshirani, R., Friedman, J., 2009. The Elements of Statistical Learning, Second Edition. Springer Series in StatisticsSpringer New York Inc., New York, NY, USA

Hill, J.L., 2011. Bayesian nonparametric modeling for causal inference. J. Comput. Graph. Stat. 20 (1), 217–240.

Hirano, K., Imbens, G.W., Ridder, G., 2003. Efficient estimation of average treatment effects using the estimated propensity score. Econometrica 71 (4), 1161–1189.

Hofman, J.M., Sharma, A., Watts, D.J., 2017. Prediction and explanation in social systems. Science 355 (6324), 486–488. https://doi.org/10.1126/science.aal3856.

Huang, Y., Gilbert, P.B., Janes, H., 2012. Assessing treatment-selection markers using a potential outcomes framework. Biometrics 68 (3), 687–696.

Imai, K., Ratkovic, M., 2013. Estimating treatment effect heterogeneity in randomized program evaluation. Ann. Appl. Stat. 7 (1), 443–470.

Imbens, G., 2004. Nonparametric estimation of average treatment effects under exogeneity: a review. Rev. Econ. Stat. 86 (1), 4–29.

Ioannidis, J.P.A., Lau, J., 1997. The impact of high-risk patients on the results of clinical trials. J. Clin. Epidemiol. 50 (10), 1089–1098.

James, G., Witten, D., Hastie, T., Tibshirani, R., 2014. An Introduction to Statistical Learning: With Applications in R. Springer Publishing Company, Incorporated ISBN 1461471370, 9781461471370.

Jaskowski, M., Jaroszewicz, S., 2012. Uplift modeling for clinical trial data. In: ICML 2012 Workshop on Clinical Data.

Johansson, F.D., Shalit, U., Sontag, D., 2018. Learning representations for counterfactual inference. In: Proceedings of the 33rd International Conference on Machine Learning.

Kent, D.M., Hayward, R.A., 2007. Limitations of applying summary results of clinical trials to individual patients: the need for risk stratification. JAMA 298 (10), 1209–1212.

Kent, D.M., Steyerberg, E., van Klaveren, D., 2018. Personalized evidence based medicine: predictive approaches to heterogeneous treatment effects. BMJ 363, k4245.

Knaus, M., Lechner, M., Strittmatter, A., 2018. Machine learning estimation of heterogeneous causal effects: Empirical Monte Carlo evidence.

Künzel, S.R., Sekhon, J.S., Bickel, P.J., Yu, B., 2019. Metalearners for estimating heterogeneous treatment effects using machine learning. Proc. Natl. Acad. Sci. U.S.A. 116 (10), 4156–4165.

Lee, C., Mastronarde, N., van der Schaar, M., 2018. Estimation of individual treatment effect in latent confounder models via adversarial learning. arXiv preprint arXiv:1811.08943.

Li, J., Zhao, L., Tian, L., Cai, T., Claggett, B., Callegaro, A., Dizier, B., Spiessens, B., Ulloa-Montoya, F., Wei, L.J., 2016. A predictive enrichment procedure to identify potential responders to a new therapy for randomized, comparative controlled clinical studies. Biometrics 72, 877–887.

Li, C., Yan, X., Deng, X., Qi, Y., Chu, W., Song, L., Qiao, J., He, J., Xiong, J., 2019. Reinforcement learning for uplift modeling. arXiv:1811.10158v2.

Louizos, C., Shalit, U., Mooij, J.M., Sontag, D., Zemel, R., Welling, M., 2017. Causal effect inference with deep latent-variable models. In: Advances in Neural Information Processing Systems6446–6456.

Lu, M., Sadiq, S., Feaster, D.J., Ishwaran, H., 2018. Estimating individual treatment effect in observational data using random forest methods. J. Comput. Graph. Stat. 27 (1), 209–219.

Morgan, S.L., Winship, C., 2007. Counterfactuals and causal inference. Cambridge University Press.

Morita, S., Müller, P., 2017. Bayesian population finding with biomarkers in a randomized clinical trial. Biometrics 73 (4), 1355–1365.

Nie, X., Wager, S., 2017. Quasi-oracle estimation of heterogeneous treatment effects. arXiv preprint arXiv:1712.04912.

Powers, S., Junyang, Q., Jung, K., Schuler, A., Shah, N.H., Hastie, T., Tibshirani, R., 2018. Some methods for heterogeneous treatment effect estimation in high dimensions. Stat. Med. 37 (11), 1767–1787.

Radcliffe, N.J., 2007. Using control groups to target on predicted lift: building and assessing uplift models. Direct Mark. Anal. J. 1, 1421.

Radcliffe, N.J., Surry, P.D., 2011. Real-world uplift modelling with significance-based uplift trees. In: White Paper TR-2011-1, Stochastic Solutions. Citeseer.

Robins, J.M., Rotnitzky, A., 1995. Semiparametric efficiency in multivariate regression models with missing data. J. Am. Stat. Assoc. 90 (429), 122–129.

Rolling, C.A., Yang, Y., 2014. Model selection for estimating treatment effects. J. R. Stat. Soc. B (Statistical Methodology) 76 (4), 749–769.

Rubin, D.B., 1974. Estimating causal effects of treatment in randomized and non-randomized studies. J. Educ. Psychol. 66 (5), 688–701.

Rubin, D.B., 1977. Assignment to treatment group on the basis of a covariate. J. Educ. Stat. 2 (1), 1–26.

Rubin, D.B., 2005. Causal inference using potential outcomes. J. Am. Stat. Assoc. 100 (469), 322–331. https://doi.org/10.1198/016214504000001880.

Schnell, P.M., Müller, P., Tang, Q., Carlin, B.P., 2018. Multiplicity-adjusted semiparametric benefiting subgroup identification in clinical trials. Clin. Trials 15 (1), 75–86.

Schuler, A., Baiocchi, M., Tibshirani, R., Shah, N., 2018. A comparison of methods for model selection when estimating individual treatment effects. arXiv:1804.05146v2.

Shalit, U., Johansson, F.D., Sontag, D., 2017. Estimating individual treatment effect: generalization bounds and algorithms. In: Proceedings of the 34th International Conference on Machine Learning, Vol. 70.

Simon, H.A., 2002. Science seeks parsimony, not simplicity: searching for pattern in phenomena. In: Zellner, A., Keuzenkamp, H.A., McAleer, M. (Eds.), Simplicity, Inference and Modelling: Keeping It Sophisticatedly Simple. Cambridge University Press, pp. 32–72.

Sugasawa, S., Noma, H., 2019. Estimating individual treatment effects by gradient boosting trees. Stat. Med. 38, 5146–5159.

Tian, L., Alizadeh, A.A., Gentles, A.J., Tibshirani, R., 2014. A simple method for estimating interactions between a treatment and a large number of covariates. J. Am. Stat. Assoc. 109 (508), 1517–1532.

van Klaveren, D., Steyerberg, E.W., Serruys, P.W., Kent, D.M., 2018. The proposed 'concordance-statistic for benefit' provided a useful metric when modeling heterogeneous treatment effects. J. Clin. Epidemiol. 94, 59–68.

Xie, Y., 2013. Population heterogeneity and causal inference. PNAS 110 (16), 6262–6268.

Yang, Y., 2007, 12. Consistency of cross validation for comparing regression procedures. Ann. Stat. 35 (6), 2450–2473. https://doi.org/10.1214/009053607000000514.

Yao, L., Li, S., Li, Y., Huai, M., Gao, J., Zhang, A., 2018. Representation learning for treatment effect estimation from observational data. In: Advances in Neural Information Processing Systems2633–2643.

Yoon, J., Jordon, J., van der Schaar, M., 2018. GANITE: estimation of individualized treatment effects using generative adversarial nets. In: International Conference on Learning Representations (ICLR).

Zhao, L., Tian, L., Cai, T., Claggett, B., Wei, L.J., 2013. Effectively selecting a target population for a future comparative study. J. Am. Stat. Assoc. 108 (502), 527–539.

Chapter 7

Nonparametric data science: Testing hypotheses in large complex data

Sunil Mathur*

College of Science and Engineering, Texas A&M University-Corpus Christi, Corpus Christi, TX, United States
*Corresponding author: e-mail: Sunil.Mathur@Tamucc.edu

Abstract

Big data contains very large, structured, or unstructured data sets, requiring novel statistical techniques to extract typically not well-defined parameters. The availability of massive amounts of complex data sets has provided challenges and opportunities to process and analyze the data, which is difficult using traditional data processing techniques. New protocols and methods are needed not only to record, store, and analyze the live streaming massive data sets but also to develop new analytical tools for testing hypotheses to gain novel insights and discoveries from systems that were previously not understood. There is a need to establish a clear path, and create and implement innovative new approaches, which are not distribution dependent to increase the understanding of complex large datasets. Data analysis is a challenge due to the lack of scalability of the underlying algorithms and the complexity of the data. Existing statistical tools, most of them are developed to draw inference from incomplete information available, have not been able to keep up with the speed of advancements in modern technologies generating a massive amount of continuous streaming data. The new approaches based on nonparametric methods have capabilities to yield transformational changes in biomedical research; integrate with next-generation technology platforms that can accelerate scientific discovery; use data ecosystems based on the data generated by researchers; and facilitate harmonization of data, methods, and technologies; and provide cutting-edge theory-based nonparametric methods in advanced computing environments. Each upgrade to a larger length scale increases variability and volume, which will eventually generate a rich data landscape that must be analyzed by cutting-edge analytical tools using both structured and general data-mining novel approaches in a continuous processing mode. The nonparametric analytical tools and concepts are needed to analyze such massive data to keep up with rapidly growing technology and which can also be used in the analysis of continuous streaming big data.

Handbook of Statistics, Vol. 44. https://doi.org/10.1016/bs.host.2020.10.004

Keywords: Nonparametric methods, Ranks, Samples, Efficient, Ranked-set, Predictions, Big data

1 Introduction

With the advances in mobile phones, satellites, imaging devices, and financial services, data is generated at an ever-increasing rate, and data is automatically generated. The vast amount of data generated needs to be analyzed for making an inference in real-time with the utmost accuracy in most of the cases. In the case of national security, data collected from satellites, drones, and other devices, misfit models and wrong inference could cost millions of lives in a matter of seconds. In many cases, the vast amount of data collected, such as in case of epidemics may not conform to traditional data structures and pose challenges in making useful interpretations. In most cases, more and more data is added to the same unit of observation over time, for example, credit card purchases by a person. Now consider millions of people using credit cards every minute issued by a company. Though the data is increasing over time for a particular person the data is also becoming complex, huge, and increasing at a high speed as the purchases are made worldwide. The volume of data has gained the most attention, however, one needs to consider the variety of data and velocity of data. If we breakdown these three attributes (Fig. 1), volume, variety, and velocity, it turns out that we are talking about large samples (MB, GB, TB, PB), large variety (attributes, tables, photo, web, audio, social, video, unstructured) and data addition rate (batch, periodic, near real-time, real-time). In addition to that, all three attributes are expanding over time.

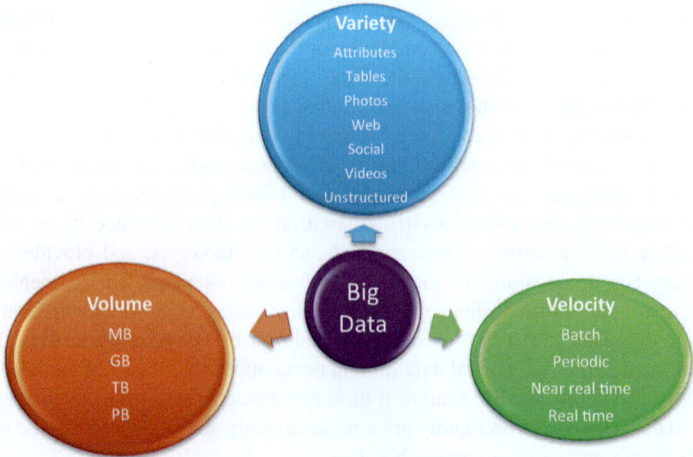

FIG. 1 Big data attributes: volume, variety, and velocity.

There is a need to develop statistical methods that can be applied to the data having big volume, variety, and velocity. Classical statistical inference methods use mean and covariance matrix to make inferences about the population parameter or prediction. These techniques require the assumption of normality. In big data with where volume, variety, and velocity are huge and expanding, it is unreasonable to assume normality of the distribution. Therefore, the classical statistical methods perform poorly in terms of their efficiency when the population distribution is heavy-tailed, skewed, and there are outliers present in the data which often the case in big data. One could argue that since the data is big, one may not need to follow the parametric form, and does not the assumption of normality and still use classical procedures. However, the dimensionality, variety, variability, and velocity of the data make the methods currently available unsuitable for analysis and interpretation of the data. Therefore, the existing methods need to be modified or new methods need to be developed for big data. Nonparametric methods can be modified easily for big data without worrying about assumptions for the statistical procedures. In this chapter, we present some of the nonparametric methods adapted/modified from existing literature or new methods that can be applied easily to big data. We call such methods as nonparametric data science to indicate that the statistical procedures presented could be applied without worrying about the parametric distribution assumptions and these methods have their origin in nonparametric statistics. Also, several of the methods applied to analyze big data require breaking down the big data into smaller groups/clusters before attempting to make sense of the data. Once they are broken down into smaller groups/cluster methods, the distributional assumptions would be required.

In Section 2, we present an overview of the nonparametric data science, its relevance, and its importance in analyzing and interpreting with respect to big data. In this chapter, we introduce nonparametric univariate, single sample, multisample, and multivariate statistical methods that can be used to make inferences about the big data. We present, one-sample nonparametric ss analysis in Section 3. In Section 4, we present methods that are useful when two samples are available and we have a univariate situation. In Section 5, we present multisample univariate procedures that can be applied to big data. Since the big data could have multi-sample and many variates to be considered before reaching a conclusion/making an interpretation of data, we present some of the methods in Section 6, which could be applied in those situations. In Section 7, we present rank-based methods which could be very useful in big data s it helps to reduce data but still preserving the distribution and other properties of data. In Section 8, we discuss detecting changepoints which has wide applications in big data related to climatology, bioinformatics, finance, oceanography, and medical imaging. In the last section, Section 9, we present our final remarks on how nonparametric data science can be used in analyzing big data and how it can be used to identify useful patterns in a decision-making process.

2 Nonparametric data science

Most statistical tests discussed so far assume that the distribution of the parent population is known. Almost all the exact tests (small sample tests) are based on the assumption that the population under consideration is having the normal distribution. The problem occurs when the distribution of the population is unknown. For example, gene expression data are not normal, socio-economic data, in general, are not normal. Similarly, the data in Psychometrics, Sociology, and Educational Statistics are seldom distributed as normal. In big data, though the sample size is too large, the data structure may not be normal. Fluctuation in data due to inherent variability in the big data is so high that it is difficult to verify the normality conditions. Big data no longer follow a defined structure and the data comes in a variety of structures. Thus, it is not possible to have a defined data structure for different applications. Moreover, big data has high velocity and that makes it unstable to verify the normality conditions. In these cases, where the parent population distribution is unknown, we nonparametric procedures work well as the nonparametric procedures make fewer and much weaker assumptions than those associated with the parametric tests.

The advantages of the nonparametric procedures over the parametric procedures are that the nonparametric procedures are readily comprehensible, very simple, and easy to apply. No assumption is made about the distribution of the parent population. If the data are mere classification, there is no parametric test procedure available to deal with the problem but there are several nonparametric procedures available. If the data are only given in terms of ranks, then only nonparametric procedures can be applied. But if all the assumptions for the parametric procedure are satisfied then the parametric procedures are more powerful than the nonparametric procedures, however, in the case of big data, it is hard to verify the normality conditions. Thus, in big data it is unreasonable to assume that the parent population is normal, therefore, we find that the nonparametric methods are very useful in analyzing the data. Since nonparametric statistical methods do not require the assumption of normal distribution and can work with big data having big volume, variety, and velocity.

3 One-sample methods

In big data, many times a single stream of data is collected on a unit of interest. The data might be collected in batches, periodic, near-real-time, or real-time. As the analysis of big data evolves, it is necessary to build models that integrate different models for developing applications of certain needs. In the case of streaming data in real-time, the processing of real-time data may be followed by batch processing. That gives rise to categories of data due to batches of data. Some of the tests available in the literature are based on the

empirical distribution function. The empirical distribution function is an estimate of the population distribution function, which works for big data. It is
defined as the proportion of sample observations that are less than or equal
to x for all real numbers x.

We consider the classical one-sample location problem with univariate
data. Let x_1, x_2, ..., x_n be an independent random sample of size n from a
continuous distribution with distribution function F. Let the hypothesized
cumulative distribution function be denoted by $F_0(x)$ and the empirical distribution function be denoted by $S_n(x)$ for all x. The hypothesis to be tested is
$H_0 : F = F_o$ vs $H_a : F \neq F_o$. If the null hypothesis is true then the difference
between $S_n(x)$ and $F_0(x)$ must be close to zero.

Thus, for large n, the test statistic

$$D_n = \sup_x |S_n(x) - F_X(x)|, \tag{1}$$

will have a value close to zero under the null hypothesis.

The test statistic, D_n, called the *Kolmogorov-Smirnov one-sample statistic*
(Gibbons and Chakraborti, 2014), does not depend on the population distribution function if the distribution function is continuous and hence D_n is a
distribution-free test statistic. The goodness-of-fit test for a sample was proposed by Kolmogorov (1933). The Kolmogorov-Smirnov test for two samples
was-proposed by Smirnov (1939).

Here we define order statistic $X_{(0)} = -\infty$ and $X_{(n+1)} = \infty$, and

$$S_n(x) = \frac{i}{n}, \quad for \ X_{(i)} \leq x \leq X_{(i+1)}, i = 0, 1, \ldots, n. \tag{2}$$

The probability distribution of the test statistic does not depend on the distribution function $F_X(X)$ for a continuous distribution function $F_X(X)$. The
asymptotic distribution of the test statistic D_n is Chi-square.

The exact sampling distribution of the Kolmogorov-Smirnov test statistic is
known while the distribution of the Chi-square goodness-of-fit test statistic is
approximately Chi-square for finite n. Moreover, the Chi-square goodness-of-
fit test requires that the expected number of observations in a cell must be greater
than five while the Kolmogorov test statistic does not require this condition. On
the other hand, the asymptotic distribution of the Chi-square goodness-of-fit test
statistic does not require that the distribution of the population must be continuous but the exact distribution of the Kolmogorov-Smirnov test statistic does
require that $F_X(X)$ must be a continuous distribution. The power of the Chi-
square distribution depends on the number of classes or groups made.

The Wilcoxon signed-rank test (Wilcoxon, 1945) requires that the parent population should be symmetric. When data is collected in batches, the data might be
symmetric at some point, particularly in the case of seasonal and periodic data.
Let us consider a random sample $X_1, X_2, \ldots, X_n$ from a continuous cdf F which
is symmetric about its median M. The null hypothesis can be stated as

$$H_0 : M = M_0 \tag{3}$$

The alternative hypotheses can be postulated accordingly. We notice that the differences $D_i = X_i - M_0$ are symmetrically distributed about zero, and hence the number of positive differences will be equal to the number of negative differences. The ranks of the differences $|D_1|, |D_2|, \ldots\ldots\ldots\ldots, |D_N|$ are denoted by Rank(.). Then, the test statistic can be defined as

$$T^+ = \sum_{i=1}^{n} a_i Rank(|D_i|) \tag{4}$$

$$T^- = \sum_{i=1}^{n} (1 - a_i) Rank(|D_i|) \tag{5}$$

where

$$a_i = \begin{cases} 1, & if D_i > 0, \\ 0, & if D_i < 0. \end{cases} \tag{6}$$

Since the indicator variables a_i are independent and identically distributed Bernoulli variates with $P(a_i = 1) = P(a_i = 0) = \frac{1}{2}$, therefore, under the null hypothesis

$$E(T^+|H_0) = \sum_{i=1}^{n} E(a_i) Rank|D_i| = \frac{n(n+1)}{4}. \tag{7}$$

and

$$Var(T^+|H_0) = \sum_{i=1}^{n} var(a_i) [Rank|D_i|]^2 = \frac{n(n+1)(2n+1)}{24} \tag{8}$$

Another common representation for the test statistic T^+ is given as follows.

$$T^+ = \sum_{1 \le i \le j \le n} \sum T_{ij} \tag{9}$$

where

$$T_{ij} = \begin{cases} 1, & if \ D_i + D_j > 0, \\ 0, & otherwise. \end{cases} \tag{10}$$

Similar expressions can be derived for T^-. The paired-samples can be defined based on the differences $X_1 - Y_1, X_2 - Y_2, \ldots, X_n - Y_n$ of a random sample of n pairs $(X_1, Y_1), (X_2, Y_2), \ldots, (X_n, Y_n)$. Now, these differences are treated as a single sample and the one-sample test procedure is applied. The null hypothesis to be tested will be

$$H_0 : M = M_0$$

where M_0 is the median of the differences $X_1 - Y_1, X_2 - Y_2,....., X_n - Y_n$. These differences can be treated as a single sample with the hypothetical median M_0. Then, the Wilcoxon signed-rank method described above for a single sample can be applied to test the null hypothesis that the median of the differences is M_0.

Since a good test must be not only fast in computing the test value but also should have the ability in finding out information hidden in big data. Wilcoxon signed-rank test fulfills that requirement, however, several other tests are available in the literature that are competitors of the Wilcoxon signed-rank test.

Chattopadhyay and Mukhopadhyay (2019) used the kernel of degree k (>1) to develop a one-sample nonparametric test. Define a kernel (k=2):

$$\psi^{(2)}(X_i, X_j) = \begin{cases} 1 & if \ \dfrac{X_i + X_j}{2} \geq 0 \\ 0 & if \ \dfrac{X_i + X_j}{2} < 0 \end{cases} \tag{11}$$

This kernel is equivalent to U-Statistic of degree 2

$$S_n^{(2)} = \binom{n}{2}^{-1} \sum_{1 \leq i_1 < i_2 \leq n} \psi^{(2)}(X_{i_1}, X_{i_2}) \tag{12}$$

Both the sign test and the Mann-Whitney test involve U-statistics with a symmetric kernel of degree one, one, and one respectively.

A general test statistic (Chattopadhyay and Mukhopadhyay, 2019) based on the kernel of k (<n) can be defined as:

$$S_n^{(k)} = \binom{n}{k}^{-1} \sum_{1 \leq i_1 < ... < i_k \leq n} \psi^{(k)}(X_{i_1}, ..., X_{i_k}) \tag{13}$$

where

$$\psi^{(k)}(X_{i_1}, ..., X_{i_k}) = \begin{cases} 1 & if \ \overline{X}_{i_k} \geq 0 \\ 0 & if \ \overline{X}_{i_k} < 0 \end{cases} \tag{14}$$

and $\overline{X}_{i_k} = \frac{1}{k} \sum_{j=1}^{k} X_{i_j}, \ 1 \leq i_1 < ... < i_k \leq n$ for $n > k$.

The null hypothesis, as given by statement (3), can be tested at a level α using the following criterion:

Reject the null hypothesis if $S_n^{(k)} > c_\alpha$, where $P_{H_o}(S_n^{(k)} > c_\alpha) \leq \alpha$, and c_α is the critical region.

In big data scenarios, face recognition has received significant attention due to increasing attention to security at public places such as airports, rail stations, and similar places. A single sample is generally received from an ID card or e-passport, captured in a very stable environment while probe images are captured in a highly unstable environment usually from surveillance cameras.

The probe images may include noise, blur, arbitrary pose, and illumination, which makes the comparison with the standard database image difficult and hence makes the recognition difficult. The performance of available computational methods based on principal component analysis, linear discriminant analysis, sparse representation, kernel-based and similar methods in face recognition, however, is heavily influenced by the number of training samples per person.

Since there is only one sample available for such problems, we try to increase the number of samples artificially using synthetic sample generation from a 3D model of the available image. The new dataset with multiple artificially generated samples can be used as a gallery set and the probe set contains images from surveillance cameras in an unconstrained environment. Now one can select 2D facial points and the landmark points in the 3D model in the gallery set and find median points at each landmark point. Similarly, select those points in the probe set and run the one-sample test, such as Eq. (9) at each landmark point. The larger similarities at landmark points will point toward similarities of probe and gallery and probe images.

The problem is generally faced when the interest is in identifying an individual at busy common places such as airports, train stations, and public meeting places. That will involve matching the gallery set with a probe set containing millions for images. In order to accomplish that task quickly and efficiently, one can set up three layers of batch processing. The first layer of processing involves eliminating the data which has more than two standard deviations of variations in major landmarks. Thus, the probe data containing a wider face or too small face will get eliminated. In the second layer, matching of finer landmarks such as ear length and width, nose length is done. The data having two standard deviations of variations in those landmarks is eliminated. Thus, the remaining data will be a lot easier to handle with the batch processing method (Fig. 2).

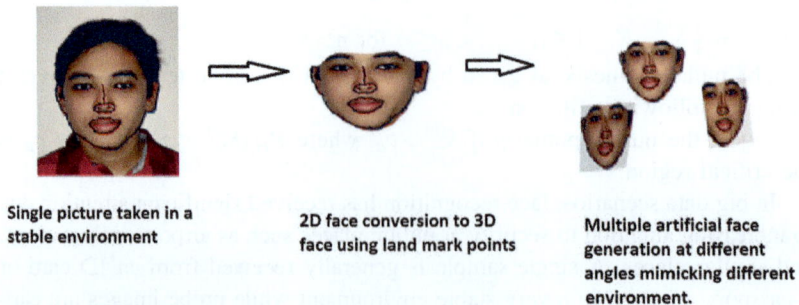

Single picture taken in a
stable environment

2D face conversion to 3D
face using land mark points

Multiple artificial face
generation at different
angles mimicking different
environment.

FIG. 2 Converting 2D facial model to 3D model and increasing virtual sample size for gallery data.

4 Two-sample methods

Let us consider two independent random samples X_1, X_2,...., X_m and Y_1, Y_2,...., Y_n from two populations X and Y respectively. Let the populations X and Y have CDFs F_X and F_Y respectively. The null hypothesis, which we would like to test, is that

$$H_0 : F_Y(x) = F_X(x) \text{ for all } x,$$

against

$$H_A : F_Y(x) = F_X(x+\theta) \text{ where } \theta \neq 0.$$

Here we assume that the populations are the same except for a possible shift in location θ. Thus, populations have the same shape and scale parameter and the difference in the population location parameter is equal to zero under the null hypothesis. We have $m + n$ random variables which can be arranged in $\binom{m+n}{m}$ ways. This sampling pattern will provide information about the difference in the location between two populations if any. Most of the tests available in the literature for detecting the shift in the location are based on some type of function of combined arrangements of two samples.

The Wilcoxon rank-sum test statistic is calculated by using the combined ordered samples. We order combined samples of size N=m+n, of X-values and Y-values from least to greatest or greatest to least value. Now we determine the ranks of X-value in the combined sample and let the rank of X_1 be R_1, X_2 be R_2, ..., X_m be R_m in the combined ordered sample. Then, the Wilcoxon rank-sum test is given by

$$W = \sum_{i=1}^{m} R_i \tag{15}$$

To test the null hypothesis that the location parameter θ is zero, we set up the null hypothesis as

$$H_0 : \theta = 0,$$

Against

$$H_A : \theta > 0.$$

We reject the null hypothesis if $W \geq w_\alpha$, where w_α is a constant chosen to make the type I error probability equal to α, the level of significance.

For the alternative hypothesis

$$H_A : \theta < 0,$$

we reject the null hypothesis if

$$W \leq n(m + n + 1) - w_\alpha \tag{16}$$

For the alternative hypothesis

$$H_A : \theta \neq 0,$$

we reject the null hypothesis if

$$W \geq w_{\frac{\alpha}{2}} \text{ or } W \leq n(m + n + 1) - w_{\alpha/2} \tag{17}$$

The direction of a vector depends on its slope. Any vector and vector obtained by rotation of 180° are not equally likely. Therefore, if the median of a population is different from the specified one then the vectors from origin having the same slope will have preferred direction. Mathur and Sepehrifar (2013) suggested a signed-ranked test by arranging the angles in ascending order with respect to the positive horizontal axis. To make the test statistic independent of correlation between two variates, they changed the slope of adjacent vectors so that angle between vectors represented by adjacent slopes is 180/n degrees. The intersection between the unit circle and the vectors is represented by the changed slope. Under the null hypothesis, the mean value of this circle, that is center of gravity, should be near to zero. If the underlying population is elliptically symmetric, then it is logical to measure the slopes of the line of each observation from the origin and to use the ranks of these slopes along with the observations' directions in forming a test statistic.

This problem arises when researchers have two sets of data and would like to test whether both samples belong to the same population. In one sample procedure, we have a photo used in an ID card or e-passport which we need to match with a set of probe images that are captured in a highly unstable environment using equipment such as surveillance cameras. In the two-sample problem, we have two sets of data from two different locations/time and we aim to investigate whether it is the same person at the two locations/time. Sample A is the gallery set captured using a surveillance camera and the probe set was captured using a surveillance camera at different locations/time, independent of the gallery set. Now use the techniques such as facial landmark detection and alignment that consider realistic facial geometric information to generate data points on the gallery set as well as probe set. Now using the 2D facial points and the landmark points in the 3D model in the gallery set and probe set, combine both gallery set and probe set data points, keeping track of each set. We order combined samples of gallery and probe set data points from least to greatest or greatest to least value. Now we determine the ranks of gallery data points value in the combined sample and let the rank of X_1 be R_1, X_2 be R_2, ..., X_m be R_m in the combined ordered sample. Then, calculate the Wilcoxon rank-sum test is given by Eq. (1). The larger similarities at landmark points will point toward similarities of probe and gallery images.

The identification of an individual at two different places/times will involve matching the gallery set with a probe set containing millions for images. As explained earlier, one can breakdown the task into batch processing, say three batches. First batch processing involves eliminating the data which has more than two standard deviations of variations in major landmarks. This may include the elimination of probe data containing a wider face or too small face. In the second batch processing, matching of finer landmarks such as ear length and width, nose length is done. In this stage, the data having two standard deviations of variations in those landmarks is eliminated. This will lead to the remaining data that will be easy to handle.

5 Multi-sample methods

In biometric recognition, the major issue is to overcome the security issues associated with traditional person recognition methods based on the ID card or passwords. Now multiple methods or samples are needed to establish the identity of a person to ensure that the right person is accessing the facility or system. Several biometric traits have been used such as hand-shape and palm-print. The multi-biometric approaches can be based on multi-modal systems that explore the combination of different traits, a multi-sensor system that uses the samples from different devices, and multi-instances which use some biometric traits present in several regions of the human body such as ten fingerprints from the two hands. This would also lead to the use of several samples as a query set instead of a single sample to increase the amount of information contained in each sample to have better efficiency and accuracy. With the ever-increasing threat of a pandemic, it is necessary to build a multi-sampling acquisition device for contactless biometric data collection. We look at the 3D face recognition in the biometric contest, we notice that 3D faces are less affected by several factors such as pose variation, lights, and surrounding which are limiting factors for the 2D image-based approaches. If we combine both 2D and 3D images, we find improvements in recognition accuracy. The 3D models are useful for both tracking and recognition purposes. For the 3D model, one can use a multi-sample approach in which a frame pair from the gallery and probe sets are arranged such that they have the same or similar pose angle using the multi-frame fusion.

In a different scenario, suppose we have several sets of data from several different locations/times and we would like to investigate whether it is the same person at those several locations/times. In this case, we denote the first sample as sample 1, the probe set captured using a surveillance camera at location 1/time1, sample 2 is the probe set captured using a surveillance camera at location 2/time2, and so on. Now use the realistic facial geometric information to generate data points on probe sets. We obtain landmark points in the 3D model in the probe sets, combine all probe set data points, keeping track of each set. We order combined samples of probe set data points from least to greatest or greatest to least value. Now we determine the ranks of each probe sets values in the combined sample and let R_j is the sum of the ranks in the jth group, k is the number of probe groups, N is the total sample size, and n_j is the sample size of the jth probe. Then, calculate the Kruskal-Wallis test given by Eq. (18). The task can be done efficiently and fast by using batch processing as described earlier.

Suppose that we have more than two independent groups to compare outcomes. The nonparametric competitor of ANOVA is the Kruskal-Wallis test. It compares medians among k comparison groups (k > 2). The null hypothesis to be tested will be

$$H_0 : M_1 = M_2 = M_3 = ...M_k$$

against

$$H_A : M_1 \neq M_2 \neq M_3 \neq ...M_k$$

where $M_1, M_2, ..., M_k$ denote the median of each sample.

Now we first order the data in the combined sample in an increasing or decreasing order, keeping track of each sample unit from where they come.

The Kruskal-Wallis test (Kruskal and Wallis, 1952) is given by

$$H = \left(\frac{12}{N(N+1)} \sum_{j=1}^{k} \frac{R_j^2}{n_j} \right) - 3(N+1) \tag{18}$$

where R_j is the sum of the ranks in the jth group, k is the number of comparison groups, N is the total sample size, and n_j is the sample size.

6   Multivariate methods

First, we look at the bivariate problem. Suppose we are interested in two characteristics (X, Y) of a population. We consider (X_{1i}, X_{2i}), $i = 1, 2,..., n$, quantitative measurements for n subjects from a population with cumulative distribution function (CDF) F(X, Y), and (X_{1j}, X_{2j}), $j = 1, 2,..., m$, quantitative measurements for m subjects from a population with CDF G(X, Y). Let F and G be continuous cumulative bivariate distributions functions. We assume variances and covariances are equal and finite, otherwise, the realignment of observations can be done to ensure that the two populations have the same variances and covariances.

We wish to test whether the two populations have the same location. In other words, we wish to test the following hypothesis concerning the location parameter $\boldsymbol{\theta} = (\theta_1, \theta_2)$ of the two unknown distributions:

$$H_0 : \boldsymbol{\theta}_F = \boldsymbol{\theta}_G \tag{19}$$

against

$$H_A : \boldsymbol{\theta}_F = \boldsymbol{\theta}_G + \delta$$

where $\boldsymbol{\delta} = (\delta_1, \delta_2) \neq (0,0)$.

We can use Hotelling's T^2 as a base to derive nonparametric tests. The Hotelling's T^2 is given by

$$T^2 = n\overline{X}^T S^{-1} \overline{X} \tag{20}$$

where $\overline{X} = mean(X_i)$, the sample mean of vector X and S is the sample mean variance-covariance matrix $= mean\left[(X_i - \overline{X})(X_i - \overline{X})^T\right]$.

We transform Hotelling's T^2 using Choleski factorization of S^{-1}. If C is a nonsingular $p \times p$ matrix such that $C^T C = S^{-1}$, then we can write

$$T^2 = n\overline{Y}^T \overline{Y} \tag{21}$$

where $Y_i = CX_i$, $i = 1, 2,..., n$. Furthermore, since $C = S^{-1/2}$, we can rewrite (21) as

$$T^2 = n\|\overline{Y}\|^2 \tag{22}$$

Thus, Hotelling's T^2 can be represented as simple size times the squared length of the average mean vector $\overline{Y}$.

The univariate sign test can be extended for the multivariate case. Oja and Randles (2004) presented decent results.

If the spatial sign function is defined as

$$S(x) = \begin{cases} \|x\|^{-1}x, & x \neq 0, \\ 0, & x = 0 \end{cases}$$

where $\|x\|$ is the L_2 norm.

The multivariate sign test (Randles, 2000) then can be defined as follows.

$$Q^2 = np\overline{S}^T \overline{S} = np\|\overline{S}\|^2 \tag{23}$$

Similarly, a multivariate median was developed by Hettmansperger and Randles (2002). Let us consider signs of transformed differences

$$S_{ij} = S(C(X_i - X_j)), \quad i, j = 1, 2, ..., n.$$

Then, the multivariate signed-rank test statistic (Oja and Randles, 2004) can be defined using Walsh's sums $x_i + x_j$ for $i \leq j$ as

$$U^2 = \frac{np}{4A^2} \|ave(S(C(X_i + X_j)))\|^2 \tag{24}$$

where $A^2 = ave\{\|R_i\|^2\}$, and $R_i = ave_j(S_{ij})$.

The test statistic (24) can be extended to develop a multi-samples multivariate test statistic.

Consider $X_1, ..., X_{n_1}; X_{n_1+1}, ..., X_{n_2}; ...; X_{n_{c-1}}, ..., X_{n_c}$ c independent samples with sample sizes $n_1, ..., n_c$, from the p-dimensional distribution $F(x - \boldsymbol{\theta}_i)$, $i = 1, 2, ..., c$, respectively. Oja and Randles (2004) presented a multivariate extension of the two-sample Wilcoxon-Mann-Whitney test and several-sample Kruskal-Wallis test as follows.

$$U^2 = \frac{p}{A^2} \sum_{i=1}^{c} n_i \|\bar{R}_i\|^2 \tag{25}$$

where $\bar{R}_i$ is the sample-wise mean vector.

Mathur and Smith (2008) proposed the following test statistic for the said problem:

$$T = T_{AB} - \frac{T_A + T_B}{2} \tag{26}$$

where

$$T_{AB} = \frac{1}{mn} \sum_{i=1}^{m} \sum_{j=1}^{n} \left[(X_{1i} - X_{2j})^2 + (Y_{1i} - Y_{2j})^2 \right],$$

$$T_A = \frac{1}{\binom{m}{2}} \sum_{1 \le i < j \le n} \sum \left[(X_{1i} - X_{1j})^2 + (Y_{1i} - Y_{1j})^2 \right]$$

and

$$T_B = \frac{1}{\binom{m}{2}} \sum_{1 \le i < j \le n} \sum \left[(X_{2i} - X_{2j})^2 + (Y_{2i} - Y_{2j})^2 \right]$$

This method can be extended to a multivariate two-sample method as given below.

$$T = T_{AB} - \frac{T_A + T_B}{2} \tag{27}$$

where

$$T_{AB} = \frac{1}{mn} \sum_{i=1}^{m} \sum_{j=1}^{n} \left[(X_{1i} - X_{2j})^2 + (Y_{1i} - Y_{2j})^2 + \ldots + (W_{1i} - W_{2j})^2 \right],$$

$$T_A = \frac{1}{\binom{m}{2}} \sum_{1 \le i < j \le n} \sum \left[(X_{1i} - X_{1j})^2 + (Y_{1i} - Y_{1j})^2 + \ldots + (W_{1i} - W_{1j})^2 \right]$$

and

$$T_B = \frac{1}{\binom{m}{2}} \sum_{1 \le i < j \le n} \sum \left[(X_{2i} - X_{2j})^2 + (Y_{2i} - Y_{2j})^2 + \ldots + (W_{2i} - W_{2j})^2 \right]$$

We denote the first sample as sample 1, the probe set captured using a surveillance camera at location 1/time1 but now we have many features of a person to compare. Faces contain a lot of information. Several features can be used to process face identities. There are generally two types of categorizations: external (e.g., head-shape) and internal (e.g., eyes) features. Sample 2 is the probe set captured using a surveillance camera at location 2/time2. Using the external and internal features along with realistic facial geometric information, generate data points on probe sets. Then, use the test statistic given by Eq. (27) to test whether the features are matching or not. Since the process could be very time-consuming to millions of data points, one can use batch processing as described earlier.

Chatterjee and Sen (1964) extended univariate Wilcoxon's rank-sum test to bivariate Wilcoxon's rank-sum test using coordinate wise Wilcoxon's statistics. There are several extensions of Wilcoxon's test for a multivariate setting (Bhapkar, 1966; Puri and Sen, 1966; Sugiura, 1965 and Tamura, 1966). Another approach to extend a univariate test to a multivariate case is to use the observations by projecting them on the unit circle. Mardia (1967) uses the center of gravity of the standardized observations to define the test statistic. The test rejects the null hypothesis for larger values of the test statistic. Peters and Randles (1991) defined a sign rank version of Mardia's test. The simulation study in Peters and Randles (1991) indicated that the Mardia's test performs better in case of heavy-tailed distributions while Chatterjee and Sen's test is better when the distribution is light-tailed.

Let $F_m(x)$ and $G_n(y)$ denote the empirical distribution function of X and Y samples, respectively. Then,

$$F_m(t) = \frac{1}{m} \sum_{i=1}^{m} I(x_{i1} \leq t_1, x_{i2} \leq t_2), \tag{28}$$

$$G_n(t) = \frac{1}{n} \sum_{i=1}^{n} I(y_{i1} \leq t_1, y_{i2} \leq t_2). \tag{29}$$

The disparity between $F_m(t)$ and $G_n(t)$ indicates the differences between population distribution functions F and G at a particular point. Hence, we consider the power divergence between $F_m(t)$ and $G_n(t)$ to define the test statistic to test the null hypothesis of no difference in locations. An important family of power divergences introduced by Cressie and Read (1984) is,

$$\phi_{(\lambda)}(x) = (\lambda(\lambda+1))^{-1}(x^{\lambda+1} - x); \quad \lambda \neq 0, \lambda \neq -1$$

$$\phi_{(0)}(x) = x \log(x) - x + 1; \quad \lambda = 0$$

$$\phi_{(-1)}(x) = \log(1/x) + x - 1; \quad \lambda = -1.$$

The power divergence between two n component probability vectors p_1 and p_2 is given by Pardo (2005),

$$D_{\phi_{(\lambda)}}(p_1,p_2) = \sum_{i=1}^{n} p_{1i}\phi\left(\frac{p_{2i}}{p_{1i}}\right). \tag{30}$$

Let t_i, $i=1, ..., N$, be i^{th} observation in the combined sample. Substituting $p_{1i}=F_m(t_i)$ and $p_{2i}=G_n(t_i)$ in Eq. (30) and using $\lambda=0$, Mathur and Sakate (2017) defined the test statistic D as

$$D = \sum_{i=1}^{N}\left[G_n(t_i)\log\left(\frac{G_n(t_i)}{F_m(t_i)}\right) - (G_n(t_i) - F_m(t_i))\right]. \tag{31}$$

If G differs with F only in location, then the power divergence between $F_m(\cdot)$ and $G_n(\cdot)$ will be large. Thus, large values of D will support the alternative hypothesis.

The exact distribution of the test statistic D defined in Eq. (31) is unknown. Hence, we use the permutation principle to implement the test. The notion of using the permutation principle for statistical hypothesis testing dates back to the origin of inferential statistics in which the earliest contribution is of Fisher (1932). The use of the permutation principle for implementing the test for any distribution in case of the two-sample location problem is advocated by Good (2013). In the following, we present the asymptotic distribution of the proposed test statistic.

Let $D_i = G_n(t_i)\log\left(\frac{G_n(t_i)}{F_m(t_i)}\right) - (G_n(t_i) - F_m(t_i))$ where t_i is the ith observation in the combined sample. Under H_0, $F=G$ so we denote the common distribution function as H.

Theorem 1 *If $m=n$,*

$$E(D) \cong 1 - \frac{1}{2n}\sum_{t}H^2(t) \tag{32}$$

and

$$Var(D) \cong \frac{1}{2n^2}\sum_{t}\left[(6n-2)H^3(t) - (4n-1)H^4(t) + (2-6n)H^2(t)\right.$$

$$\left. + (6n-2)H(t) + (1-2n)\right].$$

This test statistic is useful to test large steaming data as the batch processing can quickly determine the empirical distribution function of the samples and make a decision to accept or reject the null hypothesis. Unlike Hotelling's T^2 and Chatterjee and Sen's bivariate Wilcoxon rank-sum test, the Mathur and Sakate (2017) test doesn't require the estimation of dispersion matrix and estimation of dispersion matrix could be a problem in big data sets.

7 Ranked-set based methods

McIntyre (1952) proposed the ranked set sampling (RSS) while working on the estimation of mean pasture and forage yields, however, the term RSS was first used by Halls and Dell (1966). RSS is useful over simple random sampling (SRS) as the statistical procedures based on RSS performs better than their SRS version. RSS sampling protocol is often used to improve the cost-efficiency of an experiment (Chen et al., 2003) when the actual measurements of the sample observations are difficult but the ranking of a potential sample or auxiliary variable is relatively easy. Thus, actual measurement on observations is not taken until the judgment ranking is finished, thus saving time and cost in sampling all the units in the potential sample. Several investigators used RSS in null hypothesis testing procedures. Hettmansperger (1995) proposed the sign test based on RSS data and showed that the RSS based sign test is more efficient than the SRS based sign test. Öztürk (1999) and Öztürk and Wolfe (2000) suggested the median ranked set sampling (MRSS) selecting the median observations for quantification to test median while Wang and Zhu (2005) proposed the sign test for median using unbalanced RSS and showed analytically that the sampling allocation that maximizes the efficacy is MRSS. Wang and Zhu (2005) and Dong and Cui (2010) proposed weighted sign tests under unbalanced RSS and proved the weighted version always improves the Pitman efficiency for all distributions. Recently, Zhang et al. (2014) discussed sign test using RSS with unequal set sizes.

The procedure to select an RSS sample is given below:

Step1: Select initial SRS of k units from the population and order them according to the attribute of interest, say height.

Step 2: The smallest ordered unit is selected in the RSS and is called the first judgment order statistic and is denoted by $X_{[1]}$.

Step 3: Now select a second independent SRS of k units from the population and order them according to the same attribute of interest, say height.

Step 4: The second smallest ordered unit is selected from this second SRS and is called the second judgment order statistic. It is denoted by $X_{[2]}$.

Step 5: Now a third independent SRS is selected from the population of size k, and order them according to the same attribute, say height.

Step 6: Select the third smallest ordered unit from the third SRS and is called the third judgment order statistic. It is denoted by $X_{[3]}$,in the RSS.

Step 7: We continue the process until we have selected the kth unit, denoted by $X_{[k]}$,in the RSS.

The process of selecting k observations $X_{[1]},\ldots, X_{[k]}$ and is called a cycle. The number of units, k, in each SRS is called the set size. For selected RSS of size k, we will need a total of k^2 units from the population. The observations $X_{[1]},\ldots, X_{[k]}$ is called a balanced ranked set sample of size k, indicating that one judgment order statistic has been selected for each of the ranks 1, 2,..., k. Thus, to obtain a balanced RSS with a sample size $n=km$, we will need to repeat the entire process for m independent cycles. The process of selecting an RSS sample is explained in Figs. 3 and 4.

Let $X_{[1]j},\ldots, X_{[k]j}; j=1, \ldots, m$, be a ranked set sample (for set size k and m cycles) from X distribution. In addition, let $Y_{[1]t},\ldots, Y_{[q]t}; t=1,\ldots, n$, be a

SAMPLE 1	$X_{[1]m}$	$X_{[2]m}$	...	...	$X_{[k]m}$
SAMPLE 2	$X_{[1]m}$	$X_{[2]m}$	...	...	$X_{[k]m}$
⋮	⋮	⋮	⋮	⋮	⋮
⋮	⋮	⋮	⋮	⋮	⋮
SAMPLE k	$X_{[1]m}$	$X_{[2]m}$	...	...	$X_{[k]m}$

FIG. 3 Selecting a ranked set sample. Units shown in red are selected.

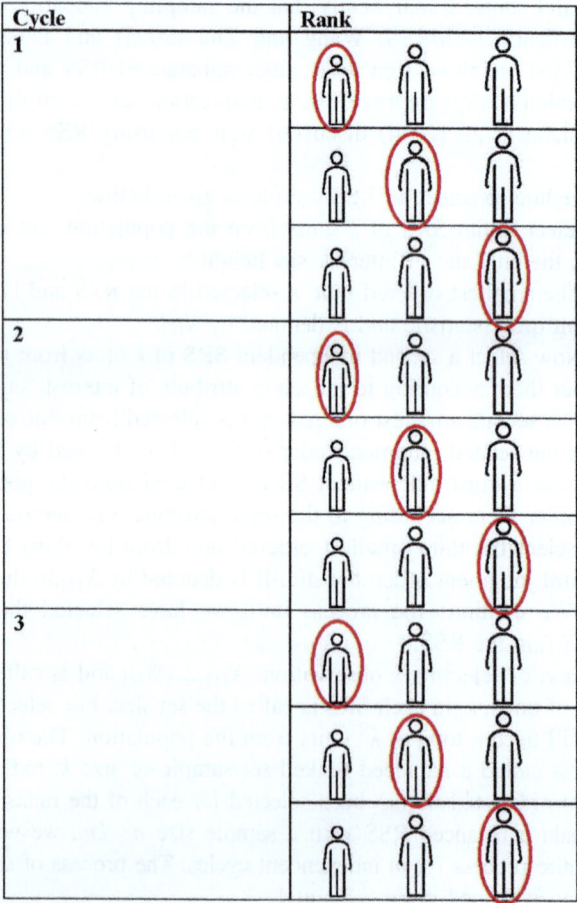

FIG. 4 Process of selecting a ranked set sample in the m=3 cycle, and k=3 units. Units selected are circled in red.

ranked set sample (for set size q and n cycles) from Y distribution. Let $F_{m,k}(x)$ and $G_{n,q}(x)$ be the empirical distribution functions (ECDF) for the X and Y ranked set samples, respectively. Several tests in literature have used ECDF to construct tests for a change in the location under the SRS scheme. For instance, Cramérs' test (Baringhaus and Franz, 2004; Cramer, 1928) and Kolmogorov-Smirnov (Kolmogoroff, 1931; Smirnov, 1948) (KS) test are based on ECDF.

The hypothesis of interest is

$$H_0 : \Delta = 0 \text{ vs } H_A : \Delta \neq 0 \tag{34}$$

Under the assumption of perfect judgment ranking for both X and Y, Bohn and Wolfe (1994) showed that the RSS version of the Mann-Whitney statistic is distribution-free. It is given by

$$BW = \sum_{s=1}^{k} \sum_{j=1}^{m} \sum_{u=1}^{q} \sum_{t=1}^{n} \psi\left(X_{[s]j}, Y_{[u]t}\right), \tag{35}$$

where, $\psi\left(X_{[s]j}, Y_{[u]t}\right) = \begin{cases} 1, & if X_{[s]j} < Y_{[u]t} \\ 0, & \text{otherwise.} \end{cases}$

Distance between two probability distributions has been used to measure the "closeness" between two distributions and this distance is also being used xzto check "how far the distributions are from each other." Different names have been used to denote the measures of distances, for example, measures of separations (Mahalanobis, 1936; Rao, 1949), measures of discriminatory information (Chernoff, 1952; Kullback, 1959), and measures of variation-distance (Kolmogorov, 1963). Several of the current tests such as likelihood tests and the Wald test are based on some distance measures from each other. The divergence measure between two probability distributions has been used to measure different statistical characteristics of distributions. In Kolmogorov distance (Kolmogoroff, 1931), the distance (or difference) between two probability distributions F_1 and F_2 (or between probability measures $P_{\theta 1}$ and $P_{\theta 2}$) is given by

$$KD(F_1, F_2) = \sup_{x \in R} |F_1(x) - F_2(x)| \tag{36}$$

Glivenko-Cantelli Theorem states that the empirical distribution function is a uniformly consistent estimate of the true distribution. Thus for any random sample of size n, from a population with distribution function F_0, for any $\epsilon > 0$, we have

$$\lim_{x \to \infty} Pr\{KD(F_n, F_0)\} = 0$$

where F_n is the empirical distribution function and it is defined as

$$F_n(x) = \frac{1}{n} \sum_{i=1}^{n} I_{(-\infty, x)}(x_i) \tag{37}$$

Therefore, it is reasonable to use empirical distribution functions as a way to measure the distance between two distribution functions. Kullback and Leibler (1951) used a divergence measure for two probability distributions $P_{\theta 1}$ and $P_{\theta 2}$. It is defined as

$$D_{KL}(\theta_1, \theta_2) = \int_x f_{\theta_1}(x) \, log \, \frac{f_{\theta_1}(x)}{f_{\theta_2}(x)} d\mu(x)$$

The natural RSS estimator for $F(x)$ considered by Stokes and Sager (1988) is the ECDF for the RSS data, namely,

$$F_{m,k}(x) = \frac{1}{mk} \sum_{i=1}^{k} \sum_{j=1}^{m} I_{(-\infty,x)}\left(X_{[i]j}\right). \tag{38}$$

Stokes and Sager (1988) show that $F_{m, k}(x)$ is an unbiased estimator of $F(x)$ and that

$$Var(F_{m,k}(x)) \leq Var\left(\widehat{F}(x)\right), \text{for all } x, \tag{39}$$

where, $\widehat{F}(x)$ is the ECDF based on a simple random sample (SRS) of size mk.

As mentioned earlier, Mahalanobis (1936) conceptualized the distance between two probability distributions to measure the closeness between two distributions. Several measures are available to measure the distance between two distributions and several of the tests such as the likelihood ratio test, Wald test, can be shown as measuring the distance between appropriate functions. The common property of increasing as two distributions get farther away from each other is called divergence. Power divergence is defined to measure the disparity between two density functions or probability mass functions.

If F_1 and F_2 are corresponding distribution functions for $P_{\theta 1}$ and $P_{\theta 2}$, then Kolmogotrov distance (Kolmogorov, 1933) between F_1 and F_2 is given by Eq. (21). Glivenko-Cantelli Theorem states that empirical distribution function is a consistent estimate of the true distribution function. In other words, if $F_n(x)$ is an empirical distribution function and $F_0(x)$ is a true population distribution function then for any $\varepsilon > 0$,

$$\lim_{n \to \infty} P\{K(F_n, F_0) > \varepsilon\} = 0.$$

We use the power divergence to measure the disparity between two ECDFs. Let t_i, $i = 1, ..., N$, be i^{th} observation in the combined sample and $N = K + Q$, $K = km$ and $Q = qn$. Substituting $p_{1i} = F_{m,k}(t_i)$ and $p_{2i} = G_{n,q}(t_i)$ in Eq. (30) and using $\lambda = 0$, Mathur and Sakate (2018) defined the test statistic D based on power divergence between the two empirical distribution functions to test the null hypothesis as given in Eq. (34).

$$D = \sum_{i=1}^{N} \left[G_{n,q}(t_i) \log \left(\frac{G_{n,q}(t_i)}{F_{m,k}(t_i)} \right) - \left(G_{n,q}(t_i) - F_{m,k}(t_i) \right) \right]. \tag{40}$$

provided $F_{m,k}(t_i) \neq 0$. If $F_{m,k}(t_i) = 0$, then contribution to D from t_i is 0.

We find that when $F_m,(t_i) > 0$ and $G_{n,q}(t_i) = 0$ should not create a problem as $0 \log \left(\frac{0}{F_{m,(t_i)}} \right) = 0$.

Let $\phi(t) = G_{n,q}(t) \log \left(\frac{G_{n,q}(t)}{F_{m,k}(t)} \right) - (G_{n,q}(t) - F_{m,k}(t))$, and $\phi \varepsilon \Phi^*$

where Φ^* is the class of all convex functions $\phi(t)$, $t \geq 0$, such that at $t = 1$, $\phi(1) = 0$, at $t = 0$, $0\phi\left(\frac{0}{0}\right) = 0$, and $0\phi\left(\frac{G}{0}\right) = \log_{v \to \infty} \frac{\phi(v)}{v}$.

For any convex function ϕ, it is known that

$$\phi(t) \leq \phi(0) + t \log_{r \to \infty} \frac{\phi(r)}{r}, \text{ for } t \geq 0.$$

The strict inequality holds when $t \epsilon (0, \infty)$.

Let $\phi(t)$ be differentiable at $t = 1$, then the function $\psi(t) \equiv \phi(t) - \phi'(1)(t - 1)$ also belongs to Φ^* and $\psi'(t) = 1$.

Using the convexity and the property $\psi'(t) = 1$, it can be shown that $\psi(t) \geq 0$, for any $t \geq 0$, and $D_\psi(t) = D_\phi(t)$. Also, for $\log \left(\frac{G_{n,q}(t)}{F_{m,k}(t)} \right) > 1$, it can be shown that

$$0 \leq D_\phi(t) \leq \phi(0) + \log_{u \to \infty} \frac{\phi(u)}{u}, \text{ where } D_\phi(t) = 0 \text{ if } G_{n,q}(t) = F_{m,k}(t).$$

If G differs with F only in location, then the power divergence between $F_{m,k}(\cdot)$ and $G_{n,q}(\cdot)$ will be large, since $F_{m,k}(x)$ and $G_{n,q}(x)$ are the empirical distribution functions for the X and Y ranked set samples. Therefore, as we deviate from the null hypothesis, the difference between $F_{m,k}(\cdot)$ and $G_{n,q}(\cdot)$ will become larger and larger, and the large values of D will support the alternative hypothesis. The SRS version of the proposed statistic in Eq. (26) is defined by Mathur and Sakate (2017). The distribution-free property of the proposed test statistic holds in the SRS case (Mathur and Sakate, 2017).

Bohn and Wolfe (1992) proposed a two-sample test based on ranked set samples using the Stokes and Sager RSS estimator of the distribution function. Their test was analogs of the Mann–Whitney version of the SRS. Bohn and Wolfe (1992) test was given as follows:

$$U_{RSS} = mnkq \int_{-\infty}^{\infty} F_{m,k}(x) \, dG_{n,q}(x)$$

$$= \sum_{s=1}^{q} \sum_{t=1}^{n} \sum_{i=1}^{k} \sum_{j=1}^{m} \Psi[Y_{[s]t} - X_{[i]j}] \tag{41}$$

$$= (\#X's \leq Y's \text{ in the RSS})$$

Bohn and Wolfe (1992) test (for RSS with perfect rankings) is shown to be distribution-free under the null hypothesis over the entire class of continuous distributions F.

Several other types of RSS schemes have been suggested such as median RSS, double ranked sampling scheme, stratified RSS, extreme ranked set sampling, moving extreme ranked set sampling, and bivariate and multivariate ranked set schemes. Al-Saleh and Zheng (2002) discussed a bivariate Bivariate ranked set sampling which is described as follows:

Step 1. To choose an RSS of size m, select a random sample of size m^4 from the population. Divide the sample randomly into m^2 pools of size m^2 each, denoting each pool as a square matrix with m rows and m columns.

Step 2. Using the first pool, rank all the value in each of the m rows with respect to the first auxiliary variable and select the minimum value with respect to the first characteristic from each of the m rows.

Step 3. Once m units are selected, choose the pair corresponding to the minimum value of the second characteristic. This pair of units, denoted by (1,1), is the first element of the ranked set sample.

Step 4. Repeat the process until you collect all the pairs (m, m).

Similarly, a multivariate version of the RSS can be constructed. The process could be faster if we use batch processing in big data sets. Using the additional information available to select the more representative sample observations and the cost-effective nature of RSS has made it more attractive than SRS and has drawn the attention of researchers in the last two decades. Even though the research in the field of RSS remained dormant since its first use, now it has gained momentum. With the increasing use of RSS protocol for sampling, there is a need for efficient and easy to apply statistical procedures in this setting. We proposed a test for a two-sample location problem based on RSS using empirical distribution functions. We used the empirical distribution functions in the power divergence measure instead of using the probability mass functions or the probability density functions in the sense of the divergence measure between two empirical distributions under two conditions. Moreover, the empirical CDF is an estimate of the true CDF by making no assumptions about the underlying distribution.

To extend Bohn and Wolfe (1992) RSS version of the Mann-Whitney Test, we can use the following procedure. Let $X_{[1]js},..., X_{[k]j}w$; $j=1,..., m$; $w=1, 2,..., p$, be a ranked set sample (for set size k, m cycles, and p-variate) from the p-variate X distribution. Let $Y_{[1]ts},..., Y_{[q]tw}$; $t=1,..., n$; $w=1,2,..., p$, be a ranked set sample (for set size q, n cycles, and p-variate) from the p-variate Y distribution. Let $F_{m,k,p}(x)$ and $G_{n,q,p}(x)$ be the empirical distribution functions (ECDF) for the X and Y multivariate ranked set samples, respectively.

$$U_{RSS} = mnkqp \int_{-\infty}^{\infty} F_{m,k,p}(x) \, dG_{n,q,p}(x)$$

$$= \sum_{w=1}^{p}\sum_{s=1}^{q}\sum_{t=1}^{n}\sum_{i=1}^{k}\sum_{j=1}^{m} \Psi\left[Y_{[s]tw} - X_{[i]jw}\right] \tag{42}$$

$$= (\#X's \leq Y's \text{ in the RSS for each variate})$$

Similarly, the test statistic proposed by Mathur and Sakate (2017) can be extended to a two-sample multivariate test statistic as follows:

$$D = \sum_{w=1}^{p} \sum_{i=1}^{N} \left[G_{n,q,w}(t_i) \log \left(\frac{G_{n,q,w}(t_i)}{F_{m,k,w}(t_i)} \right) - \left(G_{n,q,w}(t_i) - F_{m,k,w}(t_i) \right) \right]. \quad (43)$$

provided $F_{m,k,w}(t_i) \neq 0$. If $F_{m,k,w}(t_i) = 0$, then contribution to D from t_i is 0.

RSS can be used as a data reduction tool. The basic idea is discussed in Chen et al. (2003). Since there is a lot of information stored in systems, however, only parts of the information are useful in making decisions. Thus, the data reduction method should focus on discarding low information data and retain the data with high information contents. The RSS procedure focuses on selecting units with high information contents and discards no-essential information and hence reducing the data.

8 Changepoint detection

In big data, it is common that the data may change after some time and it may not be possible to use methods already in use to explain the behavior after that time point. Detecting changepoints is widely used in big data in climatology, bioinformatics applications, finance, oceanography, and medical imaging (Chen and Gupta, 2011). Detecting a change is equivalent to estimating the point at which the statistical properties of a sequence of observations change (Killick et al., 2011). The change could be in the intercept, slope, or both. We aim to detect that changepoint and also develop a flexible model. A two-line model based on F-test was proposed to detect a change in the regression coefficient (Julious, 2001). Another test was proposed based on maximum F statistics to detect the changepoint in two phases linear regression model (Lund and Reeves, 2002). Reeves et al. (2007) reviewed and compared eight undocumented changepoint detection methods. The differences in assumptions among these eight methods and guidelines for which methods work best in different situations have been discussed. Murakami (2012) used the combination of the Wilcoxon and Mood statistics to the change-point setting and introduce a nonparametric location-scale statistic for detecting a changepoint. Nosek and Szkutnik (2014) proposed a new changepoint detection test based on the likelihood ratio in the two-phase regression linear model with inequality constraints.

Since big data often comes in the form of time-stamped data, for example, credit card transactions, climate data, surveillance data, and health data, therefore, it is important to develop methods which can use the time property of the data and also detect any changes or trends in the data with respect to time. The nonparametric Mann-Kendall test (Kendall, 1955; Mann, 1945) is used to detect monotonic trends in a time series data.

If the sample units are identically and independently distributed, then the Mann-Kendall test is given by

$$S = \sum_{k=1}^{n-1} \sum_{j=k+1}^{n} sgn\,(X_j - X_k) \qquad (44)$$

where

$$sgn(x) = \begin{cases} 1 & if\ x>0 \\ 0 & if\ x=0 \\ -1 & if\ x<0 \end{cases}$$

Mann-Kendall test can be extended to multiple series. For the rth series, it can be calculated as

$$S_r = \sum_{k=1}^{n-1} \sum_{j=k+1}^{n} sgn\,(X_{jr} - X_{kr}), \quad r = 1, 2, ..., m \qquad (45)$$

For the entire series, one can calculate the Mann-Kendall statistic (Hipel and McLeod, 1994) as follows.

$$\sum_{r=1}^{m} S_r \qquad (46)$$

Dietz and Killeen (1981) proposed a modified Mann-Kendall test based on the covariance of a multivariate vector of Mann-Kendall statistic and its estimator. They used Spearman's rank correlation and Kendall's correlation between the underlying variables. Another version of the Mann-Kendall statistic was proposed by Hirsch and Slack (1984) derived a modified variance to be used in the test statistic. In these methods, there are a few problems encountered. The up- and down-trends can cancel each other and may lead to an overall indication of no trend. The tests may be sensitive to trends regardless of direction and have been shown to behave poorly for typical time series quality record lengths. Another approach is suggested by Sun et al. (2019). The simple linear regression model is given by

$$Y_i = \beta_0 + \beta_1 X_i + \varepsilon_i, \quad i = 1, ..., n. \qquad (47)$$

The two-phase linear regression model with a single regressor and single changepoint is given by

$$Y_i = \begin{cases} \alpha_1 + \beta_1 X_i + \varepsilon_i, & if\ i<c \\ \alpha_2 + \beta_2 X_i + \varepsilon_i, & if\ i>c \end{cases} \qquad (48)$$

where ε_i is the zero-mean independent random error with a constant variance σ^2. Let there be a single undocumented changepoint c, then the null hypothesis which we wish to test is given by.

$$H_0: \alpha_1 = \alpha_2 = \alpha\ and\ \beta_1 = \beta_2 = \beta \quad vs \quad H_a: \alpha_1 \neq \alpha_2\ and\ \beta_1 \neq \beta_2.$$

Under H_0, the model in Eq. (48) reduces to

$$Y_i = \alpha + \beta X_i + \varepsilon_i, \ i = 1, ..., n, \tag{49}$$

where Y is the response variable, X is the independent variable, α and β are the unknown regression coefficients, and ε_i is the error term. To test the null hypothesis H_0, assuming normality for the error terms, Lund and Reeves (2002) proposed an F-test based on least-squares estimators of the parameters involved. Sun et al. (2019) proposed a new procedure for identification of changepoint in a two-phase linear regression model with a single regressor using rank-based estimation. Let e_i denote the ith residual expressed as $y_i - (\alpha + \beta x_i)$. The rank-based estimate of regression coefficients are obtained by minimizing

$$\sum_{i=1}^{n} [rank(\widehat{e}_i)] \widehat{e}_i$$

It is equivalent to minimize

$$\sum_{i=1}^{n} \left[rank\left(y_i - x_i \widehat{\beta}\right) - \frac{n+1}{2} \right] \left(y_i - \widehat{\beta} x_i\right), \tag{50}$$

as α doesn't affect the minimizer.

The rank-based estimation assumes that the errors have a symmetric distribution. α is estimated as the median of the differences $d_i = y_i - \beta x_i$.

To test H_0, McKean and Hettmansperger (1976) proposed a nonparametric test statistic based on rank estimators of the parameters involved and is given by

$$T(c) = \frac{\left(SRWR_{reduced} - SRWR_{full}\right)}{\widehat{\tau}} \tag{51}$$

where

$$SRWR = \frac{\sqrt{12}}{n+1} \sum_{i=1}^{n} \left(rank(\widehat{e}_i) - \frac{1}{2}(n+1)\right) \widehat{e}_i; \text{ with } \widehat{e}_i = y_i - \left(\widehat{a} + \widehat{\beta} x_i\right) \tag{52}$$

$$\widehat{\tau} = f \frac{\sqrt{n} \left[A_{(k_2)} - A_{(k_1)}\right]}{2(2.1645)}, \tag{53}$$

where

$$f = \sqrt{\frac{n}{n-5}},$$

and $A_{(1)} \le A_{(2)} \le ... \le A_{(N)}$ are the following $N = \frac{n(n+1)}{2}$ numbers in increasing order,

$$A_{ij} = \frac{(\hat{e}_i + \hat{e}_j)}{2} \text{ for } 1 \leq i \leq j \leq n,$$

and

$$k_1 = \text{the closest integer to } \tfrac{1}{2} + a - (1.645)b,$$
$$k_2 = \text{the closest integer to } \tfrac{1}{2} + a + (1.645)b,$$

$$a = \frac{n(n+1)}{4},$$

$$b = \sqrt{\frac{n(n+1)(2n+1)}{24}}.$$

Sun et al. (2019) defined the changepoint as

$$\tilde{c} = arg\,max\,T(c), \tag{54}$$

provided that $arg\,max\,T$ is large enough to guarantee that it is not just due to chance.

If τ is unknown, it is estimated from the data based on the full model. Here, the full model involves two phases. We assume that τ remains the same in each phase and we estimate it using the formula

$$\hat{\tau} = \left[\frac{(n_1 - 2)\hat{\tau}_1^2 + (n_2 - 2)\hat{\tau}_2^2}{n_1 + n_2 - 4} \right]^{\frac{1}{2}} \tag{55}$$

$\hat{\tau}_1$ and $\hat{\tau}_2$ are obtained by Eq. (53) for the first and second phases separately.

$$\hat{\tau} = f \frac{\sqrt{n}\left[A_{(k_2)} - A_{(k_1)}\right]}{2(2.1645)}, \tag{56}$$

where

$$f = \sqrt{\frac{n}{n-5}},$$

and $A_{(1)} \leq A_{(2)} \leq \ldots \leq A_{(N)}$ are the following $N = \frac{n(n+1)}{2}$ numbers in increasing order,

$$A_{ij} = \frac{(\hat{e}_i + \hat{e}_j)}{2} \text{ for } 1 \leq i \leq j \leq n,$$

and

$$k_1 = \text{the closest integer to } \tfrac{1}{2} + a - (1.645)b,$$
$$k_2 = \text{the closest integer to } \tfrac{1}{2} + a + (1.645)b,$$

$$a = \frac{n(n+1)}{4},$$

$$b = \sqrt{\frac{n(n+1)(2n+1)}{24}}$$

A multivariate nonparametric approach based on Euclidean distances between sample observations was proposed by Matteson and James (2014). Since it does not use density estimation, which could be difficult in big data, it makes the calculations relatively easier. Let $Z_1, Z_2, \ldots, Z_T \in \mathbb{R}^d$ be an independent sequence of time-ordered observations. Here T denotes time (positive), and d is the dimension of the time series. Let there be a known number of change points k in the series, but with unknown locations. Thus, there exist change points $0 < \tau_1 < \tau_2, \ldots, < \tau_K < T$. Thus, it is like dividing the sequence into $k+1$ clusters, such that observations within clusters are identically distributed, and observations between adjacent clusters are not. For random variables X and Y $\in \mathbb{R}^d$, let ϕ_x and ϕ_y denote the characteristic function of X and Y respectively. For some fixed constant $\alpha \in (0,2)$, and if $E\lceil X \rceil^\alpha$, $\lceil Y \rceil^\alpha < \infty$, the divergence measure can be defined (Matteson and James, 2014) as

$$D(X, Y; \alpha) = \int_{\mathbb{R}^d} \lceil \phi_x(t) - \phi_y(t) \rceil^2 \left(\frac{2\pi^d \Gamma\left(1 - \frac{\alpha}{2}\right)}{\alpha 2^\alpha \Gamma \frac{(d+\alpha)}{2}} \lceil t \rceil^{d+\alpha} \right)^{-1} dt \qquad (57)$$

The alternative divergence measure based in Euclidean distances, defined by Szekely and Rizzo (2005) is

$$\Delta(X, Y; \alpha) = 2E\lceil X - Y \rceil^\alpha - E\lceil X - X' \rceil^\alpha - E\lceil Y - Y\prime \rceil^\alpha \qquad (58)$$

where X, X' $\overset{iid}{\sim} F_x$, and Y, Y' $\overset{iid}{\sim} F_y$. Also, X, X', Y, Y' are mutually independent with $E\lceil X \rceil^\alpha$, $\lceil Y \rceil^\alpha < \infty$.

Missing values could cause the failure of a test statistic and may lead to wrong interpretations. It is important to consider the pattern of the missing data and the amount of missing data. The distribution of missing data can provide clues and also help in dealing with the missing data. The handling of missing data can affect the reliability and accuracy of the inferences about the population of interest. In general, there are three mechanisms for missingness. They are Missing Completely at Random (MCAR), Missing at Random (MAR), and Not Missing at Random (MNAR). The missing data can be handled by several methods depending on the type of missingness, including listwise or case deletion, pairwise deletion, mean substitution, regression imputation, last observation carried forward, maximum likelihood, expectation-maximization, and multiple imputations. More details are provided by Molenberghs et al. (2014).

9 Final remarks

Nonparametric data science can be used in analyzing big data and can be used to identify useful patterns which may help in proper interpretations and decision-making process. Nonparametric data science can overcome challenges associated with big data such as are volume, variety, velocity, and validity of data. Nonparametric methods can be applied to data that is no longer restricted to structured database records but include unstructured data having no standard formatting. The complexity involved in big data is not only about a large amount of unstructured data but also about the need for rapid analysis, where interpretations/decisions need to be provided in seconds. Nonparametric methods can overcome both challenges and provide inference/decisions within a split of seconds. It can handle multivariate multi-samples big data with its innovative techniques which do not require unnecessary assumptions. There is a lot of potential in nonparametric data science and there is an urgent need to develop new nonparametric methods that can work with cloud computing at super speed to provide rapid analysis.

Since the availability of big data is increasing day-by-day due to rapid innovations in technology, there is an urgent need to provide training to students who can handle big data for finding useful patterns using cloud computing. Currently, there are nearly 70–80% of business entities do not have employees with the necessary skills to analyze big data properly and take advantage of big data to find useful information for making proper business decisions which could be critical for business expansion or survival. In healthcare, big data could provide useful information in making life-saving decisions and choosing the right treatment strategies. Big data presents challenges in data analysis, organization, retrieval, and modeling which were not present when classical methods were developed. Classical methods work well for structured data when certain assumptions are met which makes data analysis using classical methods a clear bottleneck in many applications. Due to the scalability of the underlying algorithms and due to the complexity of the data that needs to be analyzed, nonparametric data science could be very useful in big data analysis and has a lot of potential in revealing patterns and relationships in big data that would otherwise remain hidden.

Nonparametric methods have been very useful in predictive analytics that are can be helpful to determine the outcome of an event that might occur in the future. Nonparametric methods use the strength and magnitude of the associations to develop predictive models. These predictive models also use implicit dependencies on the conditions under which the past events occurred which is useful in big data as more and more information is gathered on the same variables. Nonparametric models are also robust and can handle changes in the conditions change. There is a greater need to develop more efficient predictive models that are robust to underlying conditions. Further development of nonparametric data science could provide more innovative methods and techniques for analyzing big data and help in making informed decisions that could save lives and businesses.

References

Al-Saleh, M.F., Zheng, G., 2002. Theory & methods: estimation of bivariate characteristics using ranked set sampling. Aust. N. Z. J. Stat. 44 (2), 221–232.

Baringhaus, L., Franz, C., 2004. On a new multivariate two-sample test. J. Multivar. Anal. 88 (1), 190–206.

Bhapkar, V.P., 1966. A note on the equivalence of two test criteria for hypotheses in categorical data. J. Am. Stat. Assoc. 61 (313), 228–235.

Bohn, L.L., Wolfe, D.A., 1992. Nonparametric two-sample procedures for ranked-set samples data. J. Am. Stat. Assoc. 87 (418), 552–561.

Bohn, L.L., Wolfe, D.A., 1994. The effect of imperfect judgment rankings on properties of procedures based on the ranked-set samples analog of the Mann-Whitney-Wilcoxon statistic. J. Am. Stat. Assoc. 89 (425), 168–176.

Chatterjee, S.K., Sen, P.K., 1964. Non-parametric tests for the bivariate two-sample location problem. Calcutta Stat. Assoc. Bull. 13 (1–2), 18–58.

Chattopadhyay, B., Mukhopadhyay, N., 2019. Constructions of new classes of one-and two-sample nonparametric location tests. Methodol. Comput. Appl. Probab. 21 (4), 1229–1249.

Chen, J., Gupta, A.K., 2011. Parametric Statistical Change Point Analysis: With Applications to Genetics, Medicine, and Finance. Springer Science & Business Media.

Chen, Z., Bai, Z., Sinha, B., 2003. Ranked Set Sampling: Theory and Applications. Media Springer Science & Business Media.

Chernoff, H., 1952. A measure of asymptotic efficiency for tests of a hypothesis based on the sum of observations. Ann. Math. Stat. 23 (4), 493–507.

Cramer, H., 1928. On the composition of elementary errors: second paper: statistical applications. Scand. Actuar. J. 1928 (1), 141–180.

Cressie, N., Read, T.R., 1984. Multinomial goodness-of-fit tests. J. R. Stat. Soc. B. Methodol. 46 (3), 440–464.

Dietz, E.J., Killeen, T.J., 1981. A nonparametric multivariate test for monotone trend with pharmaceutical applications. J. Am. Stat. Assoc. 76 (373), 169–174.

Dong, X., Cui, L., 2010. Optimal sign test for quantiles in ranked set samples. J. Stat. Plann. Infer. 140 (11), 2943–2951.

Fisher, R., 1932. Statistical Methods for Research Workers. Oliver and Boyd, 1925, Edinburgh. Google Scholar.

Gibbons, J.D., Chakraborti, S., 2014. Nonparametric Statistical Inference: Revised and Expanded. CRC Press.

Good, P., 2013. Permutation Tests: A Practical Guide to Resampling Methods for Testing Hypotheses. Springer Science & Business Media.

Halls, L.K., Dell, T.R., 1966. Trial of ranked-set sampling for forage yields. For. Sci. 12 (1), 22–26.

Hettmansperger, T.P., 1995. The ranked-set sample sign test. J. Nonparametr. Stat. 4 (3), 263–270.

Hettmansperger, T.P., Randles, R.H., 2002. A practical affine equivariant multivariate median. Biometrika 89 (4), 851–860.

Hipel, K.W., McLeod, A.I., 1994. Time Series Modelling of Water Resources and Environmental Systems. Elsevier.

Hirsch, R.M., Slack, J.R., 1984. A nonparametric trend test for seasonal data with serial dependence. Water Resour. Res. 20 (6), 727–732.

Julious, S.A., 2001. Inference and estimation in a changepoint regression problem. J. R. Stat. Soc. Ser. D Stat. 50 (1), 51–61.

Kendall, M.G., 1955. Rank Correlation Methods. 1955. Griffin, London.

Killick, R., Eckley, I.A., Jonathan, P., Chester, U., 2011. Efficient detection of multiple change-points within an oceano-graphic time series. In: Proceedings of the 58th World Science Congress of ISI.

Kolmogoroff, A., 1931. Über die analytischen methoden in der wahrscheinlichkeitsrechnung. Math. Ann. 104 (1), 415–458.

Kolmogorov, A.N., 1933. Sulla determinazione empírica di uma legge di distribuzione. Giornale dell' Istituto Italiano degli Attuari 4, 83–91.

Kolmogorov, A.N., 1963. On the approximations of distributions of sums of independent summands by infinitely divisible distributions. Sankhya 25, 159–174.

Kruskal, W.H., Wallis, W.A., 1952. Use of ranks in one-criterion analysis of variance. J. Am. Stat. Assoc. 47 (260), 583–621.

Kullback, S., 1959. Information Theory and Statistics. John Wiley and Sons, New York.

Kullback, S., Leibler, R.A., 1951. On information and sufficiency. Ann. Math. Stat. 22 (1), 79–86.

Lund, R., Reeves, J., 2002. Detection of undocumented changepoints: a revision of the two-phase regression model. J. Climate 15 (17), 2547–2554.

Mahalanobis, P.C., 1936. On the Generalized Distance in Statistics. National Institute of Science of India.

Mann, H.B., 1945. Nonparametric tests against trend. Econometrica 13, 245–259.

Mardia, K., 1967. A non-parametric test for the bivariate two-sample location problem. J. R. Stat. Soc. B. Methodol. 29 (2), 320–342.

Mathur, S., Sakate, D., 2017. A new test for two-sample location problem based on empirical distribution function. Commun. Stat. Theory Methods 46 (24), 12345–12355.

Mathur, S., Sakate, D., 2018. A new test for two-sample location problem based on empirical distribution functions under ranked set sampling, from data to knowledge, working for a better world. In: International Indian Statistical Association Conference. University of Florida, Gainesville.

Mathur, S., Sepehrifar, M.B., 2013. A new signed rank test based on slopes of vectors for bivariate location problems. Stat. Methodol. 10 (1), 72–84.

Mathur, S.K., Smith, P.F., 2008. An efficient nonparametric test for bivariate two-sample location problem. Stat. Methodol. 5 (2), 142–159.

Matteson, D.S., James, N.A., 2014. A nonparametric approach for multiple change point analysis of multivariate data. J. Am. Stat. Assoc. 109 (505), 334–345.

McIntyre, G., 1952. A method for unbiased selective sampling, using ranked sets. Aust. J. Agr. Res. 3 (4), 385–390.

McKean, J.W., Hettmansperger, T.P., 1976. Tests of hypotheses based on ranks in the general linear model. Commun. Stat. Theory Methods 5 (8), 693–709.

Molenberghs, G., Fitzmaurice, G., Kenward, M.G., Tsiatis, A., Verbeke, G., 2014. Handbook of Missing Data Methodology. CRC Press.

Murakami, H., 2012. A nonparametric location–scale statistic for detecting a change point. Int. J. Adv. Manuf. Technol. 61 (5–8), 449–455.

Nosek, K., Szkutnik, Z., 2014. Change-point detection in a shape-restricted regression model. Statistics 48 (3), 641–656.

Oja, H., Randles, R.H., 2004. Multivariate nonparametric tests. Stat. Sci. 19 (4), 598–605.

Öztürk, Ö., 1999. Two-sample inference based on one-sample ranked set sample sign statistics. J. Nonparametr. Stat. 10 (2), 197–212.

Öztürk, Ö., Wolfe, D.A., 2000. Alternative ranked set sampling protocols for the sign test. Stat. Probab. Lett. 47 (1), 15–23.

Pardo, L., 2005. Statistical Inference Based on Divergence Measures. CRC Press.

Peters, D., Randles, R.H., 1991. A bivariate signed rank test for the two-sample location problem. J. R. Stat. Soc. B. Methodol. 53 (2), 493–504.

Puri, M.L., Sen, P.K., 1966. On a class of multivariate multisample rank-order tests. Sankhyā, Ser. A 28, 353–376.

Randles, R.H., 2000. A simpler, affine-invariant, multivariate, distribution-free sign test. J. Am. Stat. Assoc. 95 (452), 1263–1268.

Rao, C.R., 1949. On the distance between two populations. Sankhya 9, 246–248.

Reeves, J., Chen, J., Wang, X.L., Lund, R., Lu, Q.Q., 2007. A review and comparison of change-point detection techniques for climate data. J. Appl. Meteorol. Climatol. 46 (6), 900–915.

Smirnov, N., 1939. On the estimation of the discrepancy between empirical curves of distribution for two independent samples. Mat. Sb. 48 (1), 3–26.

Smirnov, N., 1948. Table for estimating the goodness of fit of empirical distributions. Ann. Math. Stat. 19 (2), 279–281.

Stokes, S.L., Sager, T.W., 1988. Characterization of a ranked-set sample with application to estimating distribution functions. J. Am. Stat. Assoc. 83 (402), 374–381.

Sugiura, N., 1965. Multisample and multivariate nonparametric tests based on U statistics and their asymptotic efficiencies. Osaka J. Math. 2 (2), 385–426.

Sun, J., Sakate, D., Mathur, S., 2019. A nonparametric procedure for changepoint detection in linear regression. Commun. Stat. Theory Methods, 1–11. https://doi.org/10.1080/03610926.2019.1657453. In press.

Szekely, G.J., Rizzo, M.L., 2005. Hierarchical clustering via joint between-within distances: extending Ward's minimum variance method. J. Classif. 22 (2), 151–183.

Tamura, R., 1966. Multivariate nonparametric several-sample tests. Ann. Math. Stat. 37 (3), 611–618.

Wang, Y.-G., Zhu, M., 2005. Optimal sign tests for data from ranked set samples. Stat. Probab. Lett. 72 (1), 13–22.

Wilcoxon, F., 1945. Individual comparisons by ranking methods. Biometrics 1, 80–83.

Zhang, L., Dong, X., Xu, X., 2014. Sign tests using ranked set sampling with unequal set sizes. Stat. Probab. Lett. 85, 69–77.

Section IV

Network models and COVID-19 modeling

Chapter 8

Network models in epidemiology

Tae Jin Lee[a],*, Masayuki Kakehashi[b], and Arni S.R. Srinivasa Rao[c]

[a]Center for Biotechnology and Genomic Medicine, Medical College of Georgia, Augusta University, Augusta, GA, United States
[b]Department of Health Informatics, Graduate School of Biomedical and Health Sciences, Hiroshima University, Hiroshima, Japan
[c]Laboratory for Theory and Mathematical Modeling, Division of Infectious Diseases, Medical College of Georgia, Augusta, GA, United States
*Corresponding author: e-mail: talee@augusta.edu

Abstract

Communicable diseases such as flu and measles are transmitted through contacts. Given the context of modeling assumptions and disease mechanisms, various modeling frameworks have been developed. Among them, homogeneous mixing models such as the differential equation modeling approach have produced valuable results on epidemiology. However, most human interactions are not homogeneous. Network modeling is an appropriate approach utilizing contact pattern information and give a higher resolution of epidemic dynamics. In this chapter, we introduced homogeneous mixing model and the concept of networks. After the introduction, the connection of network to epidemic dynamics is covered with an example of non-homogeneous mixing with the syringe sharing model of IDU population on Hepatitis C.

Keywords: Network, Epidemiology, Branching process, Hepatitis C, Injection drug user (IDU)

1 Introduction

The discrepancy in the prediction through traditional ordinary differential equation (ODE) model can be seen at the SARS epidemic where basic reproduction number R_0 was estimated as a value between 2.2 and 3.6, but it was

Handbook of Statistics, Vol. 44. https://doi.org/10.1016/bs.host.2020.09.002

later estimated to be 1.4 (Allen et al., 2008). The inflation of the estimated transmission was due to difference in contact intensity between hospital and non-hospital setting, or heterogeneity of the population mixing. Network modeling on epidemiology is one of the approaches to improve our understanding of the disease epidemic through addressing the non-homogeneous mixing in the population.

Within the last two decades, we have experienced several epidemics including SARS, Ebola, and Zika. Concerning the impact on society through these epidemics, many mathematical models have been developed to study these disease dynamics. Most notable of these models are the compartmental model known as the SIR model. The SIR model studies the population of susceptible(S), infectious(I), and recovered or removed (R) over time, often utilizing ODE. One of the assumptions of the SIR model is that it assumes homogeneous mixture of populations, however, this is an oversimplification of the reality of epidemic process. For example, consider the case of modeling the epidemic progression among school children in a town with a child who has a flu virus. Classmates who interact with this child would likely have higher chance of infection than children from the different classrooms who are not in network with this child. Likewise, children from different school in the town would not have any chance of infection if they are assumed to be not in contact with this child network the child belongs to. Observe that susceptibility of a child in the town greatly differs, violating the assumption of a homogeneous population.

Consider the above scenario, and let S, I, and R be the number of susceptible, flu infected, and recovered children in high school respectively. Assuming the student population keep constant size over the study period and homogeneous, the standard SIR model can be obtained as

$$\frac{dS}{dt} = -\beta SI$$

$$\frac{dI}{dt} = \beta SI - \delta I \qquad (1)$$

$$\frac{dR}{dt} = \delta I$$

where β and δ represent infection and recovery rate respectively. However, student interaction may differ as courses taken by those students preparing for college and those students preparing for working differ. Thus, we need to divide both S into two groups: S_1 and S_2. Where, the subscript of 1 and 2 represent college preparing and job preparing student respectively. Infected students population is divided into I_1 and I_2. However, these two groups of students interact sometimes, as they may take the same courses or have lunch together. Let α_1 and α_2 be these mixing rates of the susceptible and infected population respectively, then the model is given as

$$\frac{dS_1}{dt} = -\beta_1 S_1 I_1 - \alpha_1 S_1 + \alpha_1 S_2$$

$$\frac{dS_2}{dt} = -\beta_2 S_2 I_2 - \alpha_1 S_2 + \alpha_1 S_1$$

$$\frac{dI_1}{dt} = \beta_1 S_1 I_1 + \alpha_2 I_2 - \alpha_2 I_1 - \delta I_1 \tag{2}$$

$$\frac{dI_2}{dt} = \beta_2 S_2 I_2 + \alpha_2 I_1 - \alpha_2 I_2 - \delta I_1$$

$$\frac{dR}{dt} = \delta(I_1 + I_2)$$

The schematic diagram of Model 1 and Model 2 is shown in Fig. 1. First, let consider the trivial case of $\beta_1 = \beta_2 = \beta$ and $\alpha_1 = \alpha_2 = \alpha$ to see the effect of non-homogeneity in population. In the school example scenario, consider the scenario that a college preparing student has a flu virus, and all other students are susceptible. If $\alpha = 0$ or student population do not mix, then flu infection occurs among college preparing students only. On the other hand student population mix well as α increases, the dynamics depend on the infectivity of infective groups. Fig. 2 shows the simulated results with various parameter combinations. The homogeneous population was shown as a solid black line in the Fig. 2, and all the population got infected within 100 days. In the same figure, no mixing scenario was shown in red line, and mixing with various β_1 and β_2 combinations was shown as other lines. Note that change in infectivity between two groups of population changed the dynamics markedly. In other words, we need to know the characteristics such as the infection rate of each group in the non-homogeneous mixing population in order to understand the disease outbreak dynamics.

In the example, we only considered two groups of students, however for better resolution, compartment sizes approach to the individual level. Note that in individual based model, homogeneous mixing assumption that states any individuals constantly interact with all other individuals would likely to be violated. In fact, some students may interact with 30 students while some

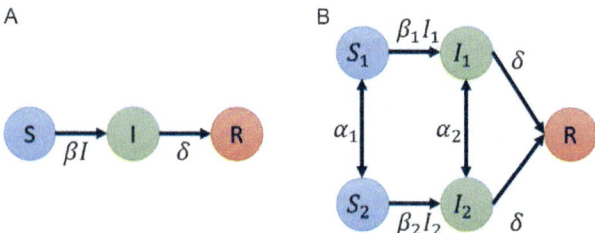

FIG. 1 Schematic diagram of the SIR model where (A) represent the well-mixed population, and (B) represent the two clustered population. The cluster 1 and 2 represented by subscript mix with α_1 and α_2 rate and have different infection rate β_1 and β_2. For this model, the recovery rate is same for both clusters.

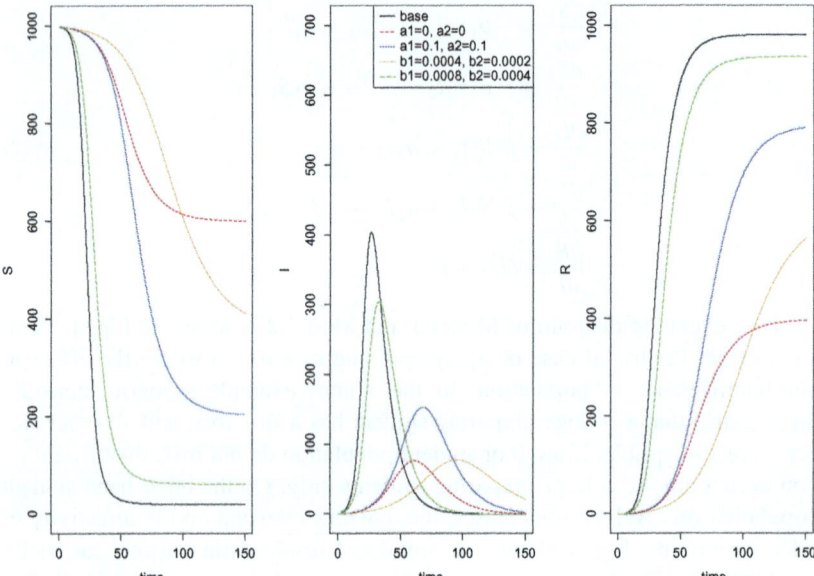

FIG. 2 Simulation for the Model 1 and Model 2. The simulation for Model 1 is shown in black line. For initial population, 999 susceptible children and 1 infected child, and infection rate of $\beta_1 = \beta_2 = 0.0004$ was used otherwise specified. Red and blue line are simulated with $S_1 = 499$ and $S_2 = 500$ population. Larger the mixing parameter α, Model 2 dynamics become closer to Model 1 dynamics.

other group of students may interact with 5 students out of entire student population. Network model incorporates this information, thus we consider network models for non-homogeneous populations.

Now consider Hepatitis C spread. More than 60% of injection drug users (IDU) in 37 countries had exposure to the Hepatitis C virus (Nelson et al., 2011), suggesting control on injection drug use or syringe sharing may have the major impact on the Hepatitis C control. In the US, 1.5 million IDUs had been exposed (Nelson et al., 2011). Since syringe sharing may weave off the population from homogeneous mixing, we need to consider the network of syringe sharing to model the Hepatitis C epidemics in the injection drug user population. However, the network structure of the syringe sharing of the injection drug users is unknown. Meanwhile, we can study the simulation of the disease network model to study the potential dynamics.

2   Network models

A network is a collection of connected dots as shown in Fig. 3. The dots and lines of the network are called nodes and vertex respectively. The network is used in broad areas including computer science, social studies, etc. Nodes in a graph represent objects and edges in a graph represent the relationship

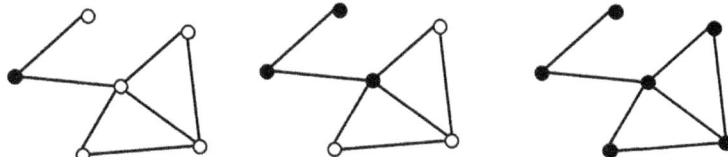

FIG. 3 Infection progression in network. White nodes represent susceptible individuals while black nodes represent infectious individuals. In this figure, assume disease transmit 100% with contact. The infection progression in the network can be seen from left network to central then right network. Here, white nodes adjacent to black nodes switched their color. That is, susceptible individuals who are in contact with infectious individuals get infected and will be spreading the disease to other susceptible individuals in contact.

between paired objects. In epidemiology, nodes represent individuals and edge represents the presence of interaction. Observe that in the SIR model scenario, each individual or node takes one of the three states (S, I, R). Nodes with S state that are adjacent to node with I state have a positive probability of being infected, while those nodes with S state that are not adjacent to a node with I state do not have positive probability of being infected. Those nodes with the state I would be recovered over time, and dynamics progresses. Meanwhile, the edges of each node are the contact of that node. When we displace contacts through graph theory notation, we mean that only those edges which are adjacent to I are participating in transmission of a disease to S.

Network modeling on epidemiology is diverse and may look very different from one discussed above. For example, Rao et al. has studied nonoverlapping path of chicken to understand the parasite spread within the chicken (Rao et al., 2015). Here, nodes are location and edges are the paths taken by chicken. Furthermore, in the gene expression networks, each node represents a gene and is associated with the expression values of the gene. Moreover, the edges in the gene expression network are directed and is clearer for modeling the dynamics. Meanwhile, in the disease transmission network, each node represents an individual and associated with disease status. Thus, the disease transmission network is a directed graph and consider the state of each node as random variables like gene expression network. However, the direction of transmission is not fixed with nodes regardless of its state because the susceptible node has a chance of becoming infected node if the edge has both infected and susceptible node.

There are several types of networks, such as the lattice network, small-world network, and scale-free network shown in Fig. 4. No node in both lattice and small-world network are isolated. The difference between them arises with distance between nodes. In Fig. 4, the lattice network requires at most four steps to reach one node to any other nodes while small-word network shrinks the step to 3. Lattice network may be used to study epidemics that is clustered in the region and slow to spread out, and small-world networks

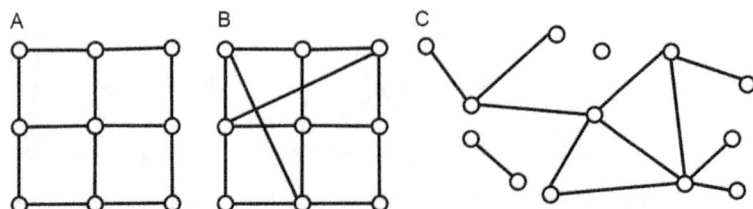

FIG. 4 Types of networks, (A) Lattice network (B) Small-world network (C) scale-free network. The rigid shape of the lattice network has larger distance than small-world network. The maximum distance between two nodes in (A) is 4 while the distance is 3 for (B). For the scale-free network, we see hubs or nodes with high number of edges.

may used to study the disease with faster spread (Moore and Newman, 2000; Rhodes and Anderson, 1997). In the other hand, the scale-free network is concerned with the connectivity of nodes or the number of edges on a node in network. We call the number of edges on a node in a network as degree of network, and if the distribution of degree of network follows power law, we call the network as a scale-free network. Because we are interested in the disease spread within injection drug users (IDU), we consider the scale-free network as it is a good representation of the social network.

2.1 Static network

When we construct the network for a disease, we assume interaction between individuals does not change over time. With the relaxation of the homogeneous mixture of the susceptible and infectious population, some of the susceptibles may not have any risk of infection. These populations are represented graphically as susceptible nodes without connection to infected nodes. To represent the presence of edges in the network mathematically, the adjacency matrix is a matrix used. Let $V = \{v_1, \cdots, v_N\}$ be individuals, and X_v be adjacency matrix represented as

$$X_v(v_i, v_j) = \begin{cases} 1, & \text{if } v_i \sim v_j, \\ 0, & \text{if } v_i \nsim v_j. \end{cases} \tag{3}$$

The row and column of the adjacency matrix represent the nodes in a network, and elements of the matrix are the indicator function of the presence of the edge between two subjects. Significant but very rare events of self-acquisition of communicable disease are not concerned, and diagonal elements of the adjacency matrix are zero. The acquisition of ebola or avian flu can be modeled expanding networks including other animals. Observe that $X_v(v_i, v_j) = X_v(v_j, v_i)$, however, direction of infection is not known as it depends on state of v_i and v_j. Note that infection happens with the contact, and the number of edges for a node called degree of node will be used for analysis. The degree of node $v \in V$ is defined as

$$deg(v) = \sum_{v_i \neq v} X_v(v, v_i) \tag{4}$$

Because the contact pattern of individuals differ from each other, the degree of nodes is not constant. We can consider the degree of epidemic network as the potential of individuals to either obtain or spread disease. Individuals with a higher degrees will have a higher risk of contracting the infection as well as a higher risk of spreading infection. Observe that degree of nodes for a network is a random variable, and we define P_k as the probability of a node with degree k. The distribution characterizes the heterogeneity of the epidemic potential: potential of individuals to either obtain or spread a disease. We will work with degree of nodes and its distribution to understand the epidemics.

2.1.1 Basic reproduction number

One of the main question for any epidemic is to predict incoming epidemic and finding the strength of the disease transmission. Basic reproduction number R_0 is the average number of secondary infection for almost all susceptible population, and is usually regarded as the strength of the disease transmission. If $R_0 > 1$, epidemic happens.

Let us examine the disease transmission network with an initially infected node. The infection happens to the adjacent node with degree k, and this node can infect $k-1$ nodes for $k = 1, 2, \ldots$ because one node is already infected. In other words, there will be $k-1$ possible secondary infection if disease spread to node with degree k. Since R_0 is the average secondary infection for almost all susceptible population, we need to obtain the distribution of infecting node with k degrees. Fig. 5 shows the $k-1$ possible secondary infections through an infecting node with the degree k.

It is tempting to consider only the nodes, however, keep in mind that we are studying the incidence of the disease here. That is, we also need to consider the edges of the network. If we randomly choose any node for the first incidence, the chance of obtaining node with degree k is p_k. Now, consider the node as a spiky ball as shown in Fig. 6. Since there are k ways to reach

FIG. 5 Node with degree 1, 3, and 4 are shown left to right. The black node represents the initial infectious individual, and the white node represents the primary infected individual who has k contacts. We can see they have we still have $k-1$ node yet to be infected.

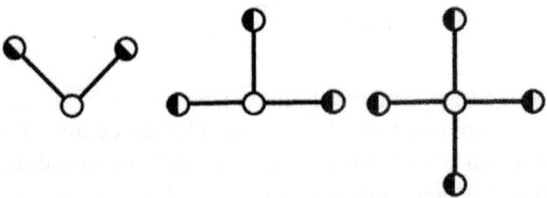

FIG. 6 Nodes with degree 2 through 4 are shown left to right. The black and white node represent the possible initially infectious individual, and the white node represent the primary infected individual who has k contacts.

to the node with degree k, the probability of the first infection to have k degree or probability of $k-1$ secondary infection is

$$q_{k-1} = \frac{kp_k}{\sum_{k=0}^{\infty} kp_k} \tag{5}$$

Let a be the total number of all edges. Then, akp_k is the total number of edges in network for all nodes with degree k. Because a is constant, kp_k is proportional to the total number of edges if primary infected node has degree k, and $\sum_{k=0}^{\infty} kp_k$ is proportional to the total number of edges. That is, kp_k represent the total possible contacts that can yield the primary infection to individual who contacts with k individuals, and $\sum_{k=0}^{\infty} kp_k$ represent total possible contacts.

Then, R_0 can be obtained as

$$
\begin{aligned}
R_0 &= \sum_{k=1}^{\infty} (k-1)q_{k-1} \\
&= \sum_{k=1}^{\infty} \frac{k(k-1)p_k}{\sum_{k=0}^{\infty} kp_k} \\
&= \frac{\sum_{k=1}^{\infty} k(k-1)p_k}{\sum_{k=0}^{\infty} kp_k} \\
&= \frac{\sum_{k=1}^{\infty} k^2 p_k}{\sum_{k=0}^{\infty} kp_k} - 1 \\
&= \frac{\mathrm{Var}(k) + \mathrm{Mean}(k)^2}{\mathrm{Mean}(k)} - 1 \\
&= \mathrm{Mean}(k) + \frac{\mathrm{Var}(k)}{\mathrm{Mean}(k)} - 1
\end{aligned}
\tag{6}
$$

One interesting observation is that the R_0 can be estimated by the mean of a line graph. A line graph of a graph G treats edges of G as a node (Fig. 7). In epidemiological sense, each node in the line graph may represents the first

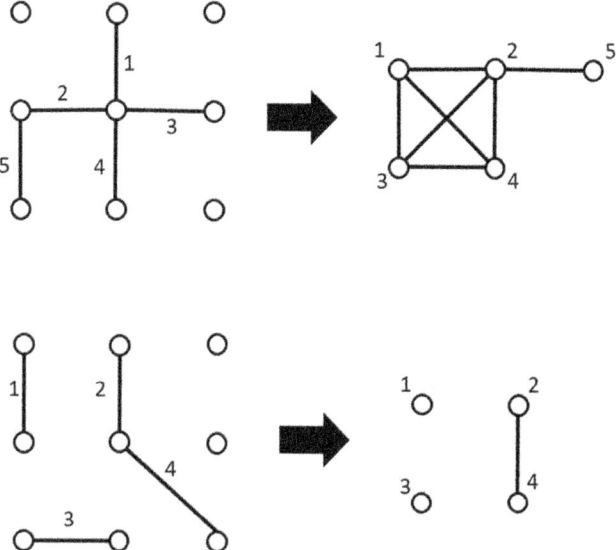

FIG. 7 Line graph. The original graphs are shown in LHS and corresponding line graphs are shown in RHS. The upper network has higher mean degree of line network and is interest to epidemiologist.

primary case of disease transmissions, and the degree of the line graph represents the secondary case of disease transmission. Thus, mean of the line network can be used to estimate R_0 (Morris and Kretzschmar, 1997).

So far, we have considered the case that every contact between individuals transmit disease. However, not all the contacts between infected and susceptible individuals transmit disease in reality. To take account of this transmissibility, we let Z to be the random variable describing the probability of successful transmission of disease between susceptible and infected individual contact. Then, the new R_0 can be obtained as

$$
\begin{aligned}
R_0 &= \sum_{k=1}^{\infty} \sum_{m=0}^{k-1} q_{k-1} m \binom{k-1}{m} Z^m Z^{k-m-1} \\
&= \sum_{k=1}^{\infty} q_{k-1} \sum_{m=0}^{k-1} m \binom{k-1}{m} Z^m Z^{k-m-1} \\
&= \sum_{k=1}^{\infty} q_{k-1} (k-1) Z \\
&= \left[\frac{\sum_{k=1}^{\infty} k^2 p_k}{\sum_{k=0}^{\infty} k p_k} - 1 \right] Z
\end{aligned}
\tag{7}
$$

Observe that the new R_0 is proportional to old R_0 with Z, and decreasing transmissibility decreases the R_0. That is "partially immune" populations are likely protected from the epidemics.

2.2 Dynamic network

We have so far seen the approaches to address the heterogeneity of finite population through the assumption that contact patterns of individuals do not change over time. However, in reality, the disease transmission network can be dynamic. Consider sexually transmitted diseases such as HIV. Partners may be reformed over course of the study period. This may also apply to the IDU group as the needle sharing group may be changed over time.

2.2.1 Correlation and dynamics

Even though the branching process discussed in the previous subsection was sufficing for the initial epidemic dynamics and estimation of R_0 value, it is not suited to study epidemic dynamics. Even when we study an edge of the infection network over time, the state of the node is not constant over time. In fact, the states of the nodes evolve over time. Consider an edge with two susceptible. Eventually, the edge may become an edge with a susceptible and infected. If an infected node was not healed before infecting the susceptible node, then the edge would have two infected nodes. Infected nodes would be healed over time, and we would have an edge with a recovered and an infected node which would progress to an edge with two recovered nodes. However, if an infected node was healed before infecting the susceptible node, then the edge would have a susceptible and a recovered node. Fig. 8 shows all possible nodes pair for an edge.

In epidemiology modeling, we consider the node as an individual. Because the infection status of an individual can change between susceptible, infectious, and recovered, we need to consider node at time t to take one of the element in set $\{S, I, R\}$ at time t. Define status of individual i at time t as $v_i(t)$ for $i = 1, 2, \ldots N$. Then, $v_i(t)$ is the random variable with state space $\{S, I, R\}$ for SIR model. When we consider the HIV transmission dynamics between couples or paired nodes, each couple undergoes a transition of states as shown in Fig. 9. For couple status, the discordant couple is the couples with the susceptible individual with an infected partner, whereas the concordant couple are the couples with both partners are either susceptible or infected. Discordant couples become infected concordant couples in monogamous STD dynamics because infected partner transmit the disease to an uninfected

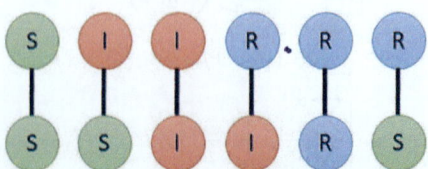

FIG. 8 All possible node pairs an edge can take. For SIR model, we have three states having $2^3 = 8$ possible contact patterns between pair of individuals (pairs of nodes). Only S-I pair in I node can transfer infection to S node in the pair.

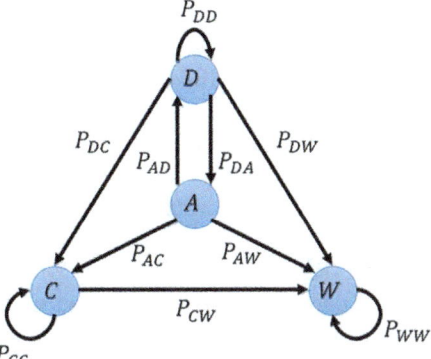

FIG. 9 Network representing Status of HIV infection for couples. D, A, C, and W represents discordant, under therapy, concordant, and widowed couple status respectively. Assuming no reformation of couples, $P_{WW} = 1$.

partner. Thus, Markov Chain model can be used to study the static monogamous STD dynamics as done in Lee (2017). We provide treatment to the discordant couple to reduce the infected concordant couples. This results in four states of couples: discordant couples without intervention, discordant couples with intervention, infected concordant couples, and singles from break-up couples. Let D, A, C and W be discordant couples without intervention, discordant couples with intervention, infected concordant couples, and singles respectively, and Y_n be the status of a couple at time n. Then, Y_n is the random variable with state space $S = \{D, A, C, W\}$ and we can model with transition matrix.

$$
P = \begin{array}{c} \\ D \\ A \\ C \\ W \end{array}
\begin{array}{cccc}
D & A & C & W
\end{array}
\left(\begin{array}{cccc}
P_{DD} & P_{DA} & P_{DC} & P_{DW} \\
P_{AD} & P_{AA} & P_{AC} & P_{AW} \\
0 & 0 & P_{CC} & P_{CW} \\
0 & 0 & 0 & 1
\end{array}\right)
\tag{8}
$$

where $\sum_{j \in S} P_{ij} = 1$ for each $i \in S$. The P_{ij} stand for probability of couple with state i to become couple with state j in one time step. This models the probability of status of a couple over time assuming infection probability is uniform for the population. However, infection depends on contact network and states of adjacent nodes. Letting $Y(t) = \{y_1(t), \cdots, y_N(t)\}$ for N couples allow us to consider contact network structure that is we are considering hierarchical structures where state transition network is incorporated within each node of the contact network. Several studies has incorporated using this setting (Pellis et al., 2015).

2.2.2 SIR model in network

Now, consider the edge of the network as contacts involving disease transmission instead of all contacts between individuals. Then, recovered with lifetime immunity subjects as observed in measle infection would be isolated from the rest of the population in the disease transmission contact network. Considering the position of nodes or individuals are fixed geographically and edges form for nearest neighbors of the nodes, this originally static network becomes dynamics with the evolution of states. A few studies have considered this scenario for modeling (Rhodes and Anderson, 1996, 1997). In addition, if an individual shifts to changes the contact pattern in the contact network over time, network structure changes over time. Let focus on a blood borne pathogen infection such as HIV and Hepatitis C among injection drug user (IDU). Then, the nodes of the network represent the state of individuals and the edges of the network represent the needle sharing between individuals. Assuming needle sharing pattern changes over time, it is convenient to separate susceptible and infectious groups in two groups as "first user" and "second user" groups. Here, the "first user" uses the needle first, thus contaminate the needle if his or her status is infectious. The "second user" uses a needle after the "first user" and may be infected if the "first user" is infectious. With pair formation rate α and pair inactivation rate ρ with degrees of each node is considered for study this scenario, and was used in STD modeling (Eames and Keeling, 2004).

The formation of the destruction of edges is a characteristic of a dynamic network. Recall individual i represented as v_i is partnered with individual j, who is represented as v_j, and element of adjacency matrix $X_v(v_i, v_j)$ indicate presence of connections between node v_i and v_j. Thus, $X_v(v_i, v_j)$ is the indicator function of edge between node v_i and v_j, and $X_v(v_i, v_j)$ changes over time for the dynamic network. Thus, we need to consider the dynamics of an adjacency matrix along with the status of each node. However, in epidemiology scenario, not all the edges participate in infection: individuals in concordant couples maintain their disease status while susceptible individuals in discordant couples may become infectious status for STD dynamics. The pair can retain the information regarding infections, and we can use pairs for the compartment of models as shown in Fig. 10. Thus, we consider couple pairs instead of individuals when we study dynamic epidemic networks such as STD network where partners may break-up and forms new partners.

To understand and simulate realistic disease transmission dynamics, we need to understand the relation between degrees of a network. Transmission of disease can be followed from infectious node to a susceptible node. For particular transmissions, let consider the sequence of disease incidence nodes $v_{I1}, v_{I2}, \ldots$ where v_{Ii} is node turned to infectious from susceptible at time i for $i = 1, 2, \ldots$. Because each node has degree, we can also show the sequence of degree for disease incidence nodes as $k_{I1}, k_{I2}, \ldots$ for corresponding degrees

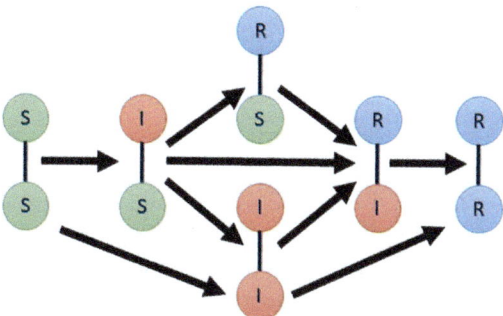

FIG. 10 Flowchart of the node pair model. Disease transmit from infected node to susceptible node, thus infection can be observed to S-I pairs. Since nodes can have multiple edges, S-S pair can be infected if adjusting nodes contains infected nodes. Infected nodes will recover over time, yielding R-S, R-I, and eventually R-R pairs.

where k_{Ii} is the degree of node turned to infectious from susceptible at time i for $i = 1, 2, \ldots$. Now, let us consider the conditional probability

$$\rho_k = P(K_{Ii}|K_{Ii-1}) \tag{9}$$

The conditional probability is also known as correlation of network, and is estimated using ratio of total number of edges on nodes. Here, correlation value close to zero means nodes are independent. Alternatively, Keeling has used considered multiplicative ideas to represent the correlation (Keeling et al., 1997). Consider correlation of edge [X-Y] or the conditional probability of edge [X-Y],

$$\rho_{XY} = P([X\text{-}Y]|[X] \text{ and } [Y] \text{ are independent}) = \frac{N}{n}\frac{[X\text{-}Y]}{[X][Y]}. \tag{10}$$

Here, N represents number of nodes, and n represents mean number of degree. We need to be careful to interpret this multiplicative correlation of network because we usually refer correlation of 0 for independent. For this correlation, value close to 1 related with independent. That is, dynamics of the epidemics would deviates with complex network structures or correlation deviate from 1.

Let [S], [I], [R], be the number of susceptible, infectious, and recovered nodes in the network. Also define [X-Y] be the number of edges with individual with status X in one node and individual with status Y in another node, where $X, Y \in \{S, I, R\}$. We can write the SIR model as

$$\frac{d[S]}{dt} = -\tau[S\text{-}I]$$
$$\frac{d[I]}{dt} = \tau[S\text{-}I] - \gamma[I] \tag{11}$$
$$\frac{d[R]}{dt} = \gamma[I]$$

Assuming independence of nodes, we can estimate the [S-I] as

$$[S\text{-}I] \approx \sum_{k=0}^{\infty} kp_k [S] \frac{[I]}{N}$$

$$= \frac{\sum_{k=0}^{\infty} kp_k}{N} [S][I]$$

(12)

Note that $\tau[S\text{-}I] \approx \frac{\tau \sum_{k=0}^{\infty} kp_k}{N} [S][I] = \beta[S][I]$, and we obtain the SIR model by assuming independence of the degree of nodes (Keeling et al., 1997).

Since every edge in the disease transmission network has the potential to transmit disease from infectious node to adjacent susceptible node, let us consider the length of all edges is equal. Then, average distance or step for a node to reach any node in the network is shorter for small-world network than lattice network which is shown in Fig. 4. We see that there are more triangles in the small-world network, and correlations seem to be present in the small-world network. Moreover, one of the properties of the small-world network is the high clustering of nodes, where clustering measures the closeness of nodes through measuring the number of triangles in network. The clustering of node i can be obtained as

$$C_i = \frac{\text{Number of Triangles containing } V_i}{\text{Number of connected tripe edges containing } V_i}$$

(13)

The clustering coefficient C_i ranges [0,1]. When $C_i=0$ means no clusters while $C_i=1$ means all nodes has degree k, the degree of nodes. Previous study (Keeling et al., 1997) has mentioned utilization of clustering to adjust the correlations in order to keep the topological information of the network. See Fig. 11 for the example of network with clusters. Identification of clusters are important for making policies regarding intervention and require complete graph structures (Rao, 2020).

To incorporate the clustering, we improve Model 11 to Model 14. The Model 14 models the number of edges and uses triplets.

$$\frac{d[S\text{-}S]}{dt} = -\tau[S\text{-}S\text{-}I] + \gamma([S\text{-}I] + [S\text{-}R])$$

$$\frac{d[S\text{-}I]}{dt} = -\tau([I\text{-}S\text{-}I] + [S\text{-}I] - [S\text{-}S]) + \gamma([I\text{-}I] + [I\text{-}R] - [S\text{-}I])$$

$$\frac{d[S\text{-}R]}{dt} = -\tau[R\text{-}S\text{-}I] + \gamma([I\text{-}R] + [R\text{-}R] - [S\text{-}R])$$

(14)

$$\frac{d[I\text{-}I]}{dt} = \tau[S\text{-}I] - \gamma[I\text{-}I]$$

$$\frac{d[I\text{-}R]}{dt} = \tau[S\text{-}R] - \gamma([I\text{-}R] - [I\text{-}I])$$

Here, number of triplets [X-Y-Z] would be estimated as

$$[X\text{-}Y\text{-}Z] = \frac{n-1}{n} \frac{[X\text{-}Y][Y\text{-}Z]}{[Y]}$$

(15)

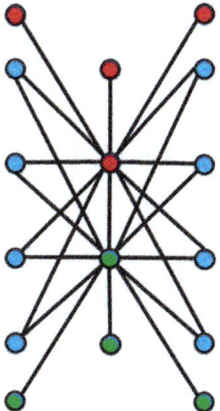

FIG. 11 Network with clustering. The blue nodes share two nodes and form a cluster. If either red or green nodes are infectious, all blue nodes have the same risk of getting infection even though they are two edges apart. Similarly, red and the green nodes has the same risk of getting infection if one of the blue nodes is infectious.

if no clustering is considered. However, behavior of dynamics changes with presence of clustering, and the Keeling et al. (1997) has used correlation between X and Z, C_{XZ}, and adjustment factor ϕ to incorporate clustering information in estimation of the [X-Y-Z]. The adjusted equation is

$$[X\text{-}Y\text{-}Z] = \frac{n-1}{n} \frac{[X\text{-}Y][Y\text{-}Z]}{[Y]} ((1-\phi) + \phi C_{XZ}) \tag{16}$$

where $C_{XZ} = \frac{N[X\text{-}Y]}{n[X][Y]}$. Note that the accuracy of the pair-wise model estimation may not match that of agent based model (ABM), it reduces complexity of model and deliver outbreak size or threshold for disease outbreak analytically.

So far, we have assumed that every edge has the same contact behaviors. However, contact behaviors between individuals are not the same. For example, a child would have frequent or long contact with his or her family or friends compared to other classmates or teachers. When we study the spreading of disease, this discrepancy in contact behaviors may result in different dynamics. For STD model, partner selection may not be distributed uniformly, and Moslonka-Lefebvre et al. has studied the effects of the preference pattern of partner preference (Moslonka-Lefebvre et al., 2012).

2.3 Multi-layered model

The other type of epidemic dynamics is the mixing of the different networks. Consider disease spreads in people who live in islands. Disease spreads within islands with contact network of each islands, and spreads to other islands through movement of infectious individuals between islands. Studying such a system is important as it can applied to model the disease transmission networks of multiple nations as well as animal to human infections. With

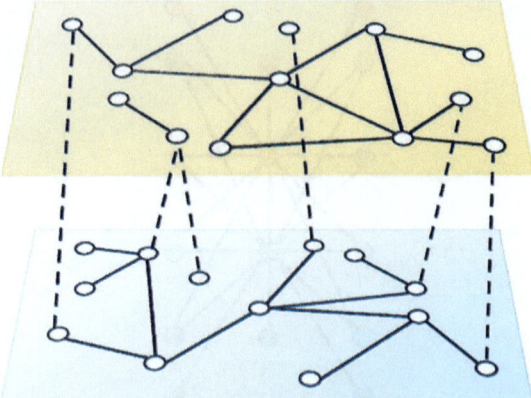

FIG. 12 Two-layered networks. Two different network may interact. For example, interaction of bird contact network and human contact network would yield better model for H1N1 model.

emergence of MERS, ebola, and H1N1 epidemics, multi-layered model is gaining importance. In case of avian flu epidemics, the disease spreads not only among human, but also spread among birds. Because human-bird interaction can spread the disease between two species, we need to consider the interaction of two networks: one for human disease transmission network and another for bird disease transmission network. Fig. 12 shows the two-layered networks. Not only for animal-human interaction network, this framework can expand to international disease mapping including multiple nations.

3 Syringe exchange on Hepatitis C

One of the interest in controlling the Hepatitis C infection among IDU is effectiveness of the needle distribution. In Vancouver study, Needle and syringe programmes (NSPs) increased cessation. That is increasing availability of needle would reduce the IDU population. This was promising results, however, we may want to see other possible scenarios before generalize the Vancouver trend to other places. Assuming no influx of IDU population, we will consider several scenarios with alteration of needle sharing behavior. We separate the groups to "sharing" and "non-sharing" group according to syringe sharing behavior of the IDU.

3.1 Syringe exchange SIRS model

For dynamics of Hepatitis C infection of IDU, we start with considering SIRS model as recovered individual may also practice old habits. Let S_i and I_i be number of susceptible and infectious individuals in group $i \in \{1, 2\}$, where $i = 1$ represent syringe sharing and $i = 2$ represent non-sharing group. At instance of drug injection contact, edges are formed only between needle sharing users. So, we have three possible way of contacts: (S_1, S_1), (S_1, I_1), and (I_1, I_1). Among them a single pair, (S_1, I_1) participate in dynamics. However, the IDU may alter their syringe sharing behavior. We let α_1 and α_2 be rate of sharing behavior transition between sharing and non-sharing group. The infected individuals would quit needle sharing practice or under goes treatment with rate δ, and return to old habits with rate γ. These are modeled as

$$\frac{dS_1}{dt} = -\beta_1 S_1 I_1 - \alpha_1 S_1 + \alpha_2 S_2 + \rho \gamma R$$

$$\frac{dS_2}{dt} = -\alpha_2 S_2 + \alpha_1 S_1 + (1 - \rho)\gamma R$$

$$\frac{dI_1}{dt} = \beta_1 S_1 I_1 + \alpha_2 I_2 - \alpha_1 I_1 - \delta I_1 \qquad (17)$$

$$\frac{dI_2}{dt} = \alpha_1 I_1 - \alpha_2 I_2 - \delta I_2$$

$$\frac{dR}{dt} = \delta(I_1 + I_2) - \gamma R$$

where ρ represents probability of returned subjects to be participate in needle sharing. From Model (IT), we can see that S_2 do not get Hepatitis C infection and I_2 do not transmit Hepatitis C infection. Thus, syringe exchange program reduces size of susceptible population, an effect similar to that of vaccine. The Fig. 13 shows the effect of syringe exchange program. Assuming

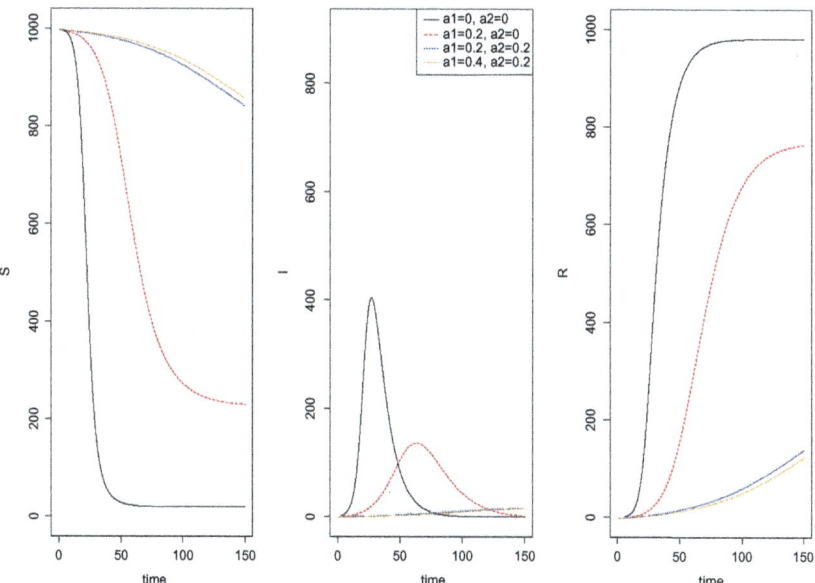

FIG. 13 Simulation results for Model 17. For initial population, $S_1 = 999$ and $I_1 = 1$ were used. If more participants of syringe exchange program quit sharing practices (larger α_1), more IDU are protected from Hepatitis C infection (larger S).

homogeneity in contact patterns, the syringe exchange program is effective in controlling the Hepatitis C infection for IDU population.

3.2 Syringe exchange SIRS model adjusted with network

Recall that R_0 was the average number of connected edges in the network, removal of some edges in the network would reduce it. Thus, we expect improved controlling result from the syringe exchange program. The effectiveness was already demonstrated through differential equation approaches as discussed earlier. However, we need to question the homogeneous mixing of population. Because syringe sharing distribution is estimated to be geometric distribution with rate 0.6, we can generate possible network for 1000 IDU as shown in Fig. 14 (Fu et al., 2016). More than half of the IDU do not share their syringes, and homogeneous contact pattern as assumed in differential equation approach fails. The dynamics can be obtained through agent based model (Fig. 15), however, we can save time and estimate rough trend through ODE estimation as done in Model (11).

FIG. 14 IDU networks. Possible IDU syringe sharing network assuming degree distribution as geometric distribution with rate 0.6. 40% of IDU do not share syringes.

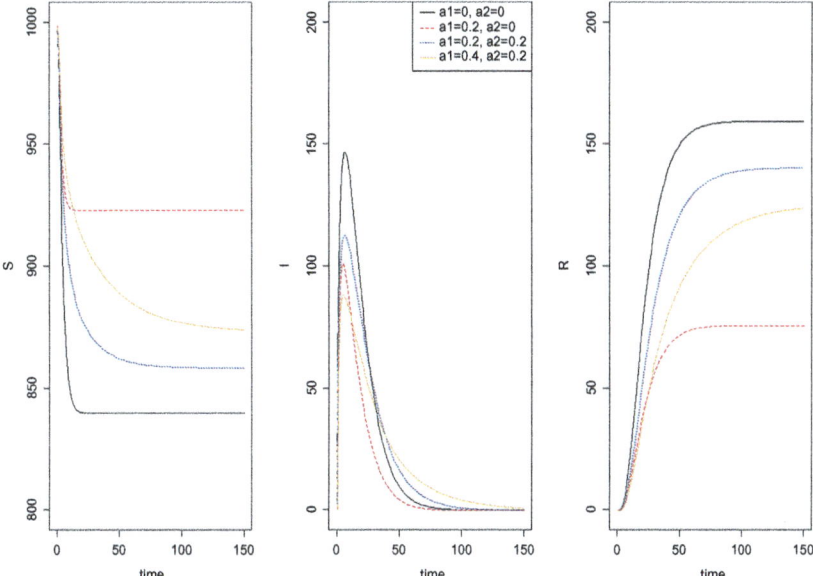

FIG. 15 Simulation results for network shown in Fig. 14. The infection rate of 0.4 and treatment rate of 0.1 were used. If more participants of syringe exchange program quit sharing practices, more IDU are protected from Hepatitis C infection.

$$\frac{d[S_1\text{-}S_1]}{dt} = -\tau[S_1\text{-}S_1] - \alpha_1[S_1\text{-}S_1] + \alpha_2[S_1\text{-}S_2]$$

$$\frac{d[S_1\text{-}I_1]}{dt} = \tau([S_1\text{-}S_1] - [S_1\text{-}I_1]) + \gamma[S_1\text{-}I_1] - \alpha_1[S_1\text{-}I_1] + \alpha_2[S_1\text{-}I_2]$$

$$\frac{d[S_1\text{-}R]}{dt} = \gamma([S_1\text{-}I_1] + [S_1\text{-}I_2])$$

$$\frac{d[I_1\text{-}I_1]}{dt} = \tau[S_1\text{-}I_1] - \gamma[I_1\text{-}I_1] - \alpha_1[I_1\text{-}I_1] + \alpha_2[I_1\text{-}I_2]$$

$$\frac{d[I_1\text{-}R]}{dt} = \gamma([I_1\text{-}I_1] - [I_1\text{-}R]) - \alpha_1[I_1\text{-}R] + \alpha_2[I_1\text{-}R]$$

$$\frac{d[S_1\text{-}S_2]}{dt} = \alpha_1([S_1\text{-}S_1] - [S_1\text{-}S_2]) - \alpha_2([S_1\text{-}S_2] - [S_2\text{-}S_2]) \qquad (18)$$

$$\frac{d[S_2\text{-}S_2]}{dt} = \alpha_1[S_1\text{-}S_2] - \alpha_2[S_2\text{-}S_2]$$

$$\frac{d[S_1\text{-}I_2]}{dt} = -\gamma[S_1\text{-}I_2] + \alpha_1[S_1\text{-}I_1] - \alpha_2[S_1\text{-}I_2]$$

$$\frac{d[I_1\text{-}I_2]}{dt} = -\gamma[I_1\text{-}I_2] + \alpha_1([I_1\text{-}I_1] - [I_1\text{-}I_2]) - \alpha_2([I_1\text{-}I_2] - [I_2\text{-}I_2])$$

$$\frac{d[I_2\text{-}I_2]}{dt} = -\gamma[I_2\text{-}I_2] + \alpha_1[I_1\text{-}I_2] - \alpha_2[I_2\text{-}I_2]$$

$$\frac{d[I_2\text{-}R]}{dt} = \gamma([I_1\text{-}I_2] + [I_2\text{-}I_2] - [I_2\text{-}R]) + \alpha_1[I_1\text{-}R] - \alpha_2[I_2\text{-}R]$$

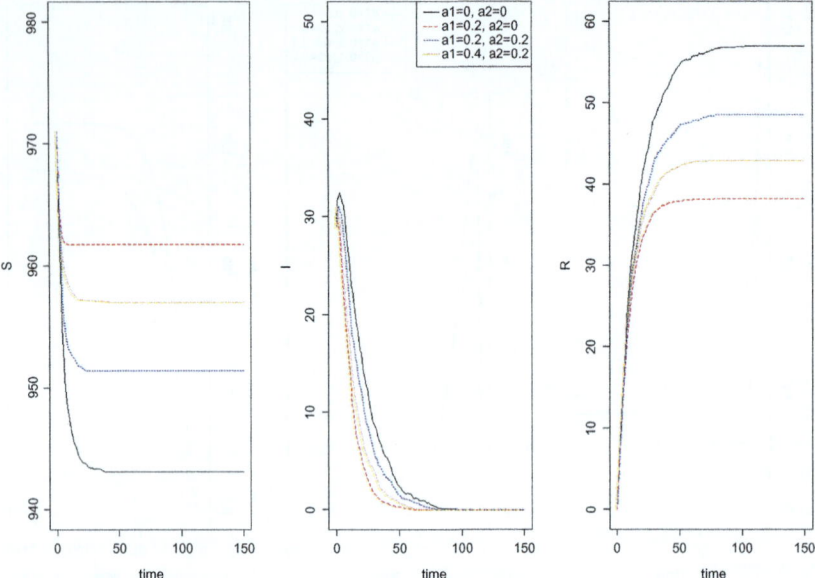

FIG. 16 Simulation results for Model 18. For initial population, $S_1 = 999$ and $I_1 = 1$ with infection rate of $\tau = 0.4$ and treatment rate of $\gamma = 0.1$ were used. If more participants of syringe exchange program quit sharing practices, more IDU are protected from Hepatitis C infection.

Simulating the model on the network obtained in Fig. 16, we observe that the syringe exchange program protect IDU people from Hepatitis C infection. However, the effect of syringe exchange program is not very impressive compared to homogeneous mixing assumptions. Because of non-homogeneous contact patters, network modeling approach is appropriate.

To confirm appropriateness in using network model on predicting efficacy of syringe exchange program for Hepatitis C, let us compare the efficacy of the syringe exchange program on HIV. Even though HIV is different disease, transmission rate of both HIV and Hepatitis C are low (<0.5%) for IDU. Moreover, they may share same network structure for IDU. Odds ratio of HIV infections on syringe exchange program was 1.5 when 28% of IDU participate in syringe exchange program (Wood et al., 2003). Assuming $\alpha_1 = 0.2$ and $\alpha_2 = 0$, odds ratio are 2.5 and 1.3 for Model (17) and Model (18) respectively while odds ratio is 1.2 for agent based model. This big deviation was introduced by heterogeneity in contact pattern and Model (17) estimated very large infection compared to Model (18) and ABM. However, ODE estimation of the network method still overestimate the infection compared to ABM and we can improve it through considering more complex features of the network such as clusters.

4 Discussion

Homogeneous mixing models have produced valuable results on epidemiology (Anderson and May, 1992). However, most human interactions are not homogenous contact, and network modeling approach is one of the alternative approach that is intuitive as well. In this chapter, we introduced the concept of networks and its application on epidemic modeling.

We demonstrated the effect of non-homogeneous mixing with syringe sharing model of IDU population on Hepatitis C. We used idealized network to study the possible dynamics, and found homogeneous mixing assumption may exaggerate the effect of syringe sharing program. The network modeling approach may yield dynamics with higher resolution, but information needed to construct the network is not readily available. The network construction is one of the major downside for the network modeling. For IDU network, we need to work with contact tracing methods, and network are very hard to obtain. However, for other diseases, the advancement of mobile technologies and GPS may give human contact patterns and bring light to potential of the network model on epidemiology (Foster et al., 2018; Moss et al., 2019; Vazquez-Prokopec et al., 2013).

References

Allen, L.J., Brauer, F., Van Den Driessche, P., Wu, J., 2008. Mathematical Epidemiology. vol. 1945 Springer.

Anderson, R.M., May, R.M., 1992. Infectious Diseases of Humans: Dynamics and Control. Oxford University Press.

Eames, K., Keeling, M., 2004. Monogamous networks and the spread of sexually transmitted diseases. Math. Biosci. 189, 115–130.

Foster, S., Adams, E., Dunn, I., Dent, A., 2018. Centers for Disease Control and Prevention Geographic Information System Data. https://www.cdc.gov/eis/field-epi-manual/chapters/GIS-data.html. Accessed: 29.06.2020.

Fu, R., Gutfraind, A., Brandeau, M.L., 2016. Modeling a dynamic bi-layer contact network of injection drug users and the spread of blood-borne infections. Math. Biosci. 273, 102–113.

Keeling, M., Rand, D., Morris, A., 1997. Correlation models for childhood epidemics. Proc. R. Soc. London, Ser. B 264 (1385), 1149–1156.

Lee, T., 2017. Mathematical and Stochastic Modeling of HIV Immunology and Epidemiology. PhD Thesis, Augusta University.

Moore, C., Newman, M.E., 2000. Epidemics and percolation in small-world networks. Phys. Rev. E 61 (5), 5678.

Morris, M., Kretzschmar, M., 1997. Concurrent partnerships and the spread of HIV. AIDS 11 (5), 641–648.

Moslonka-Lefebvre, M., Bonhoeffer, S., Alizon, S., 2012. Weighting for sex acts to understand the spread of STI on networks. J. Theor. Biol. 311, 46–53.

Moss, R., Naghizade, E., Tomko, M., Geard, N., 2019. What can urban mobility data reveal about the spatial distribution of infection in a single city? BMC Public Health 19 (1), 1–16.

Nelson, P.K., Mathers, B.M., Cowie, B., Hagan, H., Des Jarlais, D., Horyniak, D., Degenhardt, L., 2011. Global epidemiology of hepatitis B and hepatitis C in people who inject drugs: results of systematic reviews. Lancet 378 (9791), 571–583.

Pellis, M., House, T., Keeling, M., 2015. Exact and approximate moment closures for non-Markovian network epidemics. J. Theor. Biol. 382, 160–177.

Rao, A.S.S., 2020. Population network structures, graph theory, algorithms to match subgraphs may lead to better clustering of households and communities in epidemiological studies. Epidemiol. Infect. 148, e10.

Rao, A.S.S., Tomley, F., Blake, D., 2015. Understanding chicken walks on n × n grid: Hamiltonian paths, discrete dynamics, and rectifiable paths. Math. Methods Appl. Sci. 38 (15), 3346–3358.

Rhodes, C., Anderson, R., 1996. A scaling analysis of measles epidemics in a small population. Philos. Trans. R. Soc. Lond. B Biol. Sci. 351, 1679–1688.

Rhodes, C., Anderson, R.M., 1997. Epidemic thresholds and vaccination in a lattice model of disease spread. Theor. Popul. Biol. 52 (2), 101–118.

Vazquez-Prokopec, G.M., Bisanzio, D., Stoddard, S.T., Paz-Soldan, V., Morrison, A.C., Elder, J.-P., Ramirez-Paredes, J., Halsey, E.S., Kochel, T.J., Scott, T.W., et al., 2013. Using GPS technology to quantify human mobility, dynamic contacts and infectious disease dynamics in a resource-poor urban environment. PLoS One 8 (4), e58802.

Wood, E., Kerr, T., Spittal, P.M., Small, W., Tyndall, M.W., O'shaughnessy, M.V., Schechter, M.-T., 2003. An external evaluation of a peer-run "unsanctioned" syringe exchange program. J. Urban Health 80 (3), 455–464.

Chapter 9

Modeling and forecasting the spread of COVID-19 pandemic in India and significance of lockdown: A mathematical outlook

Brijesh P. Singh[*]

Department of Statistics, Institute of Science, Banaras Hindu University, Varanasi, India
[*]*Corresponding author: e-mail: brijesh@bhu.ac.in*

Abstract

A very special type of pneumonic disease that generated the COVID-19 was first identified in Wuhan, China in December 2019 and is spreading all over the world. The
ongoing outbreak presents a challenge for data scientists to model COVID-19, when
the epidemiological characteristics of the COVID-19 are yet to be fully explained.
The uncertainty around the COVID-19 with no vaccine and effective medicine available
till today create additional pressure on the epidemiologists and policy makers. In such a
crucial situation, it is very important to predict infected cases to support prevention of
the disease and aid in the preparation of healthcare service. India is fighting efficiently
against COVID-19 and facing greater challenges because of its large population and
high population density. Though the government of India is taking all needful steps
to prevent its spread but it is not enough to control and stop spread of the disease so
far, perhaps due to defiant nature of people living in India. Effective measure to control
this disease, medical professionals needs to know the estimated size of this pandemic
and pace. In this study, an attempt has been made to understand the spreading capability
of COVID-19 in India through some simple models. Findings suggest that the lockdown
strategies implemented in India are not successfully reducing the pace of the pandemic
significantly after first lockdown.

Keywords: COVID-19, Differential equations, Logistic growth, Propagation,
Joinpoint regression, Lockdown

Handbook of Statistics, Vol. 44. https://doi.org/10.1016/bs.host.2020.10.005
257

1 Background

A novel corona virus is responsible for epidemic popularly known as COVID-19 is a new strain that has not been identified previously in humans. World Health Organization (WHO) declared COVID-19 a pandemic on March 11, 2020. The virus that caused the incidence of Severe Acute Respiratory Syndrome (SARS) in 2002 in China, Middle East respiratory syndrome (MERS) in 2012 in Saudi Arabia and the virus that causes COVID-19 are genetically related to each other, but the diseases they caused are quite different (WHO). These viruses, in general, are a family of viruses that target and affect mammal's respiratory systems. The SARS corona virus spread to humans via civet cats, while the MERS virus spread via dromedaries. In case of the novel corona virus, typically happens via contact with an infected animal, perhaps the common carriers are bats initial reports from seafood market in central Wuhan, China.

The Novel Corona Virus (COVID-19) started from Wuhan, China and thus, initially known as the Wuhan virus, expanded its circle in South Korea, Japan, Italy, Iran, USA, France, Spain and finally spreading in India. It is named as novel because it is never seen before mutation of animal corona virus but certain source of this pandemic is still unidentified. It is said that the virus might be connected with a wet market (with seafood and live animals) from Wuhan that was not complying with health and safety rules and regulations.

As of July 16, 2020, with the continuously increasing global risk more than 14 million confirm positive cases and more than 0.58 million of deaths have occurred in the world. As number of cases growing day by day, in most of the countries of the world, some most populous countries like China, India, Brazil, USA, etc., are badly affected by it. In this context, the crucial role of modeling, transmission dynamics and estimating development of COVID-19 are expected. The population based mathematical model especially growth model in this scenario are the most preferable techniques to understand the epidemic future trajectory. Epidemiological characteristics like propagating dynamics, severity, susceptibility, and the effects of control measures, for COVID-19 has produced a greater concern for researchers (Cowling and Leung, 2020; Lipsitch et al., 2020).

Since preventive measures like lockdown and social distancing have immense pressure on economy of the country, quantitative estimates and predictions are necessary to learn the impact of spread that will help in plan the strategies against COVID-19. Given the paucity of such quantitative measures, the predictions on the basis of different idea given in this paper become critical and to know when the COVID-19 stops. In recent past a number of studies with various technique and tools have been carry out to understand the dynamics of propagation of disease and future course of action. For COVID-19, various models which are capable of providing worth insights for health care policy making are being continuously developed and used to explain this pandemic retrospectively as well as to project the events

(Batista, 2020; Koo et al., 2020; Kucharski et al., 2020; Tuite and Fisman, 2020; Wu et al., 2020).

Wu et al. (2020) has been done to analyzing the pace of virus transmissibility through estimating the value of R_0 with the help of stochastic Markov Chain Monte Carlo method. Another analysis with mathematical incidence decay and exponential adjustment is performed. Further to explain growth behavior of COVID-19 a statistical exponential growth model adopting the serial interval from Severe Acute Respiratory Syndrome is applied by Zhao et al. (2020). A three-parameter logistic growth function is applied and predicted for China as well as some other countries is found very satisfying (Shen, 2020). In the context of India, an early study of COVID-19 (when it started spreading in India) done by Singh and Adhikari (2020) rightly believed that countrywide lockdown on March 24 for 21 days may be insufficient for controlling the COVID-19 pandemic. Malhotra and Kashyap (2020) tried to forecast the endpoints to explain the progression of COVID-19 in Indian States, using SIR and logistic growth models and found the endpoint of COVID-19 in India is in July 23, 2020.

India with a huge population about 1.3 billion, among majority of the people are living in poor hygienic condition and the medical facilities like number of doctors and hospitals are less in India as compared to developed countries indicates that the situation of India will become very critical but comparatively better public health system and political control in India than the above developed countries. The picture of India is not so good and has more than 1 million confirm positive cases and more than 26 thousand of deaths. Although the death rate of this pandemic is low in comparison of other pandemics and diseases but its high rate of spread and no proper cure available so far is the major concern in the present time. Right now in India only 29 districts out of 739 districts have COVID-19 case more than 4000. These districts are mainly metropolitans; if we implement preventive measures properly then spread can be under control at desired level, but due to defiant nature of people living in India, political desire and rivalry, still we India society are facing problem made by COVID-19.

The first case of COVID-19 is reported in India on January 30, 2020 when a student returned from Wuhan, China (covid19india.org). The Government of India was quick to launch various levels of travel advisories beginning from February 26, 2020, with restrictions on travel to China and nonessential travel restrictions to Singapore, South Korea, Iran and Italy. The efforts to control by the Hon'ble Prime Minister Narendra Modi Ji through Janata Curfew (public curfew) on March 22, 2020, can be seen as the beginning of wide-scale public preventive measures. India has launched several social distancing measures and personal hygiene measures during the second week of March.

Symptoms of COVID-19 are reported as cough, acute onset of fever and difficulty in breathing. Out of all the cases that have been confirmed, up to 20% have been deemed to be severe. Cases vary from mild forms to severe ones that can lead to serious medical conditions or even death. It is believed that symptoms may appear in 2–14 days, as the incubation period for the COVID-19 has not yet been confirmed. However, in India 14 days minimum quarantine period is declared by Government for suspected cases. Since it is a new type of virus, there is a lot of research being carried out across the world to understand the nature of the virus, origins of its spreads to humans, the structure of it, possible cure/vaccine to treat COVID-19. India also became a part of these research efforts after the first two confirmed cases were reported here on January 31, 2020. Then in India screening of traveler at airport migrant was started, immediate Chinese visas was canceled, and who was found affected from COVID-19 kept in quarantine centers (Ministry of Home Affaires Government of India, Advisory).

For the spread of COVID-19, when disease dynamics are still unclear, mathematical modeling helps us to estimate the cumulative number of positive cases in the present scenarios. Now India is interring in the mid stages of the epidemic. It is important to predict how the virus is likely to grow among the population. The COVID-19 pandemic presents a challenge for data scientists to model it; however, the epidemiological characteristics of the COVID-19 are yet to be fully explained. The uncertainty around the COVID-19 with no vaccine and effective medicine available until today create additional pressure on the epidemiologists and policy makers. In such a crucial situation, it is very important to predict infected cases to support prevention of the disease and support in the preparation of healthcare service. A mathematical modeling approach is a suitable tool to understand the dynamics of epidemic. In the study some mathematical approach to understand the dynamics of novel COVID-19 in India has been discuss.

In absence of a definite treatment modality like vaccine, physical distancing has been accepted globally as the most efficient strategy for reducing the severity of disease and gaining control over it (Ferguson et al., 2020). Also in India it is reported that the country is well short of the WHO's recommendations of minimum threshold of 2.28 skilled health professionals per 1000 population (Anand and Fan, 2016). Therefore, on March 24, 2020, the Government of India under Prime Minister Narendra Modi Ji ordered a nationwide lockdown for 21 days, limiting movement of the entire 1.3 billion population of India as a preventive measure against the COVID-19 pandemic in India. It was ordered after a 14-h voluntary public curfew on 22 March. The lockdown was placed when the number of confirmed COVID-19 cases in India was approximately 500. On 14 April, Prime Minister of India extended the nationwide lockdown until 3 May, with a conditional relaxation after 20 April for some regions. On 4 May, the Government of India again extended the nationwide lockdown further by 2 weeks until 17 May. Also, the Government has divided the entire nation into three zones viz. green, red and orange with relaxations applied accordingly.

There are already various measures such as social distancing, lockdown masking and washing hand regularly has been implemented to prevent the spread of COVID-19, but in absence of particular medicine and vaccine it is very important to predict how the infection is likely to develop among the population that support prevention of the disease and aid in the preparation of healthcare service. This will also be helpful in estimating the health care requirements and sanction a measured allocation of resources. It is well known fact that COVID-19 has spread differently in different countries, any planning for increasing a fresh response has to be adaptable and situation-specific. Data obtained on COVID-19 outbreak have been studied by various researchers using different mathematical models (Chang et al., 2020; Srinivasa Rao Arni et al., 2020). Many other studies (Anastassopoulou et al., 2020; Corman et al., 2020; Gamero et al., 2020; Huang et al., 2020; Hui et al., 2020; Rothe et al., 2020) on this recent epidemic have been reported so many meaningful modeling results based on the different principles of mathematics.

Most of pandemics follow an exponential curve during the initial spread and eventually flatten out (Junling et al., 2014). SIR model is one of the best suited models for projecting the spread of infectious diseases like COVID-19 where a person once recovered is not likely to become susceptible to the infection again (Kermack and McKendrick, 1927). Susceptible-Infectious-Recovered (SIR) compartment model (Herbert, 2000) is used to include considerations for susceptible, infectious, and recovered or deceased individuals. These models have shown a significant predictive ability for the growth of COVID-19 in India on a day to day basis so far. A time dependent SIR models have been defined to observe the undetectable infected persons with COVID-19 (Chen et al., 2020). A recent study by Mandal et al. (2020) has shown that social distancing can reduce cases by up to 62%.

Further, time series models have been employed for predicting the incidence of COVID-19 disease. As compared to other prediction models, for instance support vector machine (SVM) and wavelet neural network (WNN), ARIMA model is more capable in the prediction of natural adversities (Zhang et al., 2019). Chatterjee et al. (2020) studied a stochastic mathematical model of the COVID-19 epidemic in India. The logistic growth regression model is used for the estimation of the final size and its peak time of the COVID-19 pandemic in many countries of the World and found similar result obtained by SIR model (Batista, 2020).

It is well known that the effects of social distancing become visible only after a few days from the lockdown. This is because the symptoms of the COVID-19 normally take some time to come out after getting infected from the COVID-19. An estimates indicates that, with hard lockdown and continued social distancing, the peak total infections in India will be 97 million and the number of infective by September is likely to be over 1100 million (Schueller et al., 2020).

2 Why mathematical modeling?

The study of infectious diseases is called epidemiology. A disease is called endemic if it persists in a population and pandemic when it occurs worldwide. The spread of an infectious disease involves not only disease related factors such as the infectious agent, mode of transmission, latent period, infectious period, susceptibility and resistance, but also social, cultural, demographic, economic and geographic factors. Mainly there are three types of models for infectious diseases that are spreading directly through person to person contact in a population. Some simple models are formulated and analyzed mathematically considering differential equations. Parameters are estimated for infectious diseases and also used to compare the vaccination levels necessary for herd immunity. The three models considered here are the simple epidemiological models and suitable for diseases which are transmitted directly from person to person. More complicated models must be used when there is transmission by insects called vectors or a reservoir of nonhuman infective. Epidemiological models are widely used to understand the pattern and policy development.

Even though vaccines are available for many infectious diseases, these diseases still cause suffering and mortality in the world, especially in developing countries. In developed countries chronic diseases such as cancer and heart disease have received more attention than infectious diseases, but infectious diseases are still a more common cause of death in the world. The transmission mechanism from an infective to susceptible is understood or nearly all infectious diseases and the spread of diseases through a chain of infections is known. However, the transmission interactions in a population are very complex so that it is difficult to comprehend the large scale dynamics of disease spread without the formal structure of a mathematical model. An epidemiological model uses a microscopic description (the role of an infectious individual) to predict the macroscopic behavior of disease spread through a population. In many sciences it is possible to conduct experiments to obtain information and test hypotheses. Experiments with infectious disease spread in human populations are often impossible, unethical or expensive. Data is sometimes available from naturally occurring epidemics or from the natural incidence of endemic; however, the data is often incomplete due to underreporting. This lack of reliable data makes accurate parameter estimation difficult so that it may only be possible to estimate a range of values for some parameters. Since repeatable experiments and accurate data are usually not available in epidemiology, mathematical models and computer simulations can be used to perform needed theoretical experiments.

Mathematical models have both limitations and capabilities that must recognized. Sometimes questions cannot be answered by using epidemiological models, but sometimes the modeler is able to find the right combination of available data, an interesting question and a mathematical model which can lead to the answer. Comparisons can lead to a better understanding of the processes of disease spread. Modeling can often be used to compare different diseases in the same population, the same disease in different populations, or the same disease at different times. Comparisons of diseases such as measles, rubella, mumps, chickenpox, whooping cough, poliomyelitis and others are made (Hethcote, 1983; Yorke and London, 1973; Yorke et al., 1979) and in the article on rubella in this volume by Hethcote (1989). Quantitative predictions of epidemiological models are always subject to some uncertainty since the models are idealized and the parameter values can only be estimated. However, predictions of the relative merits of several control methods are often robust in the sense that the same conclusions hold over a broad range of parameter values and a variety of models. Optimal strategies for vaccination can be found theoretically by using modeling. Longini et al. (1978) use an epidemic model to decide which age groups should be vaccinated first to minimize cost or deaths in an influenza epidemic. Hethcote (1988) uses a modeling approach to estimate the optimal age of vaccination for measles.

Within a short period of time, COVID-19 has traumatized the world with a greater magnitude and coercion than older pandemics. Its eventuality is grabbed by the fact that it has infected millions and killed thousands across the globe. Global markets, accessible transportation, large scale production have largely contributed to make this pandemic spread faster. This has drastically affected the social life and health mental as well as physical of human beings worldwide. The already burdened health infrastructure across the globe is virtually exposed up to an irreparable point. The WHO declared 2019–2020 corona virus outbreak a Public Health Emergency of International Concern (PHEIC) on January 30th, 2020 and a pandemic 12 days later on February 12th, 2020. With its outbreak in Wuhan, China, the pandemic seems to occupy and include all the vitals of the world thereby affecting the mechanistic processes of any nation. The countries are trying hard to combat and contain this outbreak by following suitable set of protocols that tend to alter the transmission rate effectively. In the initial phase of spread of COVID-19; Italy, Spain, France and some other European countries are one of the worst sufferers of the pandemic and the coercive measures have resulted in the disruption of all the necessary services. On the other hand, the case is virtually less severe in South Asia. India is less affected by the COVID-19, however, China is its neighboring country having border through buffer states like Nepal and Bhutan.

Being the second most populous country of the world, India is fighting hard to minimize the damage of COVID-19. As on 15th April, the total number of infected cases in India was 12,370 with 422 deaths and most recoveries (covid19india.org). India reported its first case on 30th January and entered the countrywide lockdown on March 24th, 2020 with constantly increase in number of COVID-19 cases. Indian government as well as states government has issued early guidelines and travel advisories to limit the further damage of disease. Also, the timely precautions taken by the government have contributed greatly toward combating this pandemic. The paper attempts to devise a model that would conveniently help in assessing the predictability of pandemic COVID-19 in future time period. This can be achieved by evaluating the different parameters that directly or indirectly affect the ongoing rate of pandemic. Moreover, theoretical explanation, quantitative analysis and other parameters are highly required to predict the peak and size of any pandemic.

We obtained information on cumulative number of COVID-19 confirmed cases in India from covid19india.org. All cases are laboratory confirmed following the case definition by the Government of India. Some studies modeled the epidemic curve obeying the exponential growth (De Silva et al., 2009). The nonlinear least square framework is adopted for data fitting and parameter estimation for COVID-19 at this early stage. First exponential and then logistic growth curve is used to model the COVID-19 pandemic, since epidemics grow exponentially not linearly. But it is surprising that exponential growth curve always provide increasing number of daily new cases. There is no saturation point. Another deterministic model used for understanding the dynamics of epidemic is the Susceptible-Infectious-Recovered (SIR) model, which has been used to accurately predict incidence like SARS. In the SIR model, we need to know the input parameters first the stats we feed into the model (Chatterjee et al., 2020; Mandal et al., 2020; Singh and Adhikari, 2020). The first one is R_0 called the basic reproduction number. It is essentially the number of new cases a single infected person will cause during their infectious period. It is one of the most important parameters for assessing any epidemic. Corona virus has an $R_0 \sim 2.4$. In contrast, the swine flu virus had an $R_0 \sim 1.5$ in the 2009 swine flu epidemic (Gupta, 2020). The R_0 will inform us about how many people will get infected with one infected person. Other one is the case fatality rate (CFR), which is the percentage of infected people that will die due to the infection. The CFR for corona virus has been reported between 0.5% and 4%. The lower values are more appropriate in resource better settings of medical facility. But SIR model assumes that every person is moving and has equal chance of contact with each and every other person among the population irrespective of the space or distance between different people. It is assumed that the transmission rate remains constant throughout the period of pandemic. Also this model considered to have the same transmission rate for who have been diagnosed and are in quarantine or those who have not been quarantined. The harmonic analysis methods and dynamic model (Rao Srinivasa Arni et al., 2020) estimates show that the number of COVID-19 infected would be 9225 (if there were 10 infected individuals as of March 1, 2020, who was not taking any precautions to spread), 17,986 (if there were 20) and 44,265 (if there were 50).

2.1 SIR and SEIR model

SIR model is a theoretical epidemiological model, in which, the population is categories into three component such as: susceptible (S), which is the group of people who are vulnerable to exposure with infectious people, infected (I), are those with the disease and can transmit it to the susceptible and the third component is the individuals who have recovered from the infectious disease and developed immunity and not susceptible to the same illness anymore (R). This framework enables us to understand the dynamics of any epidemic. Thus SIR model is a compartmental model in which individuals are separated into different compartments based on their status and follow the corresponding population sizes over the time. The diagrammatical representation of three-compartment model (Kermack and McKendrick, 1927) is given as

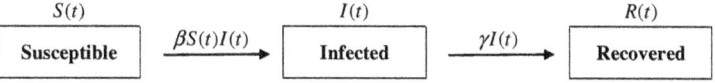

where, $S(t)$ = Proportion of individual susceptible to COVID-19 at time t, $I(t)$ = Proportion of individual who have been infected by COVID-19 and are capable of infecting others at time t, and $R(t)$ = Proportion of individual who have been infected by COVID-19 and recovered at time t, such that $S(t) + I(t) + R(t) = 1$. Here β is the transmission parameter controlling how much the disease can be transmitted. This is the average number of individuals that one infected individual will infect per unit time. It is determined by the chance of contact and the probability of disease transmission. While γ is the parameter representing the rate of recovery in a particular period. The model allows us to describe the number or proportions of persons in each compartment by solving the following ordinary differential equations,

$$\frac{dS(t)}{dt} = -\beta S(t)I(t)$$

$$\frac{dI(t)}{dt} = \beta S(t)I(t) - \gamma I(t)$$

$$\frac{dR(t)}{dt} = \gamma I(t)$$

Several assumptions have been discussed with respect to the SIR model (Brauer and Castillo-Chavez, 2012; Daley and Gani, 1999). Based on the SIR model, the basic reproduction number is defined as,

$$R_0 = \frac{\beta}{\gamma}$$

Here, R_0 is the average number of new COVID-19 cases produced by a single COVID-19 infected case over the time. In order to fit a SIR model, the parameters were obtained by minimizing the residual sum of squares between the observed active cases and the predicted active cases.

3 Formulation of SEIR model

The utilization of the SEIR model lies in the fact that it focuses on the basic processes that are directly related to this growing pandemic. In the preparation of this model, there is a need that the population is to be divided into some subdivisions which are susceptible subdivision $S(t)$, that denotes the population which is susceptible to catch the virus; exposed subdivision $E(t)$, that denotes the population which is infected but the symptoms are not visible yet; infected subdivision $I(t)$, that denotes the population which has been infected by the virus and are showing the symptoms; recovered subdivision $R(t)$, that denotes the population which has immunity to the infection. The basic assumption to formulate this model is that the recovered patients acquired permanent active immunity. It can be justified by the strong reason that none of the patients were re-infected by the COVID-19. There have been numerous cases where patients died after being discharged from the hospital but it was found that the patients were either discharged for having mild symptoms or the testing machine reported wrongly. Now we have normalized these components as $S+E+I+R=1$.

Furthermore, suppose that there are equal birth and death rates, i.e., μ and $\frac{1}{\alpha}$ is the mean latent period for the disease. $\frac{1}{\gamma}$ is the mean infectious period and recovered individuals are permanently immune. The contact rate β may or may not be a function of time. Thus the SEIR model is defined as

$$\frac{dS(t)}{dt} = \mu - \beta S(t)I(t) - \mu S(t)$$

$$\frac{dE(t)}{dt} = \beta S(t)I(t) - (\mu + \alpha)E(t)$$

$$\frac{dI(t)}{dt} = \alpha E(t) - (\mu + \gamma)I(t)$$

$$R_0 = \frac{\alpha\beta}{(\mu + \alpha)(\mu + \gamma)}$$

The variable R is determined from the other variables according to equation $S+E+I+R=1$.

3.1 Growth models

A growth curve is an empirical model of the evolution of a quantity over time. Growth curves are widely used in biology for quantities such as population size in population ecology and demography for population growth analysis, individual body height in physiology for growth analysis of individuals. Growth is also a key property of many systems such as an economic expansion, spread of an epidemic, the formation of a crystal, an adolescent's growth and the condensation of a stellar mass.

3.1.1 Linear growth

This is the simplest growth model, in which population grows at a constant rate over time. Linear growth is described by the equation

$$P_{t+1} = P_t + A \tag{1}$$

where P_t represents the numbers or size of the system at time t, P_{t+1} represents the system's numbers or size of the system one time unit later, and A is the system's (linear) growth rate. Many times this model fails to explain natural phenomenon.

3.1.2 Exponential growth (unlimited population growth)

Another simple model describes exponential growth, in which population grows at a constant proportional rate over time. The relation may be expressed in either of two forms, depending on whether reproduction is assumed to be continuous or periodic (Shryock and Siegel, 1973). Exponential growth results in a continuous curve of increase or decrease, whose slope varies in direct relation to the size of the population.

$$P_t = y = P_0 e^{rt} \tag{2}$$

where r is the constant rate of growth, P_o is the initial population size, and the variables t and P_t respectively represent time and the population at time t (Method 1). Another form of exponential curve is as follows

$$P_t = y = P_0 k^t \tag{3}$$

where $k = \left(\frac{P_n}{P_0}\right)^{1/n}$ and that therefore the growth rate in Eq. (3) does not a constant growth rate. David A. Swanson, University of California, USA used this type equation for prediction. We have used truncated information, i.e., only 30 days information (from March 4 to April 2, 2020) on number of COVID-19 cases for the prediction purpose. We have used two equations of exponential curve given below

I. $28e^{0.14t}$ up to March 31, 2020
II. $28e^{0.15t}$ from March 31, 2020 onward (adjusting faster rate of occurrence of COVID-19 cases due to Tablighi spread)

With the current incidence of the COVID-19 going on, we hear about exponential growth. In this study, an attempt has been made to understand and analyze the data through exponential growth curve. The reason for using exponential growth curve for studying the pattern of COVID-19 incidence is that epidemiologists have studied these types of happenings and it is well known that the first period of an epidemic follows exponential growth. The exponential growth function is not necessarily the perfect representation of the epidemic. I have tried to fit exponential curve first, and at the next point to study the logistic growth curve because exponential curve is only fit the epidemic at the beginning. At some point, recovered people will not spread the virus anymore and when someone is or has been infected, the growth will stop. Logistic growth is characterized by increasing growth in the beginning period, but a decreasing growth after point of inflection. For example, in the corona virus case, the maximum limit would be the total number of exposed people in India because when everybody is infected, the growth will be stopped. After that the increasing rate of curve starts to decline and reach to the minimum.

3.1.3 Logistic growth (sigmoidal)

The logistic model reveals that the growth rate of the population is determined by its biotic potential and the size of the population as modified by the natural resistance, or, in other words, by all the various effects of inherent characteristics, that are density dependence Pearl and Reed, 1920. Natural resistance increases as population size gets closer to the carrying capacity. Logistic growth is similar to exponential growth except that it assumes an essential sustainable maximum point. In exponential growth curve, the rate of growth of y per unit of time is directly proportional to y but in practice the rate of growth cannot be in the same proportion always. The logistic curve will continue up to certain level, called the level of saturation, sometimes called the carrying capacity, after reaching carrying capacity it starts declining. A system far below its carrying capacity will at first grow almost exponentially, however, this growth gradually slows as the system expands, finally bringing it to a halt specifically at the carrying capacity (Pearl and Reed, 1920; Shryock and Siegel, 1973). The logistic relationship can be expressed as

$$P_t = y_t = \frac{k}{1 + e^{a+bt}}; \ b < 0 \tag{4}$$

where a, b and k are constant and y_t is that value of the time series at the time t. The reciprocal of y_t follows modified exponential law. Hence, the given time series observation y_t will follow Logistic Law if their reciprocal $1/y_t$ follows modified exponential law. Thus in general, we may take

$$\frac{dy}{dt} = \alpha y(k - y); \alpha > 0, k > 0$$

The factor y is called the momentum factor which increases with time t and the factor $(k-y)$ is known as the retarding factor which decreases with time. When the process of growth approaches the saturation level k, the rate of growth tends to zero. Now we have

$$\frac{dy}{y(k-y)} = \alpha dt \implies \frac{1}{k}\left[\frac{1}{y} + \frac{1}{k-y}\right]dy = \alpha dt \implies \left[\frac{1}{y} - \frac{1}{k-y}\right]dy = \alpha k dt$$

Integrating, we get

$\log\left(\frac{y}{k-y}\right) = \alpha k t + \gamma$, where γ is the constant of integration.

$\frac{k}{y} = 1 + e^{-akt} . e^{-\gamma} \Rightarrow y = \frac{k}{1 + e^{-(\gamma + akt)}}$, this equation is same as Eq. (4) where $a = -\gamma$ and $b = -\alpha k$.

Logistic curve has a point of inflection at half of the carrying capacity k. This point is the critical point from where the increasing rate of curve starts to decline. The time of point of inflection can be estimate as $\frac{-a}{b}$. For the estimation of parameter of logistic curve, method of three selected point given by Pearl and Reed (1920) has been used. The estimate of the parameters can be obtained with equation given as:

$$k = \frac{y_2^2(y_1 + y_3) - 2y_1 y_2 y_3}{y_2^2 - y_1 y_3} \tag{5}$$

$$b = \frac{1}{t_2 - t_1} \ln \left[\frac{(k - y_2)y_1}{(k - y_1)y_2} \right] \tag{6}$$

$$a = \ln \left[\frac{k - y_1}{y_1} \right] - bt_1 \tag{7}$$

where $y_1 y_2$ and y_3 are the cumulative number of COVID-19 cases at a given time t_1, t_2 and t_3 respectively provided that $t_2 - t_1 = t_3 - t_2$. You may also estimate the parameter a and b by method of least square after fixing k.

To predict confirmed corona cases on different day, logistic growth curve has been also used and found very exciting results. The truncated information (means not from the beginning to the present date) on confirmed cases in India has been taken from March 13 to April 2, 2020. The estimated value of the parameters are as follows $k = 18,708.28$, $a = 5.495$ and $b = -0.174$, with these estimates predicted values has been obtained and found considerably lower values than what we observed. On April 1 and 2, 2020 the number of confirmed corona cases are drastically increasing in some part of India due to some unavoidable circumstances thus there is an earnest need to increase carrying capacity of the model, thus it is increased and considered as 22,000 and the other parameters a and b are estimated again which are $a = 5.657$ and $b = -0.173$. The predicted cumulative number of cases is very close to the observed cumulative number of cases till date. The time of point of inflection is obtained as 32.65, i.e., 35 days after beginning. We have taken data from March 13, 2020 so that the time of point of inflection should be April 14, 2020 and by May 30, 2020 there will be no new cases found in the country. Exponential growth model and model given Swanson provided natural estimate of the total infected cases by June 30, 2020 is all most all people in India. This estimate is obtained when no preventive measure would be taken by the Government of India. The testing rate is lower in India than many western countries in the month of March and April, so our absolute numbers was low, when government initiate faster testing process then we have observed more number of cases and found this logistic model fail to provide cumulative number of corona confirm cases after April 17, 2020 thus there is a need to modify this model (Fig. 1).

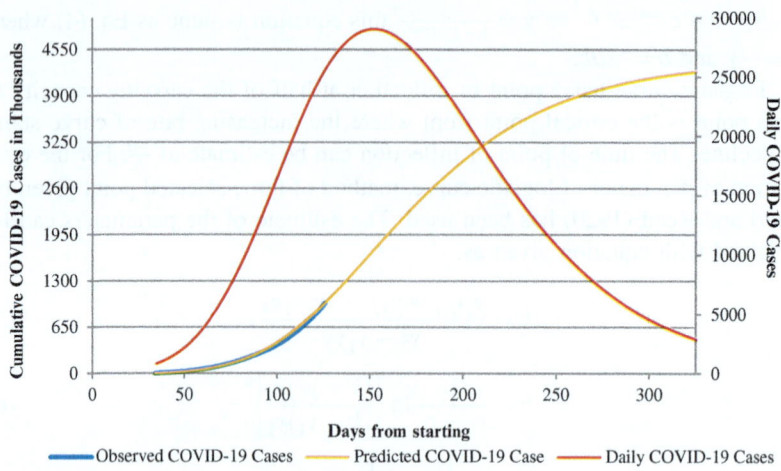

FIG. 1 Logistic growth curve of COVID-19 cases and daily cases.

In order to the modification, I have taken natural log of cumulative number of corona confirm cases instead of cumulative number of corona confirm cases as taken in the previous model. This model provides the carrying capacity is about 80,000 cases and time of point of inflection is April 30, 2020. The present model provides reasonable estimate of the cumulative number of confirmed cases and by the end of July 2020 there will be no new cases found in the country. Further, the number of COVID-19 cases increases and the model estimate does not match to the observed number of case, therefore we need to change the data period, since the logistic curve is data-driven model that provide new estimate of point of inflection and maximum number of corona positive cases by date when disease will disappear, that helps us to plan our strategies. Finally in this study we changed the data period, i.e., we have taken data from April 15th to July 16th 2020. This provides the carrying capacity is about 45 lakh cases and time of point of inflection is August 15th, 2020 with a maximum number of new cases on a day is about 30,000 per day. The model based on this data (from April 15th to July 16th 2020) provides reasonable estimate of the cumulative number of confirmed cases, and predicted value along with 95% confidence interval provided up to August 15th, 2020 (see Table 1) and by the end of March 2021 we expect there will be no new cases in the country in absence of any effective medicine of vaccine (Fig. 2).

3.2 Significance of lockdown

To know the significance of lockdown we define the COVID-19 case transmission is as $c_t = \dfrac{x_t}{\sum_{i=0}^{t} x_i}$, where x_t is the number of confirm cases on t^{th} day.

TABLE 1 Predicted value of COVID-19 cases from logistic growth model.

| Date | Predicted | 95% Confidence limit | | Date | Predicted | 95% Confidence limit | |
		Lower	Upper			Lower	Upper
15-Jul-2020	979,177	972,095	986,260	31-Jul-2020	1,414,533	1,404,319	1,424,746
16-Jul-2020	1,004,769	997,491	1,012,047	01-Aug-2020	1,443,118	1,432,717	1,453,519
17-Jul-2020	1,030,622	1,023,147	1,038,096	02-Aug-2020	1,471,804	1,461,217	1,482,391
18-Jul-2020	1,056,726	1,049,054	1,064,398	03-Aug-2020	1,500,579	1,489,809	1,511,350
19-Jul-2020	1,083,073	1,075,204	1,090,943	04-Aug-2020	1,529,434	1,518,482	1,540,386
20-Jul-2020	1,109,653	1,101,585	1,117,720	05-Aug-2020	1,558,357	1,547,226	1,569,488
21-Jul-2020	1,136,456	1,128,190	1,144,721	06-Aug-2020	1,587,339	1,576,032	1,598,647
22-Jul-2020	1,163,472	1,155,008	1,171,935	07-Aug-2020	1,616,371	1,604,889	1,627,852
23-Jul-2020	1,190,691	1,182,030	1,199,352	08-Aug-2020	1,645,441	1,633,788	1,657,093
24-Jul-2020	1,218,103	1,209,245	1,226,962	09-Aug-2020	1,674,539	1,662,718	1,686,360
25-Jul-2020	1,245,699	1,236,643	1,254,754	10-Aug-2020	1,703,657	1,691,671	1,715,644
26-Jul-2020	1,273,466	1,264,215	1,282,718	11-Aug-2020	1,732,784	1,720,636	1,744,933
27-Jul-2020	1,301,397	1,291,950	1,310,843	12-Aug-2020	1,761,912	1,749,604	1,774,219
28-Jul-2020	1,329,479	1,319,839	1,339,119	13-Aug-2020	1,791,030	1,778,567	1,803,492
29-Jul-2020	1,357,702	1,347,870	1,367,535	14-Aug-2020	1,820,129	1,807,514	1,832,743
30-Jul-2020	1,386,057	1,376,033	1,396,081	15-Aug-2020	1,849,200	1,836,437	1,861,963

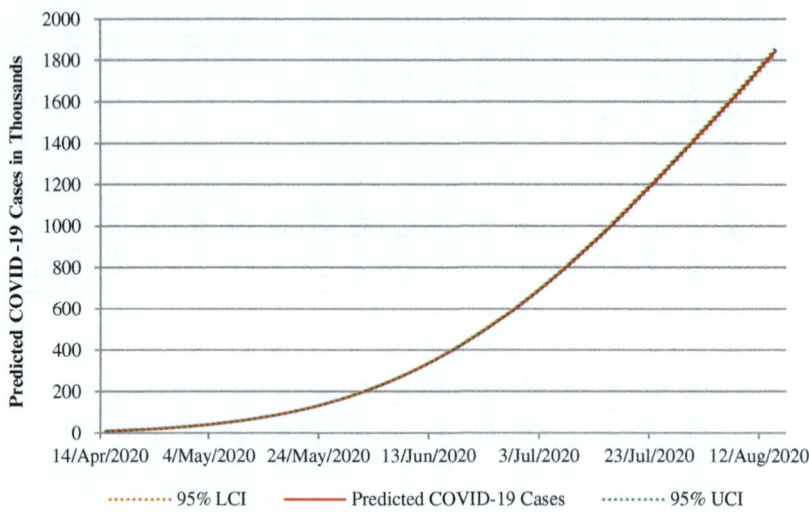

FIG. 2 95% Confidence interval of COVID-19 predicted cases (logistic growth model).

We have calculated c_t and the doubling time of the corona case transmission in India. The doubling time is calculate as $\frac{Ln2}{c_t} = \frac{0.693}{c_t}$.

We have calculated COVID-19 case transmission c_t on the basis of 5 days moving average of daily confirm cases (in the beginning the data in India is very fluctuating) and it is found gradually decreasing in India. This indicates the good sign of government attempts to combat this pandemic through implementing lockdown. These findings indicate that in future the burden of corona will be expectedly lowering down if the current status remains same. In Table 2 given below, an attempt has been made to show the summary statistics of corona case transmission c_t during various lockdown periods in India. It is observed that average COVID-19 case transmission was maximum (0.16 with standard deviation 0.033) in the period prior to the lockdown. During the first lockdown period the average COVID-19 case transmission was 0.14 with standard deviation 0.032, however, in lockdown 2 it was 0.07 with standard deviation 0.009 and in lockdown 3 COVID-19 case transmission was 0.06 with standard deviation 0.007, however, in the period of fourth lockdown the average case transmission was 0.05 with standard deviation 0.005, thus it is clear that both average transmission load and standard deviation are decreasing. Table 3 reveals the result of ANOVA for average c_t during various lockdown periods which is significant means that the average corona case transmission is significantly different is various lockdown periods considered. A group wise comparison of the average COVID-19 case transmission c_t during various lockdown periods is shown in Table 4 which reveals that first lockdown is significantly affects the spread of corona case transmission than others but second lockdown period is not significantly different than third and fourth. Same result is observed for third and fourth lockdown period.

TABLE 2 Summary of c_t during various lockdown period.

Lockdown period	N	Mean	Std. deviation	Std. error	95% confidence interval for mean	
					Lower bound	Upper bound
No	11	0.16	0.035	0.011	0.134	0.182
1	22	0.14	0.032	0.006	0.121	0.149
2	19	0.07	0.009	0.002	0.064	0.073
3	14	0.06	0.007	0.002	0.051	0.060
4	14	0.05	0.005	0.001	0.044	0.049
Total	80	0.093	0.048	0.005	0.082	0.104

TABLE 3 ANOVA test for mean of c_t during various lockdown period.

Source of variations	Sum of squares	df	Mean square	F	p value
Between groups	0.147	04	0.037	75.58	0.000
Within groups	0.036	75	0.000		
Total	0.183	79			

This indicates that the COVID-19 transmission is not under control now. Fig. 3 shows corona case transmission and doubling time in India. The corona case propagation in decreasing and doubling time is increasing day by day.

3.3 Propagation model (based on Newton's law of cooling)

Let us define a function called tempo of disease that is the first differences in natural logarithms of the cumulative corona positive cases on a day, which is as:

$$r_t = \ln\left(\frac{p_t}{p_{t-1}}\right)$$

where p_t and p_{t-1} are the number of cumulative corona positive cases for period t and $t-1$, respectively. When p_t and p_{t-1} are equal then r_t will become zero. If this value of r_t, i.e., zero will continue a week then we can assume no new corona cases will appear further. In the initial face of the disease spread, the tempo of disease increases but after sometime when some preventive measures is being taken then it decreases.

Since r_t is a function of time then the first differential is defined as

$$\frac{dr_t}{dt} = k(r_t - r_T) \qquad (8)$$

where r_t denotes the tempo that is the first differences in natural logarithms of the cumulative corona positive cases on a day, r_T is the desired level of tempo, i.e., zero in this study, t denotes the time and k is a constant of proportionality.

Eq. (8) is an example of an ordinary differential equation that can be solved by the method of separating variables. The Eq. (8) can be written as

$$\frac{dr_t}{r_t} = kdt \qquad (9)$$

Integrating Eq. (9), we get

$$\ln r_t = kt + C \qquad (10)$$

TABLE 4 Group wise comparison of means of c_t during various lockdown periods.

(I) group	(J) group	Mean difference (I − J)	Std. error	p value	95% Confidence interval	
					Lower bound	Upper bound
No	1	0.023	0.008	0.057	0.000	0.047
	2	0.090	0.008	0.000	0.066	0.114
	3	0.102	0.009	0.000	0.077	0.128
	4	0.112	0.009	0.000	0.086	0.137
1	2	0.067	0.007	0.000	0.047	0.087
	3	0.079	0.008	0.000	0.058	0.101
	4	0.089	0.008	0.000	0.067	0.110
2	3	0.013	0.008	1.000	−0.010	0.035
	4	0.022	0.008	0.059	−0.001	0.044
3	4	0.009	0.008	1.000	−0.015	0.033

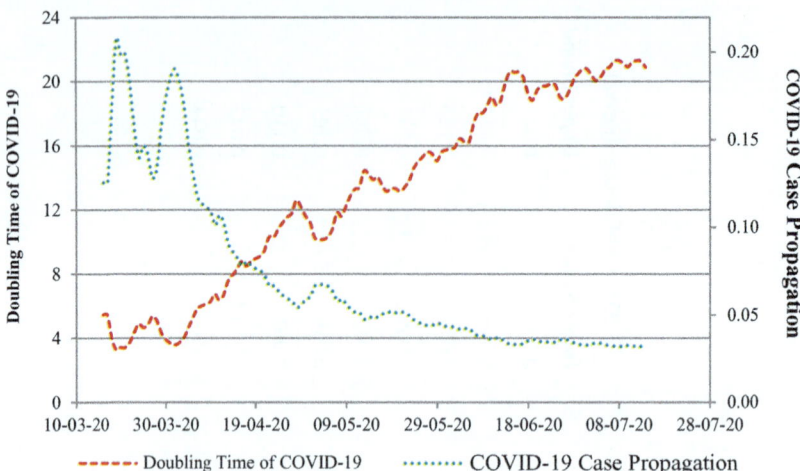

FIG. 3 Doubling time and COVID-19 propagation in India.

where C is an arbitrary constant. Taking the antilogarithms of both sides of Eq. (10) we have

$$r_t = e^{kt+C} \Rightarrow e^{kt}e^C \qquad \Rightarrow \qquad r_t = Ae^{kt} \qquad (11)$$

where $A = e^C$. This Eq. (11) is the general solution of Eq. (8). If k is less than zero, Eq. (11) tells us how the COVID-19 cases will decreases over the time until it reaches zero. Value of A and k is estimated by least square estimation procedure using the data sets.

The Government of India implemented lockdown on March 24th, 2020 and expected that the tempo of disease is decreasing. Government suggested and implemented social distancing and lockdown to control the spread of COVID-19 in the society. In Table 5, the predicted value of COVID-19 cases obtained with this method is given along with 95% confidence interval. About 21.5 lakh cases are expected by August 15th, 2020. With this model it is expected that about 45 lakh peoples will be infected in India by the end of October and after that no cases will happen since the tempo of disease r_t will become zero (Fig. 4). In Table 6 an attempt has been made to show the summary statistics of tempo of COVID-19 r_t during various lockdown periods in India. It is observed that average tempo is maximum (0.17 with standard deviation 0.062) in the period prior to the lockdown. During the first lockdown period the average tempo is 0.14 with standard deviation 0.044 and after that it is found decreasing in the various lockdowns. Table 7 indicates lockdown periods are significantly different in terms of tempo of disease spread. A group wise comparison of the average tempo of COVID-19 during

TABLE 5 Predicted confirmed cases of COVID-19 till 15 August 2020, India with propagation model.

| Date | Predicted | 95% Confidence limit | | Date | Predicted | 95% Confidence limit | |
		Lower	Upper			Lower	Upper
15-Jul-2020	1,140,781	845,770	1,568,823	31-Jul-2020	1,631,673	1,162,680	2,346,267
16-Jul-2020	1,169,628	864,891	1,613,198	01-Aug-2020	1,664,042	1,182,994	2,399,124
17-Jul-2020	1,198,757	884,133	1,658,183	02-Aug-2020	1,696,557	1,203,333	2,452,409
18-Jul-2020	1,228,163	903,487	1,703,769	03-Aug-2020	1,729,209	1,223,691	2,506,109
19-Jul-2020	1,257,836	922,949	1,749,945	04-Aug-2020	1,761,989	1,244,064	2,560,212
20-Jul-2020	1,287,768	942,513	1,796,703	05-Aug-2020	1,794,889	1,264,445	2,614,704
21-Jul-2020	1,317,951	962,171	1,844,032	06-Aug-2020	1,827,899	1,284,830	2,669,570
22-Jul-2020	1,348,376	981,919	1,891,920	07-Aug-2020	1,861,012	1,305,214	2,724,799
23-Jul-2020	1,379,035	1,001,750	1,940,359	08-Aug-2020	1,894,218	1,325,590	2,780,376
24-Jul-2020	1,409,918	1,021,658	1,989,336	09-Aug-2020	1,927,509	1,345,954	2,836,287
25-Jul-2020	1,441,019	1,041,638	2,038,840	10-Aug-2020	1,960,876	1,366,302	2,892,520
26-Jul-2020	1,472,327	1,061,682	2,088,859	11-Aug-2020	1,994,312	1,386,628	2,949,059
27-Jul-2020	1,503,833	1,081,786	2,139,382	12-Aug-2020	2,027,806	1,406,927	3,005,892
28-Jul-2020	1,535,530	1,101,944	2,190,397	13-Aug-2020	2,061,353	1,427,196	3,063,004
29-Jul-2020	1,567,409	1,122,149	2,241,891	14-Aug-2020	2,094,942	1,447,428	3,120,382
30-Jul-2020	1,599,459	1,142,396	2,293,852	15-Aug-2020	2,128,567	1,467,621	3,178,012

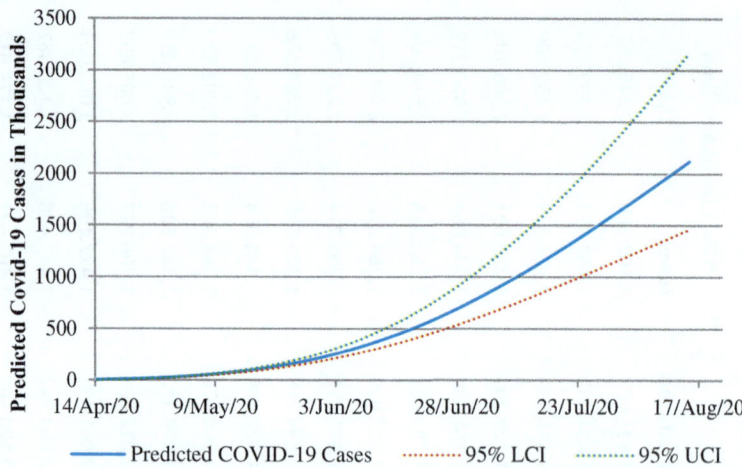

95% Confidence Interval of COVID-19 Predicted Cases (propagation model)

FIG. 4 95% Confidence interval of COVID-19 predicted cases (propagation model).

various lockdown periods is shown in Table 8 which reveals that first lock-down is significantly different than others. Consecutive mean difference shows that the decrease in disease spread has been observed but insignificant, means there is no impact of lockdown on controlling the disease spread.

3.4 Joinpoint regression model

To analyze the temporal trends and to identify important changes in the trends of the COVID-19 outbreak joinpoint regression is used in China (Al Hasan et al., 2020); here in this study we performed a joinpoint regression analysis in India to understand the pattern of COVID-19. Joinpoint regression analysis, enable us to identify time at a meaningful change in the slope of a trend is observed over the study period. The best fitting points known as joinpoints, that are chosen when the slope changes significantly in the models.

To tackle the above problem joinpoint regression analysis (Kim et al., 2000) has been employed in this study to present trend analysis. The goal of the join-point regression analysis is not only to provide the statistical model that best fits the time series data but also, the purpose is to provide that model which best summarizes the trend in the data (Marrot, 2010).

Let y_i denotes the reported COVID-19 positive cases on day t_i such that $t_1 < t_2 < \ldots < t_n$. Then the joinpoint regression model is defined as

$$\ln y_i = \alpha + \beta_1 t_1 + \delta_1 u_1 + \delta_2 u_2 + \ldots + \delta_j u_j + \varepsilon_i \qquad (12)$$

TABLE 6 Summary of r_t during various lockdown period.

Lockdown period	N	Mean	Std. deviation	Std. error	95% Confidence interval for mean	
					Lower bound	Upper bound
No	11	0.17	0.062	0.019	0.125	0.210
1	22	0.14	0.044	0.009	0.120	0.159
2	19	0.07	0.012	0.003	0.064	0.076
3	14	0.06	0.007	0.002	0.051	0.059
4	14	0.05	0.006	0.001	0.046	0.052
Total	80	0.10	0.056	0.006	0.084	0.109

TABLE 7 ANOVA test for mean of r_t during various lockdown period.

Source of variations	Sum of squares	df	Mean square	F	p value
Between groups	0.166	04	0.041	37.24	0.000
Within groups	0.083	75	0.001		
Total	0.249	79			

where $u_j = \begin{cases} (t_j - k_j) & \text{if } t_j > k_j \\ 0 & \text{otherwise} \end{cases}$ and $k_1 < k_2 ... < k_j$ are joinpoints. The details of joinpoint regression analysis are given elsewhere (Kim et al., 2004).

Joinpoint regression analysis is used when the temporal trend of an amount, like incidence, prevalence and mortality is of interest (Doucet et al., 2016). However, this method has generally been applied with the calendar year as the time scale (Akinyede and Soyemi, 2016; Chatenoud et al., 2015; Missikpode et al., 2015; Mogos et al., 2016). The joinpoint regression analysis can also be applied in epidemiological studies in which the starting date can be easily established such as the day when the disease is detected for the first time as is the case in the present analysis (Rea et al., 2017). Estimated regression coefficients (β) were calculated for the trends extracted from the joinpoint regression. Additionally, the average daily percent change (ADPC), calculated as a geometric weighted average of the daily percent changes (Clegg et al., 2009). The joinpoints are selected based on the data-driven Bayesian Information Criterion (BIC) method (Zhang and Siegmund, 2007).

The equation for computing the BIC for a k-joinpoints regression is:

$$BIC(k) = \ln \left[\frac{SSE(k)}{n} \right] + \frac{2(k+1) \times \ln(n)}{n} \qquad (13)$$

where SSE is the sum of squared errors of the k-joinpoints regression model and n is the number of observations. The model which has the minimum value of BIC(k) is selected as the final model. There are other methods also for identifying the joinpoints such as permutation test method and the weighted BIC methods. Relative merits and demerits of different methods of identifying the joinpoints are discussed elsewhere (National Institute Cancer, 2013). The permutation test method is regarded as the best method but it is computationally very intensive. It controls the error probability of selecting the wrong model at a certain level (i.e., 0.05). The BIC method, on the other hand, is less complex computationally.

TABLE 8 Group wise comparison of means of r_t during various lockdown periods.

(I) group	(J) group	Mean difference (I − J)	Std. error	p value	95% Confidence interval	
					Lower bound	Upper bound
No	1	0.027	0.012	0.306	−0.009	0.063
	2	0.097	0.013	0.000	0.061	0.134
	3	0.112	0.013	0.000	0.073	0.151
	4	0.118	0.013	0.000	0.079	0.157
1	2	0.070	0.010	0.000	0.040	0.100
	3	0.085	0.011	0.000	0.052	0.118
	4	0.091	0.011	0.000	0.058	0.124
2	3	0.015	0.012	1.000	−0.019	0.049
	4	0.021	0.012	0.790	−0.013	0.055
3	4	0.006	0.013	1.000	−0.030	0.043

In the present case, data on the reported confirmed cases of COVID-19 are available on a daily, thus the daily percent change (*DPC*) from day t to day $(t+1)$ is defined as

$$DPC = \left(\frac{y_{t+1} - y_t}{y_t}\right) \times 100 \tag{14}$$

If the trend in the daily reported confirmed cases of COVID-19 is modeled as

$$\ln(y_t) = b_0 + b_1 t + \varepsilon \tag{15}$$

then, it can be shown that the *DPC* is equal to

$$DPC = \left(e^{b_1} - 1\right) \times 100 \tag{16}$$

It is worthwhile to discuss here is that the positive value of *DPC* indicates an increasing trend while the negative value of *DPC* suggests a declining trend. The *DPC* reflects the trend in the reported COVID-19 positive cases in different time segments of the reference period observed through joinpoint regression techniques. For the entire study period, it is possible to estimate average daily percent change (*ADPC*) that is the weighted average of *DPC* of different time segments of the study period with weights equal to the length of different time segments. However, when the trend changes frequently, *ADPC* has little meaning. It assumes that the random errors are heteroscedastic (have nonconstant variance). Heteroscedasticity is handled by joinpoint regression using weighted least squares (WLS). The weights in WLS are the reciprocal of the variance and can be specified in several ways. Thus standard error is used to control heteroscedastic in the analysis during the entire period.

To observe the trend of reported cases, the moving average method has been used in this study. The daily percent change (*DPC*) in the daily reported confirmed cases of COVID-19 during the period March 14th, 2020 through July 16th, 2020 is used for forecasting the daily reported confirmed cases of COVID-19 in the immediate future under the assumption that the trend in the daily reported confirmed cases of COVID-19 remains unchanged. The number of cases increased by the rate of 6.20% per day in India; however, the rate is different in the different segment. Also Table 9 reveals that the growth rate is positive and significant (about 19%) from 16th March to 3rd April and after that the growth rate is decreasing in comparison of first segment, i.e., for 28 days (from 3rd April to 30th April). The possible reason may be lockdown imposed in India. In the third segment, i.e., from 30th April to 4th May a high increase has been observed but it is insignificant. From 4th April to 13th May the rate is although the positive but dramatically lower than the previous segments growth rate. In the next segment, i.e., 5th segment which is of 8 days, we observe a significant increase of 6.55% in COVID-19 cases. In the last and 6th segment from 20th May to 14th July, i.e., for 56 days, the growth rate is found again positive and significant (3.03% per day) in the COVID-19 cases.

TABLE 9 Results of the joinpoint regression analysis, India with sd.

Segment	Lower endpoint	Upper endpoint	Number of days[a]	Average daily percent change (ADPC)[b]	95% Confidence interval		Test statistic (t)	P>|t|
					Lower	Upper		
1	16 March	3 April	19	19.19[c]	17.20	21.20	20.74	0.00
2	3 April	30 April	28	5.27[c]	4.99	5.56	37.68	0.00
3	30 April	4 May	5	11.34	−2.48	27.12	1.61	0.11
4	4 May	13 May	10	1.78[c]	0.44	3.13	2.64	0.01
5	13 May	20 May	8	6.55[c]	3.63	9.54	4.54	0.00
6	20 May	14 July	56	3.03[c]	2.98	3.08	117.45	0.00
All	16 March	14 July	121	6.20[c]	5.63	6.78	21.82	0.00

[a]Number of days in a segment include both lower endpoint and upper endpoint.
[b]Average daily percent change.
[c]Statistically significant.

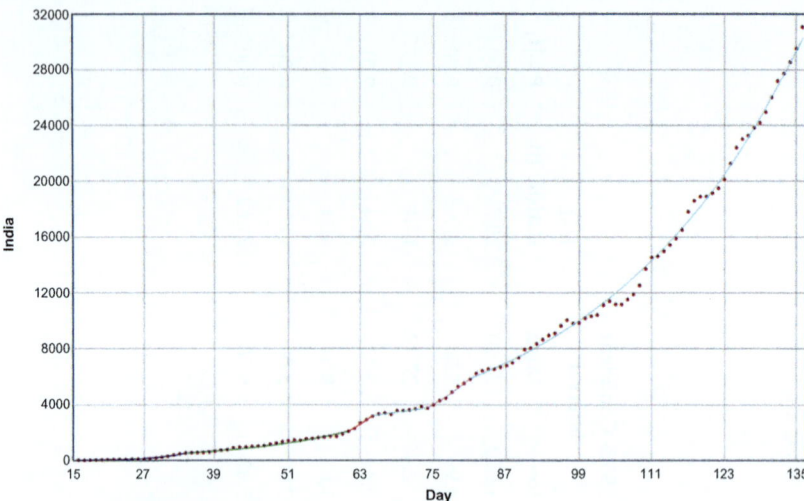

* Indicates that the Annual Percent Change (APC) is significantly different from zero at the alpha = 0.05 level.
Final Selected Model: 5 Joinpoints.

FIG. 5   Trend in daily reported confirmed cases of COVID-19 in India using joinpoint regression analysis with BIC criterion.

Fig. 5 shows that the trend increases in India still sharply and there is no hope of decline in COVID-19 cases. Fig. 2 shows the forecasted value of COVID-19 daily cases in India. The COVID-19 cases will increase further if the same trend prevailing.

Table 10 presents the forecast of the predicted cases of COVID-19 in India along with 95% confidence intervals. This exercise suggests that by August 15th, 2020, the confirmed cases of COVID-19 in India is likely to be 2,587,007 with a 95% confidence interval of 2,571,896–2,602,282 and daily reported cases will be 78,729 with 95% confidence interval of 77,516–79,961. This daily reported COVID-19 positive cases may change only when an appropriate set of new interventions are introduced to fight COVID-19 pandemic. It is observed that analysis indicates that in the month of August, India faces more than 50 thousand cases per day (Fig. 6).

4   Conclusions

India is in the comfortable zone with a lower growth rate than other countries. Logistic model shows that, the epidemic is likely to stabilize with 45 lakh cases by the end of March 2021 and peak will come in middle of the August, however, propagation model provide estimate of maximum COVID-19 case as 45 lakh but the timing is different (by end October) than the logistic model.

TABLE 10 Forecast of daily predicted confirmed cases of COVID-19 till August 15, 2020, India.

Date	Predicted	95% Confidence limit		Date	Predicted	95% Confidence limit	
		Lower	Upper			Lower	Upper
15-Jul-2020	31,208	31,193	31,223	31-Jul-2020	50,313	49,900	50,730
16-Jul-2020	32,153	32,122	32,184	01-Aug-2020	51,837	51,387	52,292
17-Jul-2020	33,127	33,079	33,176	02-Aug-2020	53,408	52,918	53,903
18-Jul-2020	34,131	34,065	34,198	03-Aug-2020	55,026	54,495	55,563
19-Jul-2020	35,165	35,080	35,251	04-Aug-2020	56,694	56,119	57,274
20-Jul-2020	36,231	36,126	36,337	05-Aug-2020	58,412	57,791	59,038
21-Jul-2020	37,329	37,202	37,456	06-Aug-2020	60,181	59,513	60,857
22-Jul-2020	38,460	38,311	38,609	07-Aug-2020	62,005	61,287	62,731
23-Jul-2020	39,625	39,452	39,799	08-Aug-2020	63,884	63,113	64,663
24-Jul-2020	40,826	40,628	41,024	09-Aug-2020	65,819	64,994	66,655
25-Jul-2020	42,063	41,839	42,288	10-Aug-2020	67,814	66,931	68,708
26-Jul-2020	43,337	43,086	43,590	11-Aug-2020	69,868	68,925	70,824
27-Jul-2020	44,650	44,370	44,933	12-Aug-2020	71,985	70,979	73,005
28-Jul-2020	46,003	45,692	46,317	13-Aug-2020	74,167	73,094	75,254
29-Jul-2020	47,397	47,053	47,743	14-Aug-2020	76,414	75,273	77,572
30-Jul-2020	48,833	48,456	49,214	15-Aug-2020	78,729	77,516	79,961

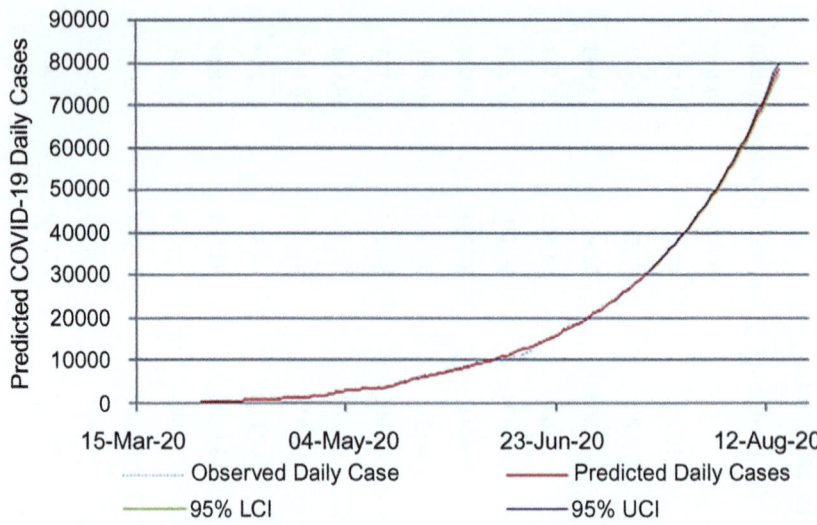

FIG. 6 Forecast of daily reported confirmed COVID-19 cases in India up to August 15, 2020.

Logistic model need to monitor the data time to time for good long term prediction. The projections produced by the model and after their validation can be used to determine the scope and scale of measures that government need to initiate. Joinpoint regression is based on the daily reported confirmed cases of COVID-19, asserts that there has virtually been little impact of the nationwide lockdown as well as relaxations in restrictions on the progress of the COVID-19 pandemic in India. The joinpoint regression analysis provides better estimate up to 15th August for the confirmed COVID-19 cases than the other two methods. To know the better understanding of the progress of the epidemic in the country may be obtained by analyzing the progress of the epidemic at the regional level. In conclusion, if the current mathematical model results can be validated within the range provided here, then the social distancing and other prevention, treatment policies that the central and various state governments and people are currently implementing should continue until new cases are not seen. The spread from urban to rural and rich to poor populations should be monitor and control is an important point of consideration. Mathematical models have certain limitations that there are many assumptions about homogeneity of population in terms of urban/rural or rich/poor that does not capture variations in population density. If several protective measures will not be taken effectively, then this rate may be changed. However, the government of India under the leadership of Modi Ji has already taken various protective measures such as lockdown in several areas, make possible quarantine facility to reduce the rate of increase of COVID-19, thus we may hopefully conclude that, country will be successful to reduce the rate of this pandemic.

References

Akinyede, O., Soyemi, K., 2016. Joinpoint regression analysis of pertussis crude incidence rates, Illinois, 1990-2014. Am. J. Infect. Control 44 (12), 1732–1733.

Al Hasan, S.M., Saulam, J., Kanda, K., Hirao, T., 2020. The novel coronavirus disease (COVID-19) outbreak trends in mainland China: a joinpoint regression analysis of the outbreak data from January 10 to February 11, 2020 (preprint). Bull. World Health Organ. https://doi.org/10.2471/BLT.20.253153.

Anand, S., Fan, V., 2016. The Health Workforce in India. WHO; Human resources for Health Observer, Geneva. Series No. 16 Available at: https://www.who.int/hrh/resources/hwindia_health-obs16/en/.

Anastassopoulou, C., Russo, L., Tsakris, A., Siettos, C., 2020. Data-based analysis, modelling and forecasting of the COVID-19 outbreak. PLoS One 15 (3), e0230405. https://doi.org/10.1371/journal.pone.0230405.

Batista, M., 2020. Estimation of the Final Size of the Second Phase of the Coronavirus COVID 19 Epidemic by the Logistic Model. medRxiv. preprint https://doi.org/10.1101/2020.03.11.20024901.

Brauer, F., Castillo-Chavez, C., 2012. Mathematical Models in Population Biology and Epidemiology, second ed. Springer, New York, NY.

Chang, S.L., Harding, N., Zachreson, C., Cliff, O.M., Prokopenko, M., 2020. Modelling Transmission and Control of the Covid-19 Pandemic in Australia. arXiv. preprint arXiv:2003.10218.

Chatenoud, L., Garavello, W., Pagan, E., Bertuccio, P., Gallus, S., La Vecchia, C., et al., 2015. Laryngeal cancer mortality trends in European countries. Int. J. Cancer 842, 833–842.

Chatterjee, K., Chatterjee, K., Kumar, A., Shankar, S., 2020. Healthcare impact of COVID-19 epidemic in India: a stochastic mathematical model. Med. J. Armed Forces India 76 (2), 147–155. https://doi.org/10.1016/j.mjafi.2020.03.022.

Chen, Y.C., Lu, P.E., Chang, C.S., Liu, T.H., 2020. A Time-Dependent SIR Model for COVID-19 With Undetectable Infected Persons. http://gibbs1.ee.nthu.edu.tw/A_Time_Dependent_SIR_Model_For_Covid 19.pdf.

Clegg, L.X., Hankey, B.F., Tiwari, R., Feuer, E.J., Edwards, B.K., 2009. Estimating average annual per cent change in trend analysis. Stat. Med. 28 (29), 3670–3682.

Corman, V.M., Landt, O., Kaiser, M., Molenkamp, R., Meijer, A., Chu, D.K., Bleicker, T., Brunink, S., Schneider, J., Schmidt, M.L., 2020. Detection of 2019 novel coronavirus (2019-ncov) by realtime RT-PCR. Euro Surveill. 25 (3), 2000045.

Cowling, B.J., Leung, G.M., 2020. Epidemiological research priorities for public health control of the ongoing global novel coronavirus (2019-nCoV) outbreak. Euro Surveill. 25 (6), 2000110.

Daley, D.J., Gani, J., 1999. Epidemic Modelling: An Introduction. Cambridge University Press, Cambridge.

De Silva, U., Warachit, J., Waicharoen, S., Chittaganpitch, M., 2009. A preliminary analysis of the epidemiology of influenza A (H1N1) v virus infection in Thailand from early outbreak data, June-July 2009. Euro Surveill. 14 (31), 19292.

Doucet, M., Rochette, L., Hamel, D., 2016. Prevalence and mortality trends in chronic obstructive pulmonary disease over 2001 to 2011: a public health point of view of the burden. Can. Respir. J., 1–10. https://doi.org/10.1155/2016/7518287.

Ferguson, N.M., Laydon, D., Nedjati-Gilani, G., Imai, N., Ainslie, K., Baguelin, M., Bhatia, S., Boonyasiri, A., Cucunubá, Z., Cuomo-Dannenburg, G., Dighe, A., 2020. Impact of Non-Pharmaceutical Interventions (NPIs) to Reduce COVID-19 Mortality and Healthcare Demand. Imperial College COVID-19 Response Team.

Gamero, J., Tamayo, J.A., Martinez-Roman, J.A., 2020. Forecast of the Evolution of the Contagious Disease Caused by Novel Corona Virus (2019-ncov) in China. arXiv. preprint arXiv: 2002. 04739.

Gupta, M., 2020. Corona Virus in India: Make or Break. Available at https://medium. com/@mohakgupta_55841/coronavirus-in-india-make-or-break-5a13dfb9646d.

Herbert, W.H., 2000. The mathematics of infectious diseases. SIAM Rev. 42 (4), 599–653.

Hethcote, H.W., 1983. Measles and rubella in the United States. Am. J. Epidemiol. 117, 2–13.

Hethcote, H.W., 1988. Optimal ages or vaccination for measles. Math. Biosci. 89, 29–52.

Hethcote, H.W., 1989. Rubella. In: Levin, S.A., Hallam, T.G., Gross, L. (Eds.), Applied Mathematical Ecology. Biomathematics. vol. 18. Springer, Berlin, Heidelberg, New York.

Huang, C., Wang, Y., Li, X., Ren, L., Zhao, J., Hu, Y., Zhang, L., Fan, G., Xu, J., Gu, X., 2020. Clinical features of patients infected with 2019 novel coronavirus in Wuhan, China. Lancet 395 (10223), 497–506.

Hui, D.S., Azhar, E.I., Madani, T.A., Ntoumi, F., Kock, R., Dar, O., Ippolito, G., Mchugh, T.D., Memish, Z.A., Drosten, C., 2020. The continuing 2019-ncov epidemic threat of novel coronaviruses to global health-the latest 2019 novel coronavirus outbreak in Wuhan, China. Int. J. Infect. Dis. 91, 264.

Junling, M., Dushoff, J., Bolker, B.M., Earn, D.J., 2014. Estimating initial epidemic growth rates. Bull. Math. Biol. 76 (1), 245–260.

Kermack, W.O., McKendrick, A.G., 1927. A contribution to the mathematical theory of epidemics. Proc. Roy. Soc. Lond. A 115 (700–721), 1927.

Kim, H.J., Fay, M.P., Feuer, E.J., Midthune, D.N., 2000. Permutation tests for joinpoint regression with applications to cancer rates. Stat. Med. 19, 335–351.

Kim, H.J., Fay, M.P., Yu, B., Barrett, M.J., Feuer, E.J., 2004. Comparability of segmented line regression models. Biometrics 60 (4), 1005–1014.

Koo, J.R., Cook, A.R., Park, M., Sun, Y., Sun, H., Lim, J.T., Tam, C., Dickens, B.L., 2020. Interventions to mitigate early spread of SARS-CoV-2 in Singapore: a modelling study. Lancet Infect. Dis. 20, 678–688. https://doi.org/10.1016/S1473-3099(20)30162-6.

Kucharski, A.J., Russell, T.W., Diamond, C., Liu, Y., Edmunds, J., Funk, S., Davies, N., 2020. Early dynamics of transmission and control of COVID-19: a mathematical modelling study. Lancet Infect. Dis. 20 (5), 553–558.

Lipsitch, M., Swerdlow, D.L., Finelli, L., 2020. Defining the epidemiology of Covid-19-studies needed. N. Engl. J. Med. 382 (13), 1194–1196.

Longini Jr., I.M., Ackerman, E., Elveback, L.R., 1978. An optimization model for influenza A epidemics. Math. Biosci. 38, 141–157.

Malhotra, B., Kashyap, V., 2020. Progression of COVID-19 in Indian States—Forecasting Endpoints Using SIR and Logistic Growth Models. https://doi.org/10.1101/2020.05.15.20103028. PPR:PPR163964.

Mandal, S., Bhatnagar, T., Arinaminpathy, N., Agarwal, A., et al., 2020. Prudent public health intervention strategies to control the corona virus disease 2019 transmission in India: a mathematical model-based approach. Indian J. Med. Res. 151, 190–199. Epub ahead of print https://doi.org/10.4103/ijmr.IJMR_504_20.

Marrot, L.D., 2010. Colorectal Cancer Network (CRCNet) User Documentation for Surveillance Analytic Software: Joinpoint. Cancer Care Ontario, pp. 1–28.

Missikpode, C., Peek-Asa, C., Young, T., Swanton, A., Leinenkugel, K., Torner, J., 2015. Trends in non-fatal agricultural injuries requiring trauma care. Inj. Epidemiol. 2 (1), 30.

Mogos, M.F., Salemi, J.L., Spooner, K.K., McFarlin, B.L., Salihu, H.M., 2016. Differences in mortality between pregnant and nonpregnant women after cardiopulmonary resuscitation. Obstet. Gynecol. 128 (4), 880–888.

National Institute Cancer, 2013. Joinpoint Regression Program. National Institutes of Health, United States Department of Health and Human Services, Bethesda, MD.

Pearl, R., Reed, L.J., 1920. On the rate of growth of the population of the United States since 1790 and its mathematical representation. Proc. Natl. Acad. Sci. U. S. A. 6 (6), 275.

Rea, F., Pagan, E., Compagnoni, M.M., Cantarutti, A., Pigni, P., Bagnardi, V., et al., 2017. Joinpoint regression analysis with time-on-study as time-scale. Application to three Italian population-based cohort studies. Epidemiol. Biostat. Public Health 14 (3), e12616.

Rothe, C., Schunk, M., Sothmann, P., Bretzel, G., Froeschl, G., Wallrauch, C., Zimmer, T., Thiel, V., Janke, C., Guggemos, W., 2020. Transmission of 2019-ncov infection from an asymptomatic contact in Germany. N. Engl. J. Med. 382 (10), 970–971.

Schueller, E., Klein, E., Tseng, K., Kapoor, G., Joshi, J., Sriram, A., Laxminarayan, R., 2020. COVID-19 in India: Potential Impact of the Lockdown and Other Longer Term Policies. Retrieved from Washington D.C. https://cddep.org/publications/covid-19-india-potential-impact-of-the-lockdown-and-other-longer-term-policies/.

Shen, C.Y., 2020. A logistic growth model for COVID-19 proliferation: experiences from China and international implications in infectious diseases. Int. J. Infect. Dis. 96, 582–589.

Shryock, H.S., Siegel, J.S., 1973. Methods and Materials of Demography. U.S. Dept. of Commerce, Bureau of the Census, Washington.

Singh, R., Adhikari, R., 2020. Age-Structured Impact of Social Distancing on the COVID-19 Epidemic in India. arXiv. 2003.12055v1 [q-bio.PE].

Srinivasa Rao Arni, S.R., Krantz, S.G., Thomas, K., Ramesh, B., 2020. Model-based retrospective estimates for COVID-19 or coronavirus in India: continued efforts required to contain the virus spread. Curr. Sci. 118 (7), 1023–1025.

Tuite, A.R., Fisman, D.N., 2020. Reporting, epidemic growth, and reproduction numbers for the 2019 novel coronavirus (2019-nCoV) epidemic. Ann. Intern. Med. 172 (8), 567–568.

Wu, J.T., Leung, K., Leung, G.M., 2020. Nowcasting and forecasting the potential domestic and international spread of the 2019-nCoV outbreak originating in Wuhan, China: a modelling study. Lancet 395 (10225), 689–697.

Yorke, J.A., London, W.P., 1973. Recurrent outbreaks of measles, chickenpox and mumps II. Am. J. Epidemiol. 98, 469–482.

Yorke, J.A., Nathanson, N., Pianigiani, G., Martin, J., 1979. Seasonality and the requirements for perpetuation and eradication of viruses in populations. Am J Epidemiol 109, 103–123.

Zhang, N.R., Siegmund, D.O., 2007. A modified Bayes information criterion with applications to the analysis of comparative genomic hybridization data. Biometrics 63 (1), 22–32.

Zhang, Y., Yang, H., Cui, H., Chen, Q., 2019. Comparison of the ability of ARIMA, WNN and SVM models for drought forecasting in the Sanjiang Plain, China. Nat. Resour. Res. 29, 1447.

Zhao, S., Lin, Q., Ran, J., Musa, S.S., Yang, G., Wang, W., Lou, Y., Gao, D., Yang, D.He, Wang, M.H., 2020. Preliminary estimation of the basic reproduction number of novel coronavirus (2019-nCoV) in China, from 2019 to 2020: a data-driven analysis in the early phase of the outbreak. Int. J. Infect. Dis. 92, 214–217.

Further reading

COVID-19, 2020. https://www.covid19india.org.

World Health Organization, 2020. World Situation Report. (Online) Available from https://www.who.int/emergencies/diseases/novelcoronavirus-2019/situation-reports.

Chapter 10

Mathematical modeling as a tool for policy decision making: Applications to the COVID-19 pandemic

J. Panovska-Griffiths[a,b,c,*], C.C. Kerr[d,e], W. Waites[f,g], and R.M. Stuart[h,i]

[a]*Department of Applied Health Research, University College London, London, United Kingdom*
[b]*Institute for Global Health, University College London, London, United Kingdom*
[c]*The Wolfson Centre for Mathematical Biology and The Queen's College, University of Oxford, Oxford, United Kingdom*
[d]*Institute for Disease Modeling, Global Health Division, Bill & Melinda Gates Foundation, Seattle, WA, United States*
[e]*School of Physics, University of Sydney, Sydney, NSW, Australia*
[f]*School of Informatics, University of Edinburgh, Edinburgh, United Kingdom*
[g]*Centre for the Mathematical Modelling of Infectious Diseases, London School of Hygiene and Tropical Medicine, London, United Kingdom*
[h]*Department of Mathematical Sciences, University of Copenhagen, Copenhagen, Denmark*
[i]*Disease Elimination Program, Burnet Institute, Melbourne, VIC, Australia*
[*]*Corresponding author: e-mail: j.panovska-griffiths@ucl.ac.uk*

Abstract

The coronavirus disease 2019 (COVID-19) pandemic highlighted the importance of mathematical modeling in advising scientific bodies and informing public policy making. Modeling allows a flexible theoretical framework to be developed in which different scenarios around spread of diseases and strategies to prevent it can be explored. This work brings together perspectives on mathematical modeling of infectious diseases, highlights the different modeling frameworks that have been used for modeling COVID-19 and illustrates some of the models that our groups have developed and applied specifically for COVID-19. We discuss three models for COVID-19 spread: the modified Susceptible-Exposed-Infected-Recovered model that incorporates contact tracing (SEIR-TTI model) and describes the spread of COVID-19 among these population cohorts, the more detailed agent-based model called Covasim describing transmission between individuals, and the Rule-Based Model (RBM) which can be thought of as a combination of both. We showcase the key methodologies of these

Handbook of Statistics, Vol. 44. https://doi.org/10.1016/bs.host.2020.12.001
291

approaches, their differences as well as the ways in which they are interlinked. We illustrate their applicability to answer pertinent questions associated with the COVID-19 pandemic such as quantifying and forecasting the impacts of different test-trace-isolate (TTI) strategies.

Keywords: Epidemiological modeling, COVID-19, SEIR models, Agent-based models, Rule-based models

1 Introduction

1.1 Overview of mathematical modeling

As we write this in October of 2020, the world remains gripped by COVID-19 pandemic caused by the spread of a severe acute respiratory syndrome coronavirus (SARS-CoV-2). Since the emergence of this new virus, mathematical sciences—particularly modeling—have been at the forefront of policy decision making around it.

Mathematical and computational models are a way to understand the processes in complex systems that underlie empirical observations and to generate possible future trajectories of these systems. Strictly speaking, mathematical models refer to the actual framework of composing a set of equations or theoretical approaches, while computational models refer to the numerical and computational approaches used to solve the mathematical framework. In practice, these terms are used interchangeably since solving practical mathematical models analytically is rarely possible; hence, we often simply refer to both as models. The overarching purpose of models is to allow a flexible framework in which different scenarios can be tested, different questions related to future behavior can be posed and evaluated, and potential future behavior can be predicted. There is a difference between explanatory models that attempt to explain current behavior and predictive models that extend this into the future, and we will discuss this further in Section 1.2.

The overall aim of mathematical modeling is to generate answers to questions we can't get from observations. The answers are then used to understand, manage and predict future behavior of complex systems and processes, for example, to inform public policy and future decision making. The statistician George Box said, "all models are wrong; some models are useful." It is important to understand that modeling is like any other technology: it can be properly applied or not, it may produce output that admits a useful interpretation, or it may not. In Section 1.3, we discuss the notion of a "correct" model; it is important to understand this notion of correctness against the background of utility of mathematical modeling.

1.2 Modeling to explain or to predict

Mathematical modeling provides a framework that, given data, facilitates understanding of how changes within the framework can affect outcomes.

Modeling combined with data can explain past behavior, predict and forecast future behavior, and evaluate how changes may alter these predictions. Explaining past and current trends and predicting future trends are two different aspects of modeling. The clear distinction between the two concepts was highlighted by a number of scientists such as Forster and Sober (1994), Forster (2002), Hitchcock and Sober (2004), and Dowe et al. (2007).

Explanatory modeling combines theory with data to test hypotheses and explain behavior. Regression modeling, or curve-fitting to data, is a type of explanatory modeling that is widely used in statistical sciences. Explanatory modeling can answer questions like "Which population cohort is at greatest risk of infection"? by looking backwards and finding the key parameters (covariates) that help explain the observed patterns in the data. These types of models have a long history in helping to explain patterns in disease and public health. While continuing the trajectory of the curve that best explained historical trends could in theory be used to predict future trends, the statistical model best capable of explaining past trends may not be well suited to forecasting future trends.

In contrast, predictive modeling (Shmueli, 2010) consists of building a mechanistic framework that is explicitly designed to be able to explain both historical patterns and future states. In contrast to explanatory modeling, within this framework different cogs within the system can be built to resemble possible future behavior. Predictive modeling could explore different scenarios and answer questions of the general form: "What would happen to X if we did Y"? Specific examples of such questions include: "What would happen to the COVID-19 epidemic under strict social distancing for 2 months"? or "If things carried on like today, how would the epidemic look in 3, 6 or 12 months"? Finally, predictive modeling can also look at trade-offs and optimise outcomes by answering questions like: "What is the best strategy to take if we want to achieve a given outcome"?

Although both explanatory and predictive models are important and have been used widely to gain a better understanding of the COVID-19 pandemic, our focus in this chapter will be on predictive models. An overview of explanatory models can be found in Shmueli (2010).

1.3 What does it mean for models to be "right"?

Building a "correct" model is a more complex concept than may initially be thought. In the previous section, we addressed the question of why we build models: we do so in order to answer questions about things that we cannot observe. But since the questions we put to models are often complex, knowing whether the answer is "right" is also not straightforward. This is best illustrated with an example. In early 2020, the widely-publicised mathematical model produced by Imperial College London (Ferguson et al., 2020) suggested that mortality from COVID-19 was around 1% and an epidemic in a susceptible population of 67 million people could cause around 670,000 deaths in the absence of any interventions to reduce the spread of the virus (Ferguson et al., 2020). In reality, numerous non-pharmeceutical interventions were deployed, and the UK had recorded around 70,000 deaths by end of November 2020. Does this mean that the model "got it wrong"? How can we assess and validate the model under these circumstances?

This example illustrates a fundamental difficulty with evaluating predictive models. If we model a policy scenario which doesn't manifest, does that invalidate the projections produced by the model? We can try to answer this by going back to the model and asking whether it would have predicted the actual trajectory of the epidemic if given information on what policies ended up being enacted. Exercises of this nature have been attempted for the Imperial model, for example (Rice et al., 2020).

Rather than asking whether a model is *right*, it is more instructive to ask whether it is *useful*. It is for this reason that a better understanding of the processes of modeling and a greater awareness of how and when models can be reliably used are important. The COVID-19 pandemic illustrated that even the often-cited maxim among modelers that "the model is as good as the data it uses" may not always be the most useful framework, because even at the onset of the pandemic where data was scarce, modeling still proved useful.

1.4 Aims and purposes of this work

This chapter aims to bring together the current thinking about mathematical modeling of infectious diseases, zooming in on the modeling of the COVID-19 pandemic that was used to guide policy. We first give a short history of mathematical modeling (Section 2) before presenting a scoping overview of the key published models until September 2020 used to capture COVID-19 transmission dynamics (Section 3). Then in Section 4, we showcase three of the models that we have applied to understanding the COVID-19 epidemic, briefly describing their methodology and illustrating their application to answer a specific policy question. Our conclusion then brings it all together, summarising the usefulness of modeling.

2 Modeling of infectious disease

2.1 The essence of modeling infectious diseases

The essence of mathematical modeling involves building a framework, often based on a system of equations, that simplifies and mimics reality in order to derive answers to real-life questions, as shown schematically in Fig. 1A. Although these systems of equations are simplifications of reality, they can nevertheless be used to investigate real-world questions, either by solving the equations directly if possible, or by simulating outputs from the system for given values of the parameters within the equations (Fig. 1B).

A

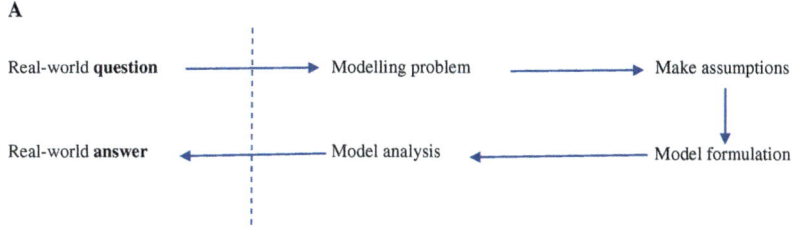

B

	Empirical	Mechanistic
Deterministic	Predicting virus spread from a **regression analysis** over time	Predicting virus spread based on known historic behaviour and based on **ordinary differential equations**
Stochastic	Analysis of **variance in regression model** of virus	Predicting virus spread based on different historic behaviour and based on **probabilistic differential equations**

FIG. 1   (A) Schematic of mathematical modeling describing broadly the steps of modeling that allow answering real world problems. (B) Broad division of mathematical models.

The parameters of the mathematical model can be refined when we compare the outputs of the model to information that we already know, for example, available data on the reported number of infections or the confirmed number of deaths due to the infection. This process of model refinement (or calibration) can be done iteratively a number of times sweeping through the parameter space, manually or in an automated way (Taylor et al., 2010), until the output of the model agrees with what we already know, e.g., about the virus spread. While simple models with only a small number of parameters can be calibrated with simple parameter sweep and sensitivity analysis, calibration of complex models require extensive complex computational techniques. A multitude of statistical approaches have been applied to this problem, some of which rely on a likelihood function and some which do not (Andrianakis et al., 2015; Kennedy and O'Hagan, 2001). In essence, these methods attempt to minimise the difference between the model and the data. More formally, this means minimising a specified objective function defined on the model by starting with an initial model parametrisation and seeing how the difference changes as the parameters change. A subset of the methods discussed in Andrianakis et al. (2015) use Bayesian approaches, in which a prior distribution is set over the parameter space, which is then updated to create a posterior distribution once the information from the data has been taken into account.

Once the model is calibrated it can be used to tell us more about future behavior of the virus spread, i.e., make predictions or forecasts. For example, in epidemiological modeling, it can predict the epidemic curve, i.e., the curve that gives the number of infections caused by the virus over time. Changes to the model parameters can mimic possible future interventions and this would allow the calibrated model to be useful in making predictions of what the epidemic curve may look like in the future.

Model discrepancy and measurement bias are two aspects that complicate model calibration. While model discrepancy explains the difference between the mathematical model and the reality as described by the data sources, the measurements bias is related to describing the measurement error, which can be caused by both the data and the calibration method. Separating the measurement bias from model discrepancy is important in calibrating models and when defining "goodness of fit." Accounting for both is important in predictive modeling.

Ideally, the system of equations within the modeling framework would be solved analytically, deriving exact solutions using historic behavior (initial or boundary conditions). But for non-trivial models it is rarely possible to obtain such analytical solutions; the complexity of modeling real-life scenarios quickly produces non-trivial models. The alternative is numerical solution of the system. Broadly speaking, the modeler has a choice when conducting a numerical simulation of the model: to formulate the model as an initial-value problem of ordinary differential equations describing entire population groups or to consider each change to the state of individuals in the population as a discrete stochastic event. For differential equations, simulating the system always gives the same result. For stochastic simulations, the inherent randomness means that each simulation results in a different evolution of the system and ascertaining how the system will behave on average requires running many such simulations and reporting the mean or the median. These two strategies are related: with suitable assumptions, the time evolution of the system described with differential equations is approximated by the average time-evolution of the stochastic system. There are important differences: the more fine-grained the model is, the less feasible it becomes to formulate differential equations for it, and the only choice is stochastic simulation. Furthermore, only some choices of distribution for the timing of events are compatible with the assumptions required for the mean trajectory of the stochastic system to coincide with the solution to the differential equations. These differences mean that the practice is that some classes of model are typically simulated using one or the other technique and this leads to two categories of widely used models: compartmental models and agent-based models (ABMs).

2.2 History of modeling infectious diseases

Mathematical modeling has a long history of being used for understanding how a virus can spread in a population.

The simplest kind of disease model is based on the concept of splitting the population in compartments, i.e., compartmental modeling and was introduced in the seminal paper by Daniel Bernoulli in 1776 (Bernoulli, 1766). This model described the spread and vaccination against smallpox and was revisited and discussed in detail in Dietz and Heesterbeek (2002). This was extended by Ross (1911) who, in work for which he was awarded the Nobel Prize in Medicine, used what we today call dynamic transmission modeling. When he published his first dynamic malaria model, Ross introduced the phrase "*a priori pathometry*" to describe the scientific process of modeling transmission dynamics (Ross, 1911). He expanded on this in his work from 1911, presenting a new set of equations for the demonstration of the dynamics of the transmission of malaria between mosquitoes and humans (Ross, 1916). The importance of Ross' work was that he believed that explaining epidemics and epidemic control quantitatively were extremely challenging and had to be combined with predictive quantitative measures; hence, giving rise to many of the processes of mathematical modeling of infectious diseases as we know them now. In his description of the work he called this "*a priori notions of observational data and exploration of patterns*" that emerge when we use observational data combined with statistical analysis; as outlined in Ross (1916). The concept of combining data with a system of theoretical equations forms the fundamental framework of mathematical modeling of infectious diseases. Further advances were made possible by Ross' collaborations with the mathematician Hilda Hudson, whose technical expertise opened the door for an understanding of the patterns that different models could produce (Ross and Hudson, 1917a,b). These developments formed the seed for the work of Kermack and McKendrick, widely considered as the conceptors of the present-day mathematical epidemiology. Their first paper from 1927 (Kermack and McKendrick, 1927), acknowledged Ross and Hudson's work and cemented the notion of "mathematical epidemiology" or "mathematical theory of epidemics" as they called it. Subsequent work by Kermack and McKendrick (1932, 1933, 1937, 1939) expanded this theory and also defined the basic reproduction number in terms of model parameters; we discuss their model in more detail in Section 2.2.2. Ross, Kermack and McKendrick influentially recognized the importance of mathematical epidemiology, and their ideas motivated the work by Macdonald (1950, 1952, 1955, 1956) and Anderson et al. (1992). This is the basis of compartmental models that are presently used in modeling infectious diseases across a number of diseases, including COVID-19; more details on this are presented in Section 3. In the next section, we give a brief description of the simplest compartmental modeling framework, while in Section 4.1 we show how we have extended this to apply it to modeling the spread of COVID-19.

2.2.1 Lotka-Volterra equations, SIR models and reproduction number

Every university mathematical epidemiology course generally starts with introducing the modeling framework that tracks the temporal evolution of populations of Susceptible to the infection (S) cohort, the subset of these that become infected (I) and the subset of these that recover from the infection (R); this describes the model of Kermack and McKendrick (1927).

But before we discuss in more details this classic Kermack-McKendrick SIR model we will introduce the classic Lotka-Volterra system of equations as a pair of first order, non-linear, differential equations that describe the dynamics of biological systems in which two species, a predator and a prey, interact (Murray, 1989). They were proposed independently by Lotka in 1925 and Volterra in 1926 just before Kermack-McKendrick SIR model was introduced in 1927. The classic Lotka-Volterra system comprises Eqs. (1)–(2) and taught modeling courses generally start with introducing this system of equations before subsequently studying them using steady-state, perturbation and bifurcation analysis exploring the existence and stability of the long-term solutions of the system; for details, see for example, Chapters 3 and 4 of Murray (1989):

$$\frac{dx}{dt} = x(a - by) \tag{1}$$

$$\frac{dy}{dt} = y(cx - d) \tag{2}$$

Assuming that $a,b,c,d > 0$, this system describes x as the predator and y as the prey with the parameters c and d representing the competition between them. The parameters a and d describe how quickly $x(t)$ and $y(t)$, respectively, grow and decay exponentially in absence of the other species; solving the system with $b = c = 0$ gives

$$x(t) = x_0 e^{at}, y(t) = y_0 e^{-dt}$$

where x_0 and y_0 are the initial values of x and y, i.e., $x_0 = x(t=0)$, $y_o = y(t=0)$.

If the parameters b and c are included, the system of Eqs. (1)–(2) can be solved in special cases as shown by Varma, Wilson or Burnside (Varma, 1977; Wilson, 1980). The numerical solutions of the system of equations, for a choice of model parameters and initial conditions, are shown in Fig. 2A and B.

The oscillatory solutions in Fig. 2A are a result of the cyclic behavior of the system confirmed in the phase plot diagram in Fig. 2B. Interestingly, using two different solvers in MATLAB to solve even a very simple system of equations such as (1)–(2) can give slightly different results; hence it is important that we discuss the robustness of numerical solutions in Section 2.3.

In fact, we can derive the cyclic solutions depicted in Fig. 2B, representing the phase trajectories, analytically by directly integrating the phase plane equation (brief introduction to phase plane analysis can be found in the Appendix A of Murray (1989))

$$\frac{dv}{du} = \alpha \frac{v}{u} \left(\frac{u - 1}{1 - v} \right) \tag{3}$$

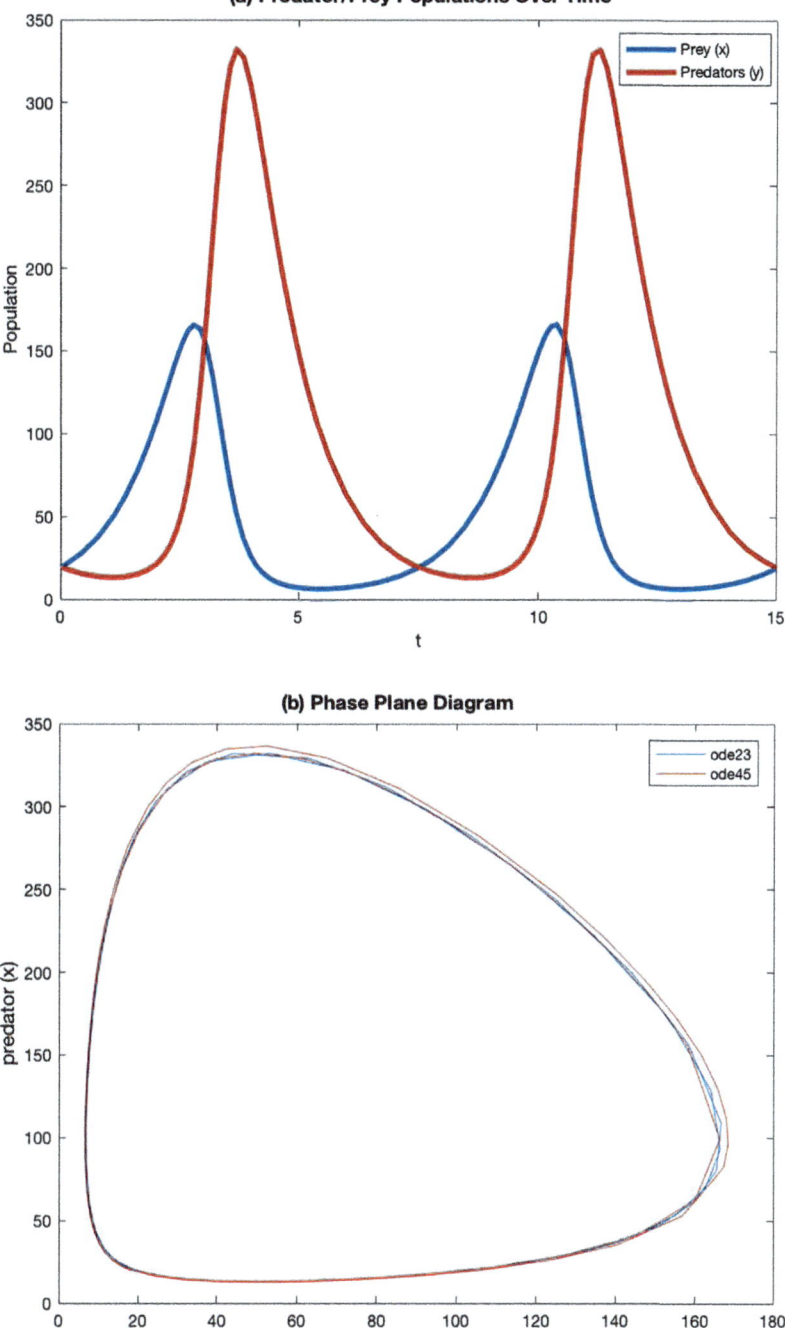

FIG. 2 Solutions to the Lotka-Volterra system of Eqs. (1)–(2) when $a=1=d$, $b=0.01$ and $c=0.02$. The initial conditions assume that $x_0=y_0=20$ in this example and we run the system for $t=[0,15]$. (A) Predator/prey populations over time. (B) Phase plane diagram. In (A) we present the temporal profiles of x and y showing the periodic solutions of the system which are confirmed in the phase plane diagram in (B). When generating the phase plane diagram, we solve the system of equations using two different numerical solvers ode45 and ode23 illustrating the subtle differences present on even the simplest set of equations as (1)–(2).

Here, u and v represent the non-dimensionalised (or rescaled) versions of the variables x and y defined as $u = \frac{cx}{d}, v = \frac{by}{a}$ and $\alpha = d/a$. Eq. (3) has singular points at $u=v=0$ and at $u=v=1$ and integrating it directly we get $\alpha u + v - \ln u^{\alpha} v = C$, where C is a constant with a minimum $1 + \alpha$ occurring when $u=v=1$. It is a straightforward mathematical conjecture to show that, for a given constant $C > 1 + \alpha$, the trajectories in the phase plane are closed; illustrated for a given parameter regime in Fig. 2B.

An important aspect to note from Fig. 2A and B is that the asymptotic solutions, i.e., the solutions of the system that the system settles to as $t \rightarrow \infty$, depend on the model parameters. Even in the simplest case, setting $a=c=1$; and the system only has two non-zero parameters that describe the strength of the competition between x and y, there are still three possible longtime solutions of the system: the coexistence steady state shown in Fig. 2A; or a limit case of this steady state where either the predator or the prey goes extinct in the long-time if the competition term from one species is much stronger (mathematically an order of magnitude larger) than the other. To determine these we need to set the derivatives in (1)–(2) to zero and solve the system of algebraic equations to determine the steady states and their stability; details on this method can be found in Appendix A of Murray (1989).

Why is the Lotka-Volterra system of equations so important when doing modeling? One of the reasons is that the classic SIR framework of Kermack-McKendrick is very similar to this system of equations.

Instead of having two competitor species, as the Lotka-Volterra system, the SIR framework splits the population cohort into three groups of Susceptible (S), Infected (I) and Recovered (R) populations. Analogous to the Lotka-Volterra system of equations, it is based on the principle of mass action that describes the rates of transition between these classes (Fig. 3A). In the simplest SIR model, the one that Kermack-McKendrick developed, births and deaths can be ignored and there is no waning of immunity, allowing only two possible transitions: spread of infection with transfer of individuals from the susceptible to the infected class (at a rate β) and recovery from the infection with transfer of individuals from the infected to the recovered class (at a rate γ). We note that different notations of the I and R compartments exist: I are sometimes called infected, infective or infectious while R are either recovered or removed; the original Kermack-McKendrick model referred to these as infected and recovered and comprised the system of Eqs. (4)–(6) below:

$$\frac{dS}{dt} = -\beta SI/N \tag{4}$$

$$\frac{dI}{dt} = \beta SI/N - \gamma I \tag{5}$$

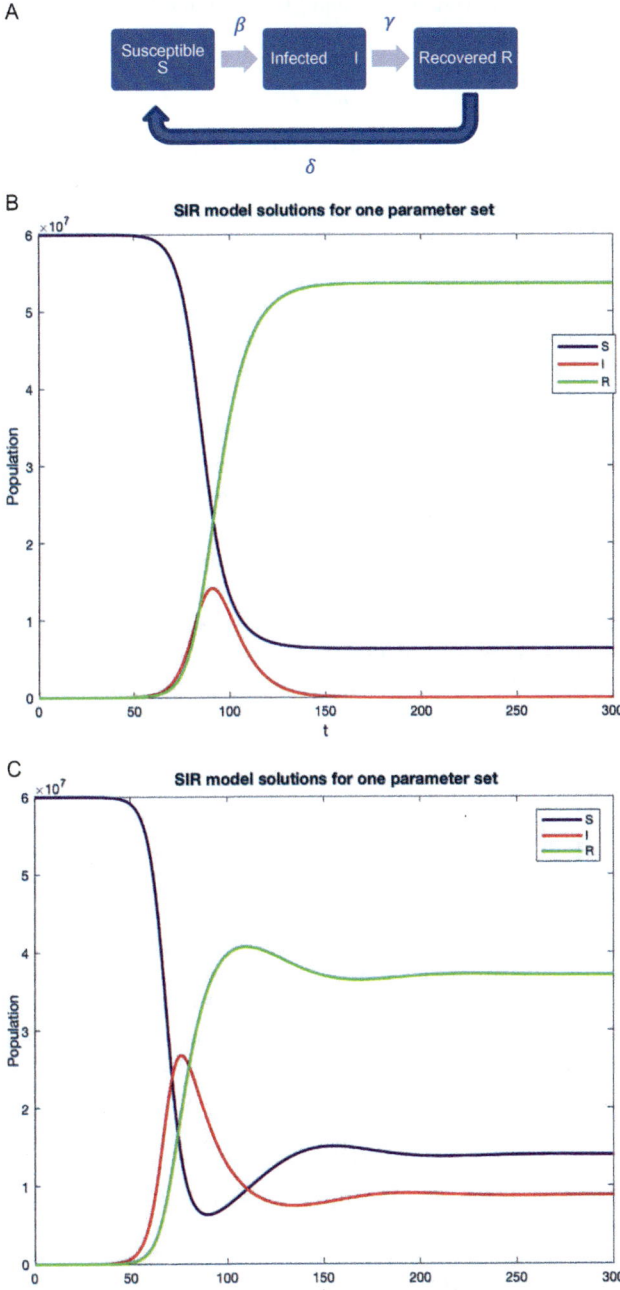

FIG. 3 (A) Schematic of SIR model. Schematic of a simple SIR model without and with waning of immunity (at a rate δ) describing the system of Eqs. (3)–(4) and numerical solutions for two set of parameters. (B and C) Model solutions for a parameter set. The model parameters are $\beta = 5 \times 10^{-9}$, $\gamma = 0.12$ in (B) and $\gamma = 0.07$ in (C) and $\delta = 0.0$ in (B) and $\delta = 1/60$ in (C). The initial conditions assume that $N(t=0) = 6 \times 10^7$, $I(t=0) = 10$, $R(t=0) = 0$, i.e., there are 10 infected people at the onset and that the basic reproduction number is $R0 = \frac{\ell}{\gamma}N(t=0) = 2.5$. We simulate the system for $t = [0,300]$. In (B) we present the temporal profiles showing the system settles to a steady state that clears the virus while in (C) the long-time solution shows the coexistence of the infected, susceptible and recovered populations (this is endemic epidemic).

$$\frac{dR}{dt} = \gamma I \tag{6}$$

Here, the parameter β represents the infectiousness rate, i.e., the rate at which the susceptible population transfers to the infected population, and γ is the rate of transfer from the infected to recovered population, with $1/\gamma$ describing the average length of the infectiousness period. It is important to note that this model assumes homogeneous mixing in the population which means that everyone interacts with equal probability with everyone else and discards situations where there is a heterogenous mixing across ages or settings. Details on the model adaptations necessary to account for such situations can be found in Chapters 3 and 7 of Keeling and Roheni (2008). Furthermore, there are some assumptions in deriving the transmission term βSI that are key for this model; details can be found for example in Chapter 2 of Keeling and Roheni (2008). Briefly, the parameter β can be defined and derived from two additional parameters c as the average number of contacts per unit of time and p as the probability of transmission per contact.

The end outcome, defined as the steady-state solution, for SIR models defined by Eqs. (4)–(6) is that the epidemic will either be cleared and the infected population eventually fully transferred to recovered population. Another possible outcome is that the epidemic will be sustained with coexistence of susceptible, infected and recovered populations (see conditions on this below, but we note that this is not possible in the case of Eqs. (4)–(6) where there is no population growth and no waning of immunity (Fig. 3). A whole theory of mathematical biology is concerned with exploring different constraints necessary for these outcomes and how dependent they are on the parameters β and γ; both Keeling and Roheni (2008) and Murray (1989) Murray (1989) have details of this. Next we revisit this briefly.

The key question when modeling epidemics is, given the values of β and γ and the initial conditions on the populations, will the infection spread or not, how will it behave over time and when will it start to decline and diminish. These questions can be formulated mathematically as the set of Eqs. (4)–(6) plus some initial conditions:

$$S(0) = S_0 > 0, I(0) = I_0 > 0, R(0) = 0$$

Considering Eq. (4), since $\frac{dS}{dt} < 0, S \leq S_0$. Hence we can see that if $S_0 < \frac{\gamma}{\beta}$ then

$$\frac{dI}{dt} = I(\beta S - \gamma) \leq 0, \forall t \geq 0$$

This suggests that $I_0 > I(t)$ as $t \to \infty$ and hence infection will die out in the longtime. But if $S_0 > \frac{\gamma}{\beta}$, then similarly we conclude that $I_0 < I(t)$ as $t \to \infty$, so an epidemic occurs and starts growing. In this sense an epidemic occurs if the infected population increases above its' initial value, i.e., if $I_0 < I(t)$ for some

$t>0$. In modeling, this gives rise to a threshold phenomenon, i.e., depending on the threshold value of the ratio $\frac{\gamma}{\beta}$ we get two behavioral phenomena: epidemic or not. The inverse ratio $\frac{1}{\frac{\gamma}{\beta}} = \frac{\beta}{\gamma}$ represents the infection contact rate. We note that as epidemiological modeling has been developing over the years, so has the notation. For example, different modelers refer to the contact rate differently and sometime writing $\beta = cp$ or $\beta = -c\,log\,(1-p)$, where as described above, p represents the probability of risk of transmission upon contact and is related to viral load of the virus, while c represents the number of contacts per unit of time; hence in this case only c is referred to as contact rate.

Whichever notation we use, the interplay between the parameters β and γ leads to the notion of the basic reproduction number which is defined as $R_0 = \frac{\beta}{\gamma}$ that simply represents the number of secondary infections that emerge from one primary infection in a wholly susceptible population. By the definition, if more than one secondary infection is produced from the primary infection this simply means $R_0 > 1$. The basic reproduction number and the condition $R_0 > 1$ has been widely used and discussed in the media as a metric for describing the state of the COVID-19 epidemic and we will revisit this again in our COVID-19 specific case studies in Section 4.

While the basic reproduction number is an important metric to consider in a growing epidemic, this simple example shows that its value is most relevant at the onset of the epidemic, as it depends on the initial pool of susceptible people. As the epidemic develops and the pool of susceptible people changes, the basic reproduction number R_0 becomes an effective reproduction number, often written as Reff, Re, Rt or just R. A detailed discussion of the differences between these highlighting what the effective reproduction number can tell us and what it can't in terms of a growing epidemic can be found in Vegvari et al. (n.d.).

It is also important to note that non-pharmaceutical interventions such imposing social distancing rules to reduce the number of contacts (reducing c) or wearing masks to reduce the risk of transmission during a contact (to reduce p) would reduce the transmission rate β. This can in turn reduce the effective reproduction number—we illustrate this in our case studies in Sections 4.1 and 4.2.

2.2.2 Stochastic models: Adding stochasticity to the SIR framework, branching processes and individual- or agent-based models (IBMs/ABMs)

Compartmental models with fixed parameters, such as the SIR framework in Section 2.2.2, can answer simple questions around initial epidemic growth and give parameter constraints on expected long-term solutions. But such models have a crucial limitation: since everyone in each compartment is assumed to be the same, these models ignore important aspects of social interactions and heterogeneous behavior patterns. Various generalisations of the structure of the simple SIR model have been made over the years to address this limitation, which have included modifications of the model structure to include sub cohorts of the population (e.g., different age or risk-groups) or the addition of a mixing matrix to the infection rate beta to account for the interaction between such sub cohorts (see for example Kiss et al., 2006; Lloyd and May, 2001; Rohani et al., 2010). But an alternative means of addressing the assumption of heterogeneity within compartments is to use probability distribution functions for the flow rates between compartments, rather than assuming a fixed value. This allows for uncertainty and variability in the model parameters to be included in compartmental models. For example, the duration of the infectiousness period (described by the parameter gamma in Section 2.2.2) need not be treated as a fixed number but rather as a varying parameter, which takes different values according to a probability distribution. For large populations, this variability averages out, but for small populations, a stochastic treatment is required. Stochastic Differential Equations (SDEs) are one way to capture this variability where there is a natural limiting behavior of the mean field that yields a corresponding ODE system for large populations. Another, more flexible way to capture this variability is by explicitly modeling individuals rather than the population as a whole. This can be done using Agent- or Individual-Based models (ABMs/IBMs).

In ABMs, epidemics are modeled by creating a set of autonomous agents or individuals that follow certain rules and/or decisions and interact with each other in certain defined ways. ABMs allow individual contacts within a network to be modeled, and infectious disease spread within realistic synthetic population to be simulated (Eubank et al., 2004). In addition, agent-based models can more easily allow for uncertainty analysis by allowing the incorporation of stochastic effects, as well as the potential to have well-defined variation between individuals. While compartmental models encourage modelers to simplify the problem at hand, ABMs encourage modelers to think more deeply about the full system. This can pair well with model-driven data collection: rather than building a model simply to use the available data, this approach involves building the model first and then identifying what data gaps remain to be filled. When applied appropriately, this approach can have enormous benefits, since "one can only understand what one is able to build" (Dudai and Evers, 2014). While still a less common approach than compartmental modeling, there are still many examples of where it has been used successfully (Heesterbeek et al., 2015).

There are several disadvantages to using ABMs as well. First, vast stochasticity within ABMs can be a disadvantage: the results from ABMs almost always converge to those of compartmental models if averaged over a large enough numbers of individuals and/or over a large number of trajectories, and in cases where this is the desired outcome, a compartmental model may be preferable because of its lower computational cost. Second, compared to compartmental models, ABMs are typically much more data-hungry, as they often have many more parameters to inform the behavior of individual agents. It is often hard enough to find sufficient data to parameterise compartmental models appropriately, much less ABMs. In such cases, the better approach is often to opt for a simple model based on available data rather than create a complex model based on many assumptions.

2.3   Challenges of modeling infectious diseases

Developing and using models to inform decision making has challenges. Some of these were highlighted in a series of challenge papers published as a special issue of Epidemics in 2015 led by a consortium of scientists (Lloyd-Smith et al., 2015). A follow-on collection of challenge papers focused on the challenges related to modeling the COVID-19 pandemic and planning for future pandemics is under preparation in late 2020 by another consortium of scientists within the Isaac Newton Institute.

The challenges of modeling infectious diseases can be grouped into four broad categories.

First, it is necessary to find good and reliable data to parametrise the model. Formulating mathematical models requires defining model variables and model parameters. The variables represent the key dynamic quantities that change in time (for example, the number of infected people or the number of disease-related deaths), while the parameters govern the ways in which the state of the model evolves over time. Defining the parameters of a model is one of the main tasks of the modeler, and it often involves a trade-off between simplicity and detail. For example, there are just two parameters (β and γ) in the SIR model presented in Section 2.2.2, while the more detailed ABM we describe in Section 4.2 includes over 500 parameters. Often, the values of the parameters will be informed by data, so a crucial first step in determining the structure and parameterisation of a model is to evaluate the extent of data available. But there are stark limitations on data that can inform the model parameters, especially when modeling new diseases or outbreaks such as COVID-19. This is one of the main challenges of good modeling. In the absence of reliable data sources, the modeler has various options: (a) use a less data-rich model (for example, a simpler SIR model instead of an ABM); (b) propagate the uncertainty stemming from the lack of data through the model to produce distributions of possible predictive outcomes, either by conducting a sensitivity analysis in which the model is simulated numerous times for different parameter values, or by conducting a more formalized statistical analysis.

The second challenge is around having a robust and efficient numerical algorithms to solve the model and produce reliable predictive outcomes. It is the responsibility of the modeler to assure that technically correct and robust numerical techniques are used for calibration and prediction, and that uncertainty within the model is clearly communicated.

Third, there are challenges associated with developing simulation code that can withstand the tests of time and be reproducible by other users. The modeler needs to have good code etiquette with the code available as open source so it can be understood, reproduced, expanded, adapted and used by other users.

The fourth and final challenge is that models should be able to answer questions from policy decision makers in a timely and informative fashion. This has rarely been more pertinent than in the worldwide pandemic of COVID-19, when mathematical modeling was brought to the forefront of policy making and communication. In the next section, we provide an overview of these models.

3 Models for COVID-19

Since the beginning of the COVID-19 pandemic, mathematical modeling was widely used to help make decisions around the control of COVID-19 spread. COVID-19 presented a unique modeling challenge to the community due to (a) the extreme urgency of generating accurate predictions; (b) the quickly-evolving data; and (c) the large uncertainties, especially early on in the epidemic, around even basic aspects of transmission such as the reproduction number, latent period, and proportion of people who are asymptomatic.

3.1 Compartmental modeling and the SEIR framework: Overview and examples of COVID-19 models

The vast majority of the models that have proliferated in response to the COVID-19 pandemic have been compartmental models, due to their relatively simple requirements for development and the long-standing body of work using them, making them most accessible to epidemiologists. For example, Walker et al. (2020) adopted the basic SIR framework from Section 2.2.2 and used an age-structured stochastic "Susceptible, Exposed, Infectious, Recovered" (SEIR) model to determine the global impact of COVID-19 and the effect of various social distancing interventions to control transmission and reduce health system burden. Read et al. (2020) developed an SEIR model to estimate the basic reproduction number in Wuhan. Keeling et al. (2020) used one to look at the efficacy of contact tracing as a containment measure; and Dehning et al. (2020) used an SIR model to quantify the impact of intervention measures in Germany. In models such as those by Giordano et al. (2020) and Zhao and Chen (2020), compartments are further divided to provide more nuance in simulating progression through different disease states, and have been deployed to study the effects of various population-wide interventions such as social distancing and testing on COVID-19 transmission.

3.2 Agent-based models: Overview and examples of COVID-19 models

ABMs shot to public prominence on the basis of the dire predictions of the Ferguson et al. model, later named CovidSim (Ferguson et al., 2020). At the time, there had been only 6000 COVID-19 deaths globally, but this model predicted half a million deaths in the UK and over 2 million deaths in the US if strong interventions were not implemented. Like many COVID models, CovidSim was based on an earlier influenza model (Ferguson et al., 2006).

Another influential early COVID-19 ABM was that of Koo et al. (2020), who adapted an existing H1N1 model by Chao et al. (2010), in order to explore the impact of interventions on COVID transmission in Singapore. Other ABMs were developed to simulate the spread of COVID-19 transmission and the impact of social distancing measures in Australia (Chang et al., 2020) and the United States (Chao et al., 2020). Due to their flexibility, ABMs can be used to evaluate micro-level policies much more accurately than compartmental models, such as to evaluate the impact of social distancing and contact tracing (Aleta et al., 2020; Kretzschmar et al., 2020; Kucharski et al., 2020) and super-spreading (Lau et al., 2020). Since these models can account for the number of household and non-household contacts (Chao et al., 2020; Kretzschmar et al., 2020; Kucharski et al., 2020); the age and clustering of contacts within households (Aleta et al., 2020; Chao et al., 2020; Kucharski et al., 2020); and the microstructure in schools and workplace settings informed by census and time-use data (Aleta et al., 2020) they can be used to investigate detailed interventions with maximal realism.

3.3 Branching process models: Overview and examples of COVID-19 models

A third type of model, in some senses halfway in between a compartmental model and an ABM, is a branching process model. Although not widely used in the epidemiology community compared to the other two model types, branching process models saw considerable use during the COVID-19 epidemic since in some ways they combine with simplicity of a compartmental model with the detail and stochasticity of an ABM. Branching process models have also been used to investigate the impact of non-pharmaceutical intervention strategies (Hellewell et al., 2020; Peak et al., 2017). Compared to ABMs, which represent both infected and susceptible individuals and their interactions, branching models consider only infected individuals, and use probabilistic algorithms for determining how many new infections each infected individual causes. Although this allows for the incorporation of properties specific to the infected individual, it does not allow for a full treatment of the interactions between infected and susceptible individuals as in an ABM.

4 Applications of modeling of COVID-19: Three case studies

4.1 Case study 1: Application of SEIR-TTI model to the UK COVID-19 epidemic

4.1.1 Overview of SEIR-TTI

For this case study, we illustrate a new method for including the effects of Testing, contact-Tracing and Isolation (TTI) strategies in classic Susceptible-Exposed-Infected-Removed (SEIR) models. The SEIR-TTI model is a direct extension of the SEIR modeling framework that incorporates a probabilistic argument to show how contact tracing at the individual level can be reflected in aggregate on the population level.

4.1.2 SEIR-TTI methodology

Details of the mathematical framework behind the SEIR-TTI model can be found in Sturniolo et al. (2020). Briefly, the SEIR set-up extends the classic SIR model from Section 2.2.1 to include a cohort of individuals exposed (E) to the virus that have been infected with the virus but not yet infectious. A key parameter than describes the latent period when the pathogen reproduces within the host, but the viral load is too low to be categorised as susceptible (S) or infected (I). Hence an intermittent cohort E, and respective mathematical equation is needed to link these. Assuming the average length of the latency period is $1/\alpha$ the system of equations becomes:

$$\frac{dS}{dt} = -\beta c S I / N \tag{7}$$

$$\frac{dE}{dt} = \beta c S I / N - \alpha E \tag{8}$$

$$\frac{dI}{dt} = \alpha E - \gamma I \tag{9}$$

$$\frac{dR}{dt} = \gamma I \tag{10}$$

An important thing to note is that although SIR and SEIR models behave very similarly at steady state, i.e., have the same longtime solutions with the E cohort emerging as an intermediate cohort, the SEIR models have a slower growth rate. This is a consequence of the delayed process of development of infectiousness following the virus/pathogen entering the host system.

The SEIR model, we developed in Sturniolo et al. (2020) has similar structure to Eqs. (7)–(10). The novelty of our work is how we then layer this model with a model for tracing the population that is either exposed or infectious—the E and I cohorts. Existing models that have attempted to do this have incorporated this by asserting that a proportion of exposed individuals become quarantine or via reducing the transmission, sometime with a possible delay.

But neither of these concepts account for finding and tracing both exposed and infectious individuals that have different probabilities of risk of onward transmission. Our SEIR-TTI framework allows us to do that; further details can be found in Sturniolo et al. (2020). Importantly, we validated our SEIR-TTI ODEs based model against two mechanistic ABMs and showed good agreement at far less computational cost.

There are two key concepts in our SEIR-TTI framework:

a) the introduction of overlapping compartments within the SEIR modeling framework;
b) defining the transition rates for people who are traced.

The first of these is achieved by defining overlapping compartments to represent model states that are not mutually exclusive like the S, E, I and R compartments are. Hence, we allow for an individual within our SEIR model to belong in more than one category, e.g., be infected and contact-traced, or exposed and tested. Specifically, as illustrated in Fig. 4A we represent unconfined and isolated individuals simply by doubling the number of states, labeling S_U, E_U, I_U, R_U respectively the undiagnosed S, E, I and R compartments; and similarly, S_D, E_D, I_D, R_D the ones who have been diagnosed or otherwise distanced from the rest of the population, by for example home isolation or hospitalization.

The second point is more complex and explicitly explained in Sturniolo et al. (2020), so we will not go into details here. The key aspect is that we derive transition rates among overlapping compartments, by considering that individuals will be traced proportionally to how quickly the infectious individuals who originally infected them are, themselves, identified. People can be identified via testing, at a rate $\frac{1}{\theta}$, or via tracing at a rate η and success χ. We define a global tracing rate that depends on the probability of an individual of being traced.

In the next section, we give an illustration of how the SEIT-TTI model can be applied to predict the effective reproduction number as a combination of different test-trace strategies and how this changes the epidemic curve.

4.1.3 Application of SEIR-TTI to estimate R

We applied the SEIR-TTI model to simulate the spread of COVID-19 in the UK. We began by creating a total population of 67 million individuals, with 100,000 individuals initially infected (i.e., $I(0) = 100,000$). To model the dynamics of transmission, we define the probability of transmission per contact per day to be $P = 0.033$, the number of contacts per day to be $c = 13$, the rate of exposed people becoming infectious to be $\alpha = 0.2$ days^{-1}; and the length of the infectiousness period (recovery time) to be 7 days. These values result in a basic reproduction number of $R_0 = 3$ at the onset.

Assuming the success of tracing $\chi = 0.5$, we then simulate different combinations of tracing and testing levels and project the value of R after 30 days

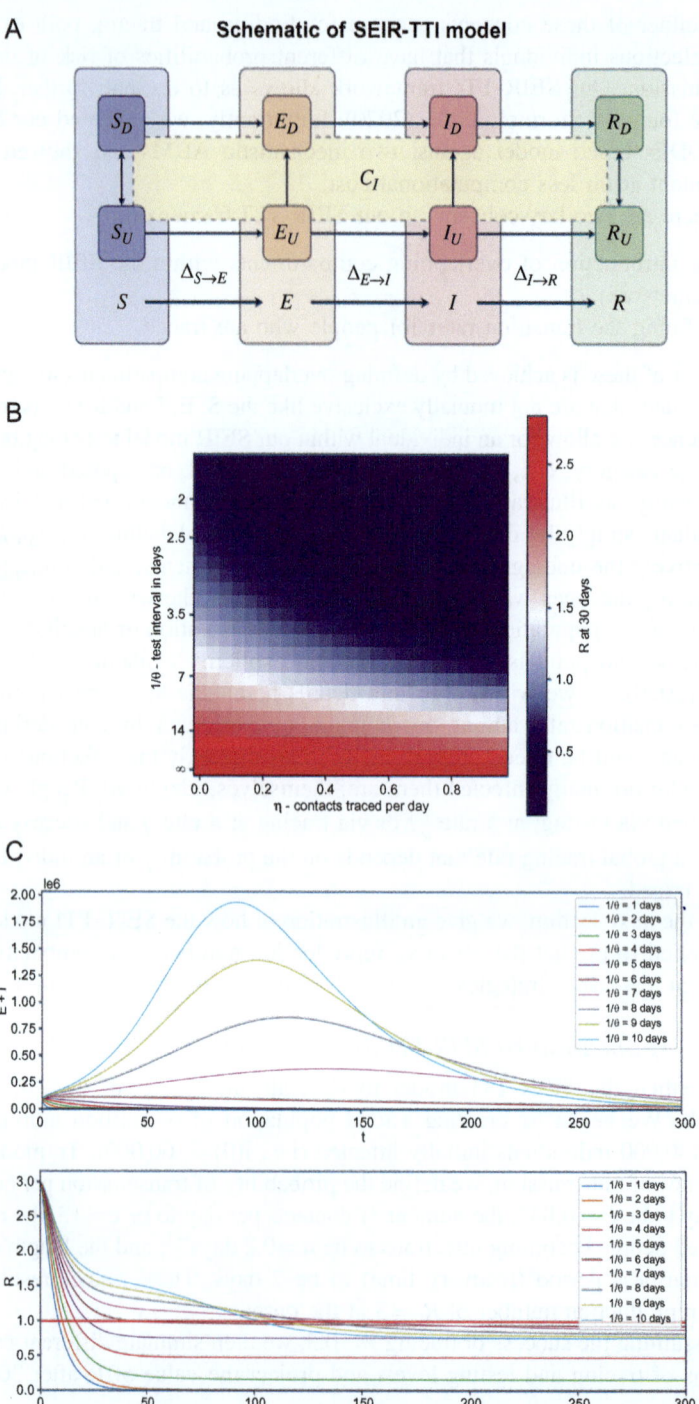

A   **Schematic of SEIR-TTI model**

B

C

FIG. 4 See figure legend on opposite page.

(Fig. 4B). Our results suggest that sufficiently effective tracing and sufficiently frequent tracing are necessary to control the virus and keep $R < 1$ and within the blue region in Fig. 4B. But we also found a non-linear relationship between testing and tracing, implying it would be possible to control the virus with less effective tracing and more frequest testing, or vice versa (Fig. 4C). The epidemic trajectory looks different for different combinations of test-trace levels; in Fig. 4C we illustrate these for different testing levels assuming a tracing rate of $\eta = 0.5$.

4.2 Case study 2: Application of Covasim to the UK COVID-19 epidemic

4.2.1 Overview of Covasim

For this case study, we illustrate the application of an open-source agent-based model called Covasim (COVID-19 Agent-based Simulator) developed by the Institute for Disease Modeling. The methodology of the model is contained in Kerr et al. (2020a) with further development and implementation details available at http:/docs.covasim.org. Since the onset of the pandemic, Covasim has been applied across a number of studies (Cohen et al., 2020; Kerr et al., 2020b; Panovska-Griffiths et al., 2020a,b; Stuart et al., 2020). Here, we illustrate an application of Covasim to answer questions around the COVID-19 epidemic in the UK.

4.2.2 Covasim methodology

Covasim is an agent-based model with individuals modeled at different stages of their infectiousness, as susceptible to the virus, exposed to it, infected, recovered, or dead. Infectious individuals are additionally categorized as asymptomatic, presymptomatic (before the viral shedding has begun) or with mild, severe or critical symptoms, as illustrated in Fig. 5A.

Covasim is coded in Python with default parameters that are regularly updated based on ongoing literature reviews. It is equipped with demographic data on population age structures and household sizes for different countries,

---

FIG. 4—CONT'D (A) Schematic of the SEIR-TTI model. Application of the SEIR-TTI model. (B) R phase plane for test-trace levels. Phase plane plot of the effective reproduction number R after 30 days of running the model for different tracing (x-axis) and frequency of testing (y-axis) levels. Larger R, implying higher numbers of new infections is shown in red, while lower values in white, and the blue region representing area where the resurgence of COVID-19 is controlled ($R < 1$) with combinations of adequate test-trace strategy. (C and D) Model predictions for different testing levels. SEIR-TTI model predictions of the Exposed and Infected populations (C) and R (D) over time when tracing level is 50% and frequency of testing increases. *Panel A schematic of the SEIR-TTI model reproduced from Sturniolo S, Waites W, Colbourn T, Manheim D, Panovska-Griffiths J. Testing, tracing and isolation in compartmental models." medRxiv. (2020) preprint doi: https:/doi.org/10.1101/2020.05.14.20101808, with permissions.*

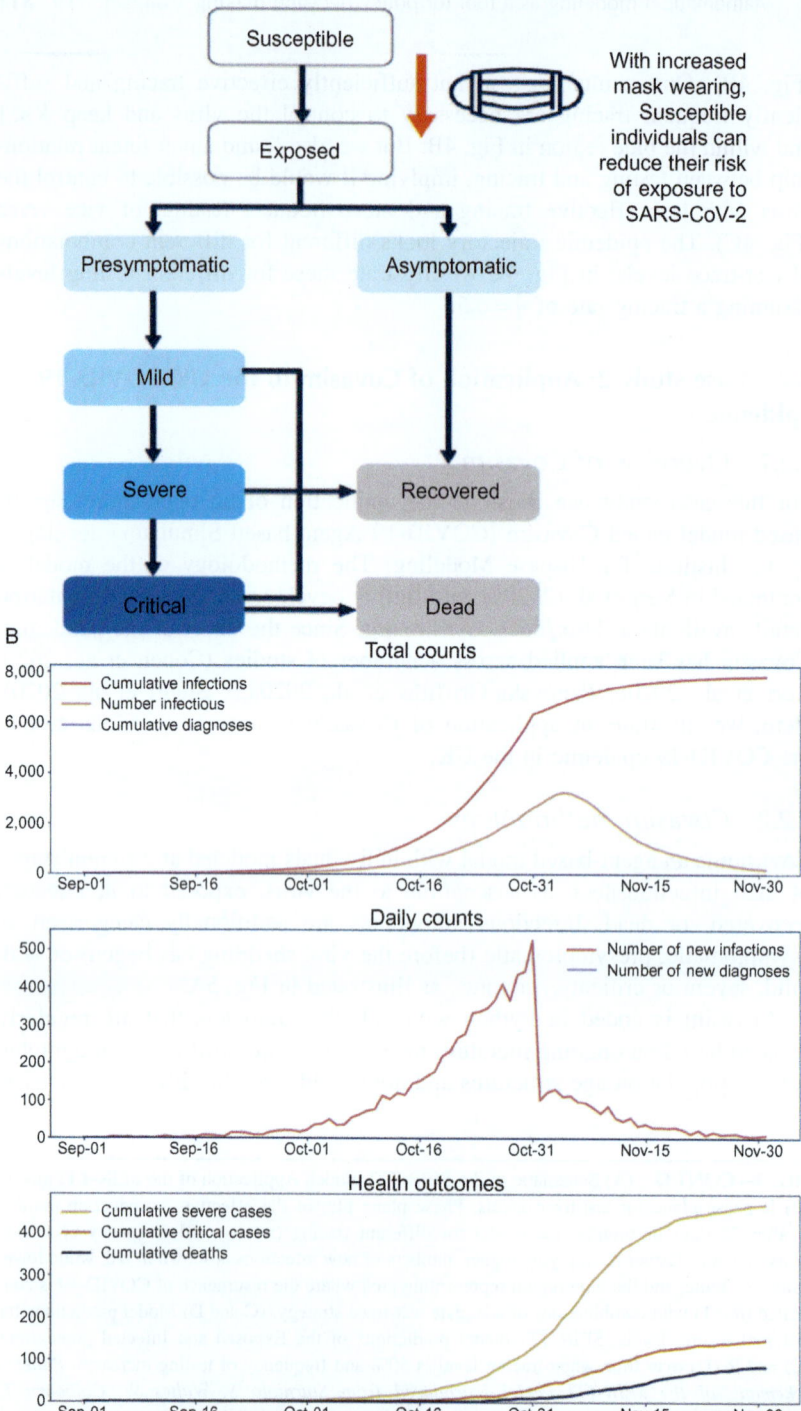

A   Schematic of Covasim model with mask wear

FIG. 5 Illustration of the Covasim model. (A) Schematic of Covasim model with mask usage showing the disease-related states that individuals can be in. (B) Typical model outcomes from Covasim. Model outcomes from a simple example of running Covasim with the defaults values

(Continued)

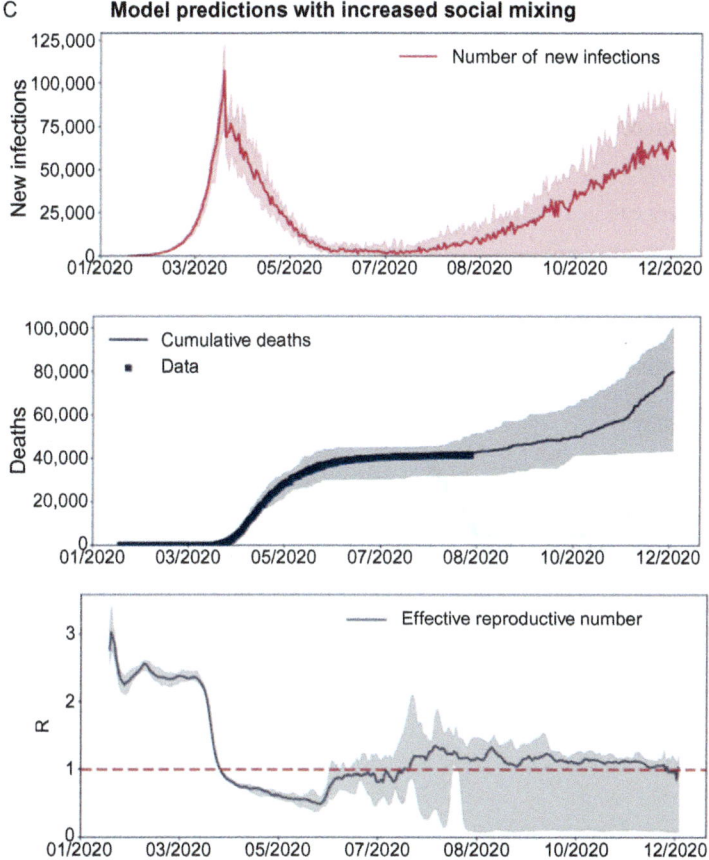

FIG. 5—CONT'D and projecting total and daily counts as well as health outcomes. (C and D) Model predictions with increased social mixing and with reduced social mixing. Outcomes from an application of Covasim to the UK epidemic, calibrated to the reported COVID-19 infections and deaths until August 28, 2020 with model parameters as per (Panovska-Griffits et al., 2020). Medians across 12 simulations are indicated by solid lines and 10% and 90% quantiles by shading.

with individuals interacting across four contact network layers for schools, workplaces, households and community settings. Different epidemics can be modeled by adjusting context-specific parameters, including rates of testing, tracing, isolation compliance, and other non-pharmeceutical interventions. Transmission occurs during contacts between infectious and susceptible individuals, according to a parameter β, which is comparable to the infection rate β in the simple SIR from Section 2.2.2, but stratified across contact network layers and across different risk groups.

Covasim has a flexible framework that can be adapted across settings and for specific conditions. To run a simulation, we can use the default parameters

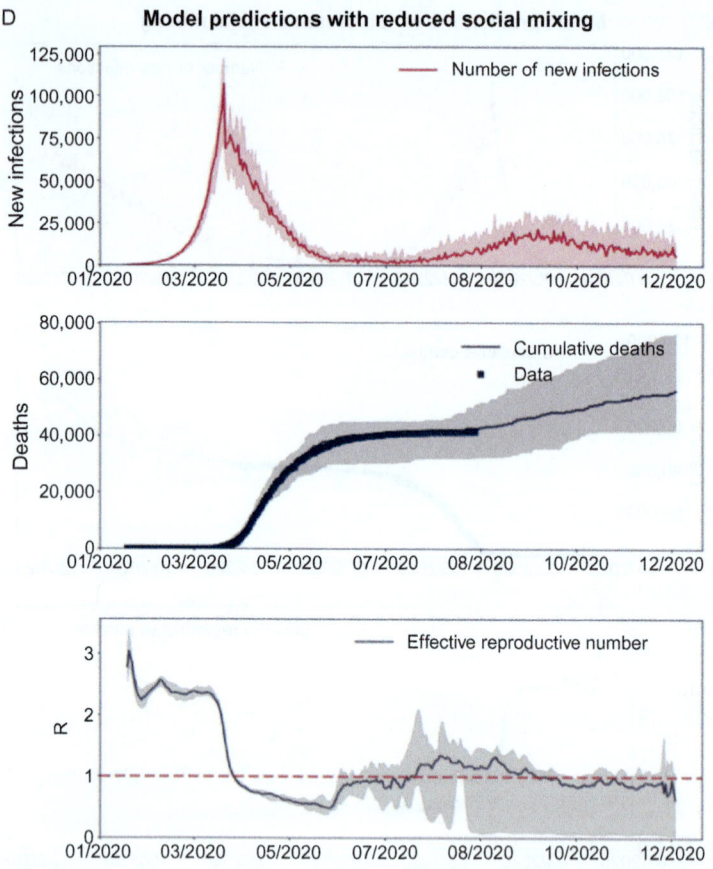

FIG. 5—CONT'D

to generate a population of agents who interact over the four contact networks layers and then change model parameters to calibrate the model to fit the specific epidemic. A simple modification would be to specify the time for which we are interested to run the simulation, or change a model parameter.

In Fig. 5B we show the outcomes of a simple simulation in Covasim with the simulation ran between September 1, 2020 and December 1, 2020 with a change in the infection rate β from 0.016 to 0.2 from November 1, 2020. The simulation shows an increase in infection with increased β, i.e., showcases the impact of a single parameter on the predicted outcomes from November 1, 2020 when β was changed (Fig. 5B).

This example can be built on and expanded within Covasim to simulate and evaluate a large number of interventions. These include non-pharmaceutical interventions, such as reduced social mixing, hygiene measures and use of face coverings (all of which can be simulated by respective changes in the contact rate c or the infection rate β), different testing interventions, such as testing people with COVID-19 symptoms or asymptomatic testing (for which specific parameters are defined within Covasim), as well as contact tracing and isolation strategies (again changing specific parameters for these within Covasim). It is also possible to simulate different vaccination strategies with Covasim.

The example illustrated above simulates the most basic intervention in Covasim—reducing transmissibility starting on a given day. Transmissibility can mean a reduction in transmissibility per contact (such as through wearing face coverings or maintaining social distance), or a reduction in the number of contacts at households, school, work, or community settings. For example, school closures can be modeled either by setting both of these to 0, while partial closures can be modeled by scaled reductions in either transmissibility per contact or the number of contacts. Our published work (Panovska-Griffiths et al., 2020a) has modeled the impact of reopening school and society on the UK COVID-19 epidemic in a similar, but more complex way to this simple example.

Importantly, Test-Trace-Isolate (TTI) interventions can also be modeled in Covasim. Testing can be modeled either by specifying the probabilities of receiving a test on each day for people with different risk factors and levels of symptoms; or by specifying the number of tests performed on each day directly. Tracing assigns a probability that a contact of a person testing positive can be traced, and allocates a time that it takes to identify and notify contacts. Isolation of persons testing positive and their contacts is the key aspect by which TTI interventions can reduce transmission. In Covasim, people diagnosed with COVID-19 are isolated with an assigned adherence level and a period duration of isolation. In the next section, we illustrate a specific application of Covasim on the UK epidemic during 2020.

4.2.3 Application of Covasim to the UK epidemic

To apply Covasim to the UK epidemic (Panovska-Griffiths et al., 2020a,b), we used the default parameters and generated a population of 100,000 agents across the household, schools, workplaces and community networks. We then calibrated the model outcomes by performing an automated search for the optimal values of the number of infected people on 21 January 2020 (when the UK first COVID-19 case was confirmed), the per-contact transmission probability and the daily testing probabilities for individuals with and without COVID-19 symptoms during May, June, July, August and September (until September 26, 2020). The optimal values determined were the ones that minimised the sum of squared differences between the model's estimates of confirmed cases and deaths and data on these same two indicators between January 21, 2020 and August 28, 2020. The data we compared against was collated from the UK government's COVID-19 dashboard (https:/coronavirus.data.gov. uk). These particular parameters were selected as the most important to estimate because of the considerable uncertainties around them. Details of the exact methodology can be found in Kerr et al. (2020a) and Panovska-Griffiths et al. (2020a,b) and the code used to run all simulations contained in here is available from https:/github.com/Jasminapg/Covid-19-Analysis.

We included policies around mask usage and TTI interventions that were part of the policy recommendations in the UK at the time of writing. Specifically, under the policy on masks in September of 2020, face coverings were mandatory in parts of community, such as public transport or in shops, and were recommended in secondary schools from September 1, 2020, but were not mandatory in workplaces. To simulate the impact of the masks policy, we defined effective coverage (a measure for effectiveness of masks) as the product of efficacy of masks (efficacy) and adherence to wearing them (coverage). A systematic review of the efficacy of face coverings suggested that taking in consideration different types of masks, their average efficacy is within the range of 11%–60% (Panovska-Griffits et al., 2020b). To simulate different face masks policy we then reduced the transmission probability of relevant contact network layers by the amount of effective coverage. For example, if masks are worn in 50% of the community settings with efficacy of 60%, then the effective coverage is 30%. To explore the impact of this policy in the model, we then reduced the transmission probability in the community contact network layer by 0.30. To account for masks additionally worn by 50% of those in school (i.e., only secondary school students) we also reduced the transmission risk in this layer by 30% (assuming again efficacy of 60%). We assumed that 60% of workforce were returning to work in September 2020 with the rest of the workforce working from home.

In addition to masks policy, since May 28, 2020 the key non-pharmaceutical intervention in the UK has been the Test-Trace-Isolate (TTI) strategy. For the testing part, the daily testing probabilities for symptomatic and asymptomatic people were fitted between May and September 26, 2020. For the level of contact tracing we collated the publicly available weekly data from NHS Test and Trace reports (NHS, 2020) between May 28, 2020 (when the program started) and September 26, 2020. To generate an average number, we multiplied the percentage of people testing positive that were interviewed, the percentage of those reporting contacts and the percentage of contacts that were traced to generate an overall percentage for contacts of those tested positive that were traced. We derived average monthly levels of contract tracing to be 43% for June, 47% for July, 45% for August and 50% for September (until September 26, 2020). For the isolation part of the TTI strategy, we assumed that 90% of people who are required to isolate do so.

Under these assumptions, the model predicted new infections, cumulative deaths and the effective reproduction number R, as shown in Fig. 5C. To compare these, we additionally simulated a scenario where only 40% of the workforce goes back to work from September 1, 2020 (instead of 60%) and there is additional 20% reduction in transmission in community (i.e., β for this layer was set to 0.4 instead of 0.6 since September 1, 2020). The results are shown in Fig. 5D.

This scenario analysis suggested that allowing fewer people to go back to work and reducing the transmission risk in the community, while keeping schools open, would result in a smaller resurgence of the COVID-19 in the latter part of 2020 than if social mixing was higher (comparing Fig. 5C and D). Fig. 5C and D show the median projections as solid lines, and the range across 12 simulations in shaded area. Although there is an obvious difference in the pattern of the projected epidemic—increasing with less stringent assumption in Fig. 5C and able to be controlled under more stringent assumptions in Fig. 5D—the shaded area highlights the wide range of possibilities across the simulations. Therefore it is important to note a level of uncertainty within the model projections.

4.3 Case study 3: Application of rule-based modeling

4.3.1 Overview

For this case study, we show how Rule-Based modeling (RBM) as a technique from computational molecular biology (Danos and Laneve, 2004) can be applied to the COVID-19 pandemic. Details of the methodology illustrated here can be found in Waites et al. (2020). This technique is based on chemical equations, and hence is similar to compartmental models of infectious disease but generalises reactions in two important ways: it allows arbitrary subsets of the population to be specified in rules, and it allows bonds to be formed between individuals.

4.3.2 Rule-based modeling methodology

One of the characteristics of compartmental models is that they experience rapid combinatorial explosion when adding features. For example, when considering the effect of face masks in the classic SIR model, this can be done via changes in the parameter β, to assume that the effect of masks is to reduce the transmission risk of susceptible people become infected; or to double the number of compartments, e.g., susceptible individuals with and without masks, infectious ones the same, and so on. Introducing vaccines, where an individual can be vaccinated or not, means that another doubling of compartments is required. Adding two features has thus caused the number of compartments to expand from the original 3–12. This explosion in the number of compartments clearly implies an explosion in the number of transitions: one $I \rightarrow R$ transition is now three, the quadratic $S+I \rightarrow I+I$ transition now requires no less than 16 to completely specify. Each of these rates needs data so using large-scale compartmental models in scarce data scenarios can lead to a large number of assumptions and hence large uncertainty of the predictive modeling.

RBM gives an alternative to this by defining an agent. This is not the kind of agent found in agent-based modeling, but agent by analogy with reagent. It has three internal states: disease progression state, wearing a mask or not, and a binding site for a vaccine:

```
%agent: Person(covid{s i r} mask{y n} vax{y n})
```

Now, if we want to refer to *any* person, we simply write, Person(). If we want to refer to infectious people, we write, Person(covid{i}). If we want to refer to those people who are susceptible and wearing a mask, we can write, Person(covid{s}, mask{y}). We can write the recovery, or removal rule as,

```
Person(covid{i}) -> Person(covid{r}) @ gamma
```

We note that not only is this representation simple, but it is the model, verbatim, meaning that precisely what is written above is provided to the simulator.

The infection rules are somewhat more complicated. Infection involves an infectious person and an unvaccinated susceptible person and infection happens at different rates depending on whether masks are worn or not. So, we have four rules:

```
Person(covid{i}, mask{n}), Person(covid{s}, mask{n}, vax[.]) ->
    Person(covid{i}, mask{n}), Person(covid{i}, mask(n), vax[.]) @
beta_nn
    Person(covid{i}, mask{y}), Person(covid{s}, mask{n}, vax[.]) ->
    Person(covid{i}, mask{y}), Person(covid{i}, mask(n), vax[.]) @
beta_yn
    Person(covid{i}, mask{n}), Person(covid{s}, mask{y}, vax[.]) ->
    Person(covid{i}, mask{n}). Person(covid{i}, mask{y}, vax[.]) @
beta_ny
    Person(covid{i}, mask{y}), Person(covid{s}, mask{y}, vax[.]) ->
    Person(covid{i}, mask{y}), Person(covid{i}, mask(y), vax[.]) @
beta_yy
```

These rules are all nearly identical: an infectious person and a susceptible one come in, two infectious ones come out, for different combinations of masks or no.

The notation `vax[.]` means that there is no vaccine bound to the person's vaccine receptor. What does it mean for a vaccine to become bound? To make use of this, we require another kind of agent. This is the second feature of RBM that does not have an equivalent in compartmental models. This agent is not a compartment, it does not correspond to a subset of the population but individuals become vaccinated by forming a bond with an available vaccine:

```
Person(vax[.]), Vaccine(p[.]) ->
Person(vax[1]), Vaccine(p[1]) @ vaxrate
```

Within the rule-based modeling framework we also need to specify initial conditions, can model wearing masks and introduce social dynamics, can model different testing and vaccinating strategies. Details of how we do this and simple examples can be found in Waites et al. (2020).

The simulation can in principle be done in a deterministic way and it is possible to produce a system of ordinary differential equations (ODEs) approximated by the mean trajectory of the model when simulated using a stochastic technique such as Gillespie's algorithm. Different software can be used to run the simulations, e.g., KaDE can produce Matlab or Octave code for the differential equations, and KaSim simulates the model stochastically. The expressiveness of the rule-based modeling language makes it very easy to write a rule-based model that results in an unreasonably large (possibly infinite) system of ODEs and for this reason stochastic simulation is normally used.

Next we apply this technique to a question related to testing in resource constraint settings.

4.3.3 Application of RBM to LMICs

We consider a scenario of modeling testing strategies against COVID-19 transmission to answer the question: how do we optimally allocate tests? We focus narrowly on surveillance testing at a relatively high rate and a relatively low rate, aiming to explore if non-pharmaceutical severe social distancing (lockdown)-like intervention is triggered on the rate of positive tests, how much surveillance is required, and how accurate the tests should be. The answer that we get is that this style of outbreak management is relatively insensitive to the accuracy of the tests, and even a low level of surveillance testing will suffice for these purposes.

The setup is as follows. We augment a transmission model with testing. Tests are discrete entities, much like the vaccines in the toy example above. Individuals in the population have two internal states related to testing: a state indicating the correct result for a perfect test, and a state indicating the actual result, for the modeled tests have a certain sensitivity and specificity. The relevant parts of the agent definitions are,

```
%agent: Person(test{p n} result{x p n})
%agent: Test(used{y n})
```

The rules relating to testing are straightforward. If an individual is not in a possession of a result, and their test site is not bound, an unused test may bind to it:

```
Test(used[.]{n}), Person(test[.], result{x}) ->
Test(used[1]{y}), Person(test[1], result{x}) @ testRate
```

The individual in possession of a test receives a result after some time,

```
Person(test[_]{p}, result{x}) -> Person(test[_]{p}, result{p}) @
r*resultRate
Person(test[_]{p}, result{x}) -> Person(test[_]{p}, result{n}) @
(1-r)*resultRate
Person(test[_]{n}, result{x}) -> Person(test[_]{n}, result{n}) @
s*resultRate
Person(test[_]{n}, result{x}) -> Person(test[_]{n}, result{p}) @
(1-s)*resultRate
```

And the result that they receive is correct or incorrect according to the characteristic sensitivity (recall) and specificity of the test. Finally, an individual becomes eligible for retesting at some rate,

```
Test(used[1]), Person(test[1]) -> Test(used[.]), Person(test[.]) @
retest
```

From this, we compute the rate of positive tests occurring in a window of time given by the retest rate. This is simply the ratio of the number of individuals with positive results bound to a test to the total number of individuals bound to a test with any result:

```
%obs: TPR
|Person(test[_], result{p})| /
  ( |Person(test[_], result{p})| + |Person(test[_], result{n})| )
```

We use this quantity in a perturbation: when it increases above 2%, a non-pharmaceutical intervention occurs that reduces the contact rate by half. When it falls again below 1%, this intervention is removed.

Our results shown in Fig. 6A suggest that, with a high level of surveillance testing—perhaps unreasonably high where each individual in the population could expect to be tested on average once a year—we see a clearly defined cycle of applying and releasing the non-pharmaceutical intervention. In contrast, with testing an order of magnitude lower as shown in Fig. 6B, where only 10% of the population can expect to be tested in a year, our results suggest this may not be necessary.

As with the previous model, it is useful to show both the mean (solids lines in Figs. 6A and B) and the range of the simulations, hence highlighting the uncertainty of the predictions.

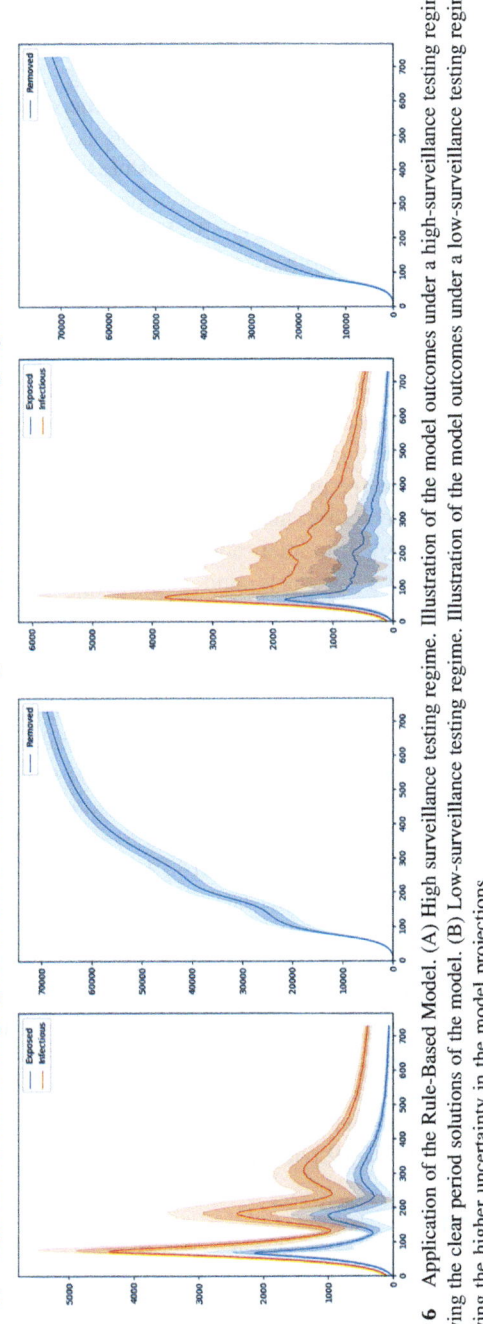

FIG. 6 Application of the Rule-Based Model. (A) High surveillance testing regime. Illustration of the model outcomes under a high-surveillance testing regime showing the clear period solutions of the model. (B) Low-surveillance testing regime. Illustration of the model outcomes under a low-surveillance testing regime showing the higher uncertainty in the model projections.

5 Conclusions

In this chapter, we aimed to highlight the need for and importance of mathematical modeling of infectious diseases, illustrating this with three case studies of how we modeled the COVID-19 pandemic during 2020. We discussed a number of issues related to what modeling is, what it can and cannot do, addressing the issues of model validity, robustness, calibration and different frameworks. By illustrating three conceptually different models applied for the same purpose—to better understand COVID-19 transmission during 2020—we showcased the power of cross-disciplinary and cross-methodologies research in the midst of a pandemic.

As discussed in greater depth in Section 1.2, mathematical modeling can be very useful in evaluating possible interventions and predicting future epidemic curves, but models need to be built well and designed to answer specific questions. There are no models that fit all the questions or can give all the answers; rather, each individual model, assuming it is fit for purpose, can contribute to a collective goal of decision support.

An important development in modeling in infectious diseases over the last 30 years has been the growth in inter- and multi-disciplinary collaborations of mathematicians, computer scientists, and data analysts with biological and medical researchers and policy decision makers. For future applicability and ability of modeling to inform decision making, this cross-disciplinary work needs to continue.

Addressing heterogeneity in both the model framework and within the data is crucial for real-time decision modeling. But behavioral heterogeneities that play an important role in describing infectious disease dynamics and designing and validating infectious diseases models remain challenging. Different modeling frameworks, as illustrated in Section 4, are amenable to modeling infectious diseases, and in this case COVID-19 spread and different interventions. Each has strengths and limitations, as outlined in Section 2.3 and it is important to remain aware of these when modeling real-life epidemics. It is especially important to identify the limits of predictability and potential uncertainty of the model predictions when discussing with policy decision makers.

In summary, modeling remains a very useful technique to support policy decision making and this chapter has included perspectives on modeling of infectious diseases, highlighting different frameworks for modeling COVID-19 and illustrating some of the models that our groups have developed and applied during the COVID-19 pandemic.

Acknowledgments

We thank all contributors to the development and application of the three models we discussed here. For the SEIR-TTI model we thank Simone Sturniolo, David Manheim and Tim Colbourn. For the Covasim model we thank a large team of collaborators: at the Institute for Disease Modeling, Daniel J. Klein, Dina Mistry, Brittany Hagedorn, Katherine

Rosenfeld, Prashanth Selvaraj, Rafael Núñez, Gregory Hart, Carrie Bennette, Marita Zimmermann, Assaf Oron, Dennis Chao, Michael Famulare, and Lauren George; at GitHub, Michał Jastrzębski, Will Fitzgerald, Cory Gwin, Julian Nadeau, Hamel Husain, Rasmus Wriedt Larsen, Aditya Sharad, and Oege de Moor; at Microsoft, William Chen, Scott Ayers, and Rolf Harms; and at the Burnet Institute, Romesh Abeysuriya, Nick Scott, Anna Palmer, Dominic Delport, and Sherrie Kelly. For the RBM model, we thank Matteo Cavalieri and Vincent Danos.

References

Aleta, A., Martin-Corral, D., Pastore y Piontti, A., Ajelli, M., Litvinova, M., Chinazzi, M., Dean, N.E., Halloran, M.E., Longini Jr., I.M., Merler, S., Pentland, A., Vespignani, A., Moro, E., Moreno, Y, 2020. Modelling the impact of social distancing, testing, contact tracing and household quarantine on second-wave scenarios of the COVID-19 epidemic. Nat. Hum. Behav. 4 (9), 964–971. https://doi.org/10.1038/s41562-020-0931-9.

Anderson, R.M., May, R.M., Anderson, B., 1992. Infectious Diseases of Humans: Dynamics and Control. Wiley Online Library. vol. 28.

Andrianakis, I., Vernon, I.R., McCreesh, N., McKinley, T.J., Oakley, J.E., Nsubuga, R.N., et al., 2015. Bayesian History Matching of Complex Infectious Disease Models Using Emulation: A Tutorial and a Case Study on HIV in Uganda. PLoS Comput. Biol. 11 (1), e1003968.

Bernoulli, D., 1766. Essai d'une nouvelle analyse de la mortalite causee par la petite verole. Mem. Math. Phys. Acad. Roy. Sci., Paris 1 (Reprinted in: L.P. Bouckaert, B.L. van der Waerden (Eds.), Die Werke von Daniel Bernoulli, Bd. 2 Analysis und Wahrscheinlichkeitsrechnung, Birkh€auser, Basel, 1982, p. 235. English translation entitled 'An attempt at a new analysis of the mortality caused by smallpox and of the advantages of inoculation to prevent it' in: L. Bradley, Smallpox Inoculation: An Eighteenth Century Mathematical Controversy, Adult Education Department, Nottingham, 1971, p. 21. Reprinted in: S. Haberman, T.A. Sibbett (Eds.) History of Actuarial Science, vol. VIII, Multiple Decrement and Multiple State Models, William Pickering, London, 1995, p. 1.).

Chang, S.L., Harding, N., Zachreson, C., Cliff, O.M., Prokopenko, M., 2020. Modelling Transmission and Control of the COVID-19 Pandemic in Australia. ArXiv 2003. 10218 [Cs, q-Bio], May http://arxiv.org/abs/2003.10218.

Chao, D.L., Halloran, M.E., Obenchain, V.J., Longini Jr., I.M., 2010. FluTE, a Publicly Available Stochastic Influenza Epidemic Simulation Model. PLoS Comput. Biol. 6 (1), e1000656. https://doi.org/10.1371/journal.pcbi.1000656.

Chao, D.L., Oron, A.P., Srikrishna, D., Famulare, M., 2020. Modeling layered non-pharmaceutical interventions against SARS-CoV-2 in the United States with Corvid. Epidemiology. medRxiv.

Cohen, J.A., Mistry, D., Kerr, C.C., Klein, D.J., 2020. Schools are not islands: Balancing COVID-19 risk and educational benefits using structural and temporal countermeasures. medRxiv. preprint. https://doi.org/10.1101/2020.09.08.20190942.

Danos, V., Laneve, C., 2004. Formal molecular biology. Theoretical Computer Science 325 (1), 69–110. https://doi.org/10.1016/j.tcs.2004.03.065.

Dehning, J., Zierenberg, J., Spitzner, F.P., Wibral, M., Neto, J.P., Wilczek, M., Priesemann, V., 2020. Inferring change points in the spread of COVID-19 reveals the effectiveness of interventions. Science 369 (6500), eabb9789. https://doi.org/10.1126/science.abb9789.

Dietz, K., Heesterbeek, J.A.P., 2002. Danie; Bernoulli's epidemiological model revisited. Math. Biosci. 180, 1–21.

Dowe, D.L., Gardner, S., Oppy, G., 2007. Bayes not Bust! Why Simplicity is no problem for Bayesians. Br. J. Philos. Sci 58 (4), 709–754.

Dudai, Y., Evers, K., 2014. To simulate or not to simulate: what are the questions? Neuron 84 (2), 254–261. https://doi.org/10.1016/j.neuron.2014.09.031.

Eubank, S., Guclu, H., Anil Kumar, V.S., Marathe, M.V., Srinivasan, A., Toroczkai, Z., Wang, N., 2004. Modelling disease outbreaks in realistic urban social networks. Nature 429, 180–184. https://doi.org/10.1038/nature02541.

Ferguson, N.M., Cummings, D.A.T., Fraser, C., Cajka, J.C., Cooley, P.C., Burke, D.S., 2006. Strategies for mitigating an influenza pandemic. Nature 442 (7101), 448–452.

Ferguson, N.M., Laydon, D., Nedjati-Gilani, G., Imai, N., Ainslie, K., Baguelin, M., Bhatia, S., et al., 2020. Impact of Non-Pharmaceutical Interventions (NPIs) to Reduce COVID-19 Mortality and Healthcare Demand. Imperial College COVID-19 Response Team, London, p. 16.

Forster, M., 2002. Predictive accuracy as an achievable goal of science. Philos. Sci. 69, S124–S134.

Forster, M., Sober, E., 1994. How to tell when simpler, more unified, or less ad-hoc theories will provide more accurate predictions. Br. J. Philos. Sci. 45, 1–35. MR1277464.

Giordano, G., Blanchini, F., Bruno, R., Colaneri, P., Di Filippo, A., Di Matteo, A., Colaneri, M., 2020. Modelling the COVID-19 Epidemic and Implementation of Population-Wide Interventions in Italy. Nat. Med., 1–6. https://doi.org/10.1038/s41591-020-0883-7.

Heesterbeek, H., Anderson, R.M., Andreasen, V., et al., 2015. Modeling infectious disease dynamics in the complex landscape of global health. Science 347 (6227), aaa4339. https://doi.org/10.1126/science.aaa4339.

Hellewell, J., Abbott, S., Gimma, A., Bosse, N.I., Jarvis, C.I., Russell, T.W., Munday, J.D., Kucharski, A.J., Edmunds, W.J., 2020. Centre for the Mathematical Modelling of Infectious Diseases COVID-19 Working Group, Funk, S., Eggo, R.M., Feasibility of controlling COVID-19 outbreaks by isolation of cases and contacts. Lancet Glob. Health 8 (4), e488–e496. https://doi.org/10.1016/S2214-109X(20)30074-7. (Erratum in: Lancet Glob. Health., 2020.).

Hitchcock, C., Sober, E., 2004. Prediction versus accommodation and the risk of overfitting. Br. J. Philos. Sci. 55, 1–34.

Keeling, M.J., Roheni, P., 2008. Modelling Infectious Diseases in Humans and Animals. Princeton University Press. ISBN: 978-0-691-116174.

Keeling, M.J., Hollingsworth, T.D., Read, J.M., 2020. Efficacy of contact tracing for the containment of the 2019 novel coronavirus (COVID-19). J. Epidemiol. Community Health 74 (10), 861–866. https://doi.org/10.1136/jech-2020-214051.

Kennedy, M.C., O'Hagan, A., 2001. Bayesian calibration of computer models. J. R.Stat. Soc. Ser. B Stat. Methodol. 63 (3), 425–464.

Kermack, W., McKendrick, A., 1927. A contribution to the mathematical theory of epidemics. Philos. Trans. R. Soc. Lond. A 115, 13–23.

Kermack, K.O., McKendrick, A.G., 1932. Contributions to the mathematical theory of epidemics - ii. The problem of endemicity. Philos. Trans. R. Soc. Lond. A 138, 55–83.

Kermack, K.O., McKendrick, A.G., 1933. Contributions to the mathematical theory of epidemics - iii. Further studies of the problem of endemicity. Philos. Trans. R. Soc. Lond. A.

Kermack, W.O., McKendrick, A.G., 1937. Contributions to the mathematical theory of epidemics: IV. Analysis of experimental epidemics of the virus disease mouse ectromelia. J. Hyg. 37, 172–187.

Kermack, W.O., McKendrick, A.G., 1939. Contributions to the mathematical theory of epidemics: V. Analysis of experimental epidemics of mouse-typhoid; a bacterial disease conferring incomplete immunity. J. Hyg. 39, 271–288.

Kerr, C.C., Stuart, R.M., Mistry, D., et al., 2020a. Covasim: an agent-based model of COVID-19 dynamics and interventions. medRxiv. https://doi.org/10.1101/2020.05.10.20097469.

Kerr, C.C., Mistry, D., Stuart, R.M., et al., 2020b. Controling COVID-19 via test-trace-quarantine. medRxiv. preprint doi https://doi.org/10.1101/2020.07.15.20154765.

Kiss, I.Z., Green, D.M., Kao, R.R., 2006. The network of sheep movements within Great Britain: network properties and their implications for infectious disease spread. J. R. Soc. Interface 3, 669–677. https://doi.org/10.1098/rsif.2006.0129.

Koo, J.R., Cook, A.R., Park, M., Sun, Y., Sun, H., Lim, J.T., Tam, C., Dickens, B.L., 2020. Interventions to mitigate early spread of SARS-CoV-2 in Singapore: a modelling study. Lancet Infect. Dis. 20 (6), 678–688. https://doi.org/10.1016/S1473-3099(20)30162-6. (Erratum in: Lancet Infect. Dis. 2020 May;20(5):e79.).

Kretzschmar, M.E., Rozhnova, G., Bootsma, M.C.J., van Boven, M., van de Wijgert, J.H.H.M., Bonten, M.J.M., 2020. Impact of Delays on Effectiveness of Contact Tracing Strategies for COVID-19: A Modelling Study. Lancet Public Health 5 (8), e452–e459. https://doi.org/10.1016/S2468-2667(20)30157-2.

Kucharski, A.J., Klepac, P., Conlan, A.J.K., Kissler, S.M., Tang, M.L., Fry, H., Gog, J.R., et al., 2020. Effectiveness of Isolation, Testing, Contact Tracing, and Physical Distancing on Reducing Transmission of SARS-CoV-2 in Different Settings: A Mathematical Modelling Study. Lancet Infect. Dis. 20 (10), 1151–1160. https://doi.org/10.1016/S1473-3099(20)30457-6.

Lau, M.S.Y., Grenfell, B., Thomas, M., Bryan, M., Nelson, K., Lopman, B., 2020. Characterizing Superspreading Events and Age-Specific Infectiousness of SARS-CoV-2 Transmission in Georgia, USA. Proc. Natl. Acad. Sci. 117 (36), 22430–22435. https://doi.org/10.1073/pnas.2011802117.

Lloyd, A.L., May, R.M., 2001. How viruses spread among computers and people. Science 292, 1316–1317. https://doi.org/10.1126/science.1061076.

Lloyd-Smith, J., Mollison, D., Metcalf, J., Klepac, P., Heesterbeek, H., 2015. Challenges in Modelling Infectious Disease Dynamics. Epidemics 10, 1–108. special issue.

Macdonald, G., 1950. The analysis of infection rates in diseases in which superinfection occurs. Trop. Dis. Bull. 47, 907–915.

Macdonald, G., 1952. The analysis of the sporozoite rate. Trop. Dis. Bull. 49, 569–586.

Macdonald, G., 1955. The measurement of malaria transmission. Proc. R. Soc. Med. 48, 295–301.

Macdonald, G., 1956. Epidemiological basis of malaria control. Bull. World Health Organ. 15, 613–626.

Murray, J.D., 1989. Mathematical Biology. Springer-Verlag, Berlin Heidelberg. Second, corrected edition. ISBN 0-387-57204-X.

NHS, 2020. NHS test and trace reports. https://www.gov.uk/government/collections/nhs-test-and-trace-statistics-england-weekly-reports.

Panovska-Griffiths, J., Kerr, C.C., Stuart, R.M., et al., 2020. Determining the optimal strategy for reopening schools, the impact of test and trace interventions, and the risk of occurrence of a second COVID-19 epidemic wave in the UK: a modelling study [published online ahead of print, 2020 Aug 3]. Lancet Child Adolesc Health. https://doi.org/10.1016/S2352-4642(20)30250-9. S2352-4642(20)30250-9.

Panovska-Griffits, J., Kerr, C.C., Waited, W., et al., 2020. Modelling the potential impact of mask use in schools and society on COVID-19 control in the UK. medRxiv. preprint doi https://doi.org/10.1101/2020.09.28.20202937.

Peak, C.M., Childs, L.M., Grad, Y.H., Buckee, C.O., 2017. Comparing Nonpharmaceutical Interventions for Containing Emerging Epidemics. Proc. Natl. Acad. Sci. U. S. A. 114 (15), 4023–4028.

Read, J.M., Bridgen, J.R.E., Cummings, D.A.T., Ho, A., Jewell, C.P., 2020. Novel Coronavirus 2019-NCoV: Early Estimation of Epidemiological Parameters and Epidemic Predictions. medRxiv. Infectious Diseases (except HIV/AIDS).

Rice, K., Wynne, B., Martin, V., Ackland, G.J., 2020. Effect of school closures on mortality from coronavirus disease 2019: old and new predictions. BMJ 371. m3588.

Rohani, P., Zhong, X., King, A.A., 2010. Contact network structure explains the changing epidemiology of pertussis. Science 330, 982–985. https://doi.org/10.1126/science.1194134.

Ross, R., 1911. Some quantitative studies in epidemiology. Nature 87, 466–467.

Ross, R., 1916. An application of the theory of probabilities to the study of *a priori* pathometry. Part I. Philos. Trans. R. Soc. Lond. A 92, 204–230.

Ross, R., Hudson, H.P., 1917a. An application of the theory of probabilities to the study of *a priori* pathometry. Part II. Philos. Trans. R. Soc. Lond. A 93, 212–225.

Ross, R., Hudson, H.P., 1917b. An application of the theory of probabilities to the study of *a priori* pathometry. Part III. Philos. Trans. R. Soc. Lond. A 93, 225–240.

Shmueli, G., 2010. To Explain or to Predict. Statistical Science 25 (3), 289–310. https://doi.org/10.1214/10-ST330.

Stuart, R.M., Abeysuriya, R.G., Kerr, C.C., et al., 2020. Robust test and trace strategies can prevent COVID-19 resurgences: a case study from New South Wales, Australia. medRxiv. preprint. https://doi.org/10.1101/2020.10.09.20209429.

Sturniolo S, Waites W, Colbourn T, Manheim D, Panovska-Griffiths J. Testing, tracing and isolation in compartmental models." medRxiv. (2020) preprint. https://doi.org/10.1101/2020.05.14.20101808.

Taylor, D.C.A., Pawar, V., Kruzikas, D., et al., 2010. Methods of Model Calibration. Pharmacoeconomics 28, 995–1000. https://doi.org/10.2165/11538660-000000000-00000.

Varma, V.S., 1977. Exact solutions for a special pre-predator or competing species system. Bull. Math. Biol. 39, 619–622.

Vegvari C. et al. Commentary on the use of the reproduction number R during the COVID-19 pandemic. Statistical Methods in Medical Research. (In submission).

Waites, W., Cavaliere, M., Manheim, D., Panovska-Griffiths, J., Danos, V., 2020. Scaling up epidemiological models with rule-based modelling. arXiv. 12077v3.

Walker, P.G.T., Whittaker, C., Watson, O.J., Baguelin, M., Winskill, P., Hamlet, A., Djafaara, B.A., Cucunubá, Z., Olivera Mesa, D., Green, W., Thompson, H., Nayagam, S., Ainslie, K.E.C., Bhatia, S., Bhatt, S., Boonyasiri, A., Boyd, O., Brazeau, N.F., Cattarino, L., Cuomo-Dannenburg, G., Dighe, A., Donnelly, C.A., Dorigatti, I., van Elsland, S.L., FitzJohn, R., Fu, H., Gaythorpe, K.A.M., Geidelberg, L., Grassly, N., Haw, D., Hayes, S., Hinsley, W., Imai, N., Jorgensen, D., Knock, E., Laydon, D., Mishra, S., Nedjati-Gilani, G., Okell, L.C., Unwin, H.J., Verity, R., Vollmer, M., Walters, C.E., Wang, H., Wang, Y., Xi, X., Lalloo, D.G., Ferguson, N.M., Ghani, A.C., 2020. The impact of COVID-19 and strategies for mitigation and suppression in low- and middle-income countries. Science 369 (6502), 413–422. https://doi.org/10.1126/science.abc0035.

Wilson, A.J., 1980. On Varma's prey-predator problem Bull. Math. Biol. 42, 599–600.

Zhao, S., Chen, H., 2020. Modeling the epidemic dynamics and control of COVID-19 outbreak in China. Quant. Biol., 1–9. https://doi.org/10.1007/s40484-020-0199-0.

Index

Note: Page numbers followed by "*f*" indicate figures and "*t*" indicate tables.

CPI Antony Rowe
Eastbourne, UK
October 28, 2023